AF903354

VOLUME ONE HUNDRED AND TWENTY SIX

CURRENT TOPICS IN DEVELOPMENTAL BIOLOGY

Myogenesis in Development and Disease

CURRENT TOPICS IN DEVELOPMENTAL BIOLOGY

"A meeting-ground for critical review and discussion of developmental processes"
A.A. Moscona and Alberto Monroy (Volume 1, 1966)

SERIES EDITOR

Paul M. Wassarman
Department of Developmental and Regenerative Biology
Icahn School of Medicine at Mount Sinai
New York, NY, USA

CURRENT ADVISORY BOARD

Blanche Capel
Wolfgang Driever
Denis Duboule
Anne Ephrussi
Susan Mango
Philippe Soriano
Cliff Tabin
Magdalena Zernicka-Goetz

FOUNDING EDITORS

A.A. Moscona and Alberto Monroy

FOUNDING ADVISORY BOARD

Vincent G. Allfrey
Jean Brachet
Seymour S. Cohen
Bernard D. Davis
James D. Ebert
Mac V. Edds, Jr.
Dame Honor B. Fell
John C. Kendrew
S. Spiegelman
Hewson W. Swift
E.N. Willmer
Etienne Wolff

VOLUME ONE HUNDRED AND TWENTY SIX

Current Topics in DEVELOPMENTAL BIOLOGY

Myogenesis in Development and Disease

Edited by

DAVID SASSOON

Myology Group, INSERM, Hopital Pitié Salpêtrière
University of Pierre and Marie Curie
Paris, France

ELSEVIER

ACADEMIC PRESS
An imprint of Elsevier

Academic Press is an imprint of Elsevier
50 Hampshire Street, 5th Floor, Cambridge, MA 02139, United States
525 B Street, Suite 1800, San Diego, CA 92101-4495, United States
The Boulevard, Langford Lane, Kidlington, Oxford OX5 1GB, United Kingdom
125 London Wall, London, EC2Y 5AS, United Kingdom

First edition 2018

Copyright © 2018 Elsevier Inc. All rights reserved.

No part of this publication may be reproduced or transmitted in any form or by any means, electronic or mechanical, including photocopying, recording, or any information storage and retrieval system, without permission in writing from the publisher. Details on how to seek permission, further information about the Publisher's permissions policies and our arrangements with organizations such as the Copyright Clearance Center and the Copyright Licensing Agency, can be found at our website: www.elsevier.com/permissions.

This book and the individual contributions contained in it are protected under copyright by the Publisher (other than as may be noted herein).

Notices
Knowledge and best practice in this field are constantly changing. As new research and experience broaden our understanding, changes in research methods, professional practices, or medical treatment may become necessary.

Practitioners and researchers must always rely on their own experience and knowledge in evaluating and using any information, methods, compounds, or experiments described herein. In using such information or methods they should be mindful of their own safety and the safety of others, including parties for whom they have a professional responsibility.

To the fullest extent of the law, neither the Publisher nor the authors, contributors, or editors, assume any liability for any injury and/or damage to persons or property as a matter of products liability, negligence or otherwise, or from any use or operation of any methods, products, instructions, or ideas contained in the material herein.

ISBN: 978-0-12-809215-6
ISSN: 0070-2153

For information on all Academic Press publications
visit our website at https://www.elsevier.com/books-and-journals

Working together to grow libraries in developing countries

www.elsevier.com • www.bookaid.org

Publisher: Zoe Kruze
Acquisition Editor: Zoe Kruze
Editorial Project Manager: Shellie Bryant
Production Project Manager: Denny Mansingh
Cover Designer: Greg Harris

Typeset by SPi Global, India

CONTENTS

CONTRIBUTORS

C. Florian Bentzinger
Faculté de médecine et des sciences de la santé, Université de Sherbrooke, Sherbrooke, QC, Canada

Andrew S. Brack
Broad Center of Regeneration Medicine and Stem Cell Research, University of California, San Francisco, San Francisco, CA, USA

Joe V. Chakkalakal
Center for Musculoskeletal Research; Stem Cell and Regenerative Medicine Institute; The Rochester Aging Research Center, University of Rochester Medical Center, Rochester, NY, United States

DDW Cornelison
Christopher S. Bond Life Sciences Center, University of Missouri, Columbia, MO, United States

Colin Crist
Lady Davis Institute for Medical Research, Jewish General Hospital, McGill University, Montreal, QC, Canada

Jonathan Dando
Echino Limited, England, United Kingdom

Francis J. Dilworth
Sprott Centre for Stem Cell Research, Regenerative Medicine Program, Ottawa Hospital Research Institute; University of Ottawa, Ottawa, ON, Canada

Ryo Fujita
Lady Davis Institute for Medical Research, Jewish General Hospital, McGill University, Montreal, QC, Canada

Penney M. Gilbert
Institute of Biomaterials and Biomedical Engineering, University of Toronto; Terrence Donnelly Centre for Cellular and Biomolecular Research, Toronto, ON, Canada

Lorenzo Giordani
Sorbonne Universités, UPMC Univ Paris 06, INSERM UMRS974, Center for Research in Myology, Paris, France

Denis C. Guttridge
Arthur G. James Comprehensive Cancer Center, The Ohio State University, Columbus, OH, United States

Ara B. Hwang
Broad Center of Regeneration Medicine and Stem Cell Research, University of California, San Francisco, San Francisco, CA, USA

Fabien Le Grand
Sorbonne Universités, UPMC Univ Paris 06, INSERM UMRS974, Center for Research in Myology, Paris, France

Emmeran Le Moal
Faculté de médecine et des sciences de la santé, Université de Sherbrooke, Sherbrooke, QC, Canada

Edward W. Li
Institute of Biomaterials and Biomedical Engineering, University of Toronto, Toronto, ON, Canada

Wenxuan Liu
Center for Musculoskeletal Research, University of Rochester Medical Center, Rochester, NY, United States

Omid Mashinchian
Nestlé Institute of Health Sciences; École Polytechnique Fédérale de Lausanne, Doctoral Program in Biotechnology and Bioengineering, Lausanne, Switzerland

Olivia C. McKee-Muir
Institute of Biomaterials and Biomedical Engineering, University of Toronto, Toronto, ON, Canada

Alice Parisi
Nestlé Institute of Health Sciences (NIHS), Campus École Polytechnique Fédérale de Lausanne, École Polytechnique Fédérale de Lausanne Innovation Park, Lausanne, Switzerland

Addolorata Pisconti
Institute of Integrative Biology, University of Liverpool, Liverpool, United Kingdom

Daniel C.L. Robinson
Sprott Centre for Stem Cell Research, Regenerative Medicine Program, Ottawa Hospital Research Institute; University of Ottawa, Ottawa, ON, Canada

Peter Y. Yu
Arthur G. James Comprehensive Cancer Center; College of Medicine, The Ohio State University, Columbus, OH, United States

PREFACE

When asked to be the managing editor for this volume on Myogenesis in Development and Disease, I realized that I had spent the greater part of three decades working in this field. While my research efforts have moved in many directions including the study of the heart, cancer, endocrine disruption, and basic embryology, these forays have all served to enrich an interest in skeletal muscle biology that has been a lodestar since my days as a PhD student. My years as a postdoctoral fellow saw the discovery of the myogenic factors (MyoD, myf5, MRF4, and myogenin) and a rebirth of embryology that came together to form a cornerstone of modern embryology, as we know it today.

While research aimed at ultimately curing cancer, heart disease, diabetes, and even obesity appears to occupy the public's attention, research into muscle diseases has occupied less time in the spotlight. The early years of the field had been focused upon the most fundamental aspects of muscle biology, and it is only in recent years that the clinical and fundamental aspects of the field have truly begun to articulate. While the field of myogenesis is now beginning to deliver on its promise to alleviate muscle diseases, it has delivered in a remarkable manner to the advancement of other fields including those focused on seemingly unrelated issues. One example of this is the discovery of the myogenic factors that led to a similar quest for tissue-specifying factors in the heart, fat tissue, and brain. The astonishing observation that the myogenic factors could phenotypically convert almost any cell into a muscle cell was one of the most dramatic illustrations of cellular reprogramming and set the ground for research in other fields including Induced Pluripotent Stem Cells (iPS). While cancer is a devastating disease, the profound muscle wasting that accompanies most cancers underlies much of the suffering and even death that ultimately occurs. Simply stated, skeletal muscle is one of the largest tissues by mass in the body and our understanding of how this tissue functions, maintains this function, and repairs itself following injury is an important topic for biology and moreover for society.

This volume consists of 10 chapters. Nine of these chapters review selected topics of active research in the field of myogenesis by recognized researchers whose interests range from the neuromuscular junction to biomaterials to epigenetics to aging. The authors have made significant contributions to the field and are poised to bring our understanding of myogenesis into a new phase in the next two decades. It should be noted that the authors are early enough in their career such that they will continue for at least that long-as editor I made this choice to try to make the relevance of this volume long lived. The first chapter is more unusual for this volume as it addresses the reality of how muscle research is driven in the framework of today's funding mechanisms. As many academic researchers know, the pressures to justify one's collective curiosity about how nature works can, at times, deter from work that might better provide a groundwork for cures in the future. This is not to say that such pressures do not also serve to focus research proposals; however, the present paradigm is in need of reevaluation as argued by the author who has managed multiple international consortiums focused on advancing biomedical research including those on skeletal muscle disease and translational research.

To those in the field, to the student thinking of what topics to pursue in this exciting field, for the established scientist outside of the field who wishes to learn more about myogenesis, and to the layperson who wishes to become familiar with what questions are driving the field today, I hope this collection will be of use and serve as a marker for where the field is presently and the remarkable promise it holds for the future.

Lastly, I wish to thank the contributors to these chapters with whom I have had the honor to work with, as well as the many young scientists, too many to be named, that have worked together with me. Of course, I would be remiss not to mention my deepest gratitude to my thesis advisor, Dr. Darcy Kelley who allowed me to stray from her laboratory's central interest in behavior to muscle research and who introduced me to the late Dr. Alex Mauro, the discoverer of the muscle stem (satellite cell) as well as Dr. Margaret Buckingham who mentored me during my postdoctoral studies and allowed me to explore the intersection of embryology and myogenesis. Lastly, I thank Dr. Giovanna Marazzi, wife, best friend, and colleague who has provided the most encouragement of all.

Dr. David Sassoon
Myology Group, INSERM, Hopital Pitié Salpêtrière
University of Pierre and Marie Curie
Paris, France

CHAPTER ONE

"What Did Maxwell's Equations Really Have to Do With Edison's Invention?": Addressing the Complexity of Developing Clinical Interventions for Skeletal Muscle Disease☆

Jonathan Dando[1]
Echino Limited, England, United Kingdom
[1]Corresponding author: e-mail address: dandojonathan@hotmail.com

Contents

Abstract

To reach the healthcare market and have a medical intervention reimbursed in any format carries high risk and very low success rates. Even when all regulatory hurdles have been surpassed, there is no guarantee that the product will be purchased; a different body makes that decision using criteria typically unknown to early-stage innovators and intervention developers.

In the context of skeletal muscle diseases, the field is at a crossroads; accurate diagnoses are difficult to obtain, patient management and monitoring are equally difficult, cures are evasive, and disease progression is not well enough understood in the human to identify clear targets (irrespective of whether the specific muscle disease is rare or frequent because the progression is slow and the tissue large). Additionally, the

☆The quote in the title is from the article "A bitter pill," written by the late Professor Paolo Bianco in 2012, who is missed as both a friend and as a professional medic/researcher who never lost sight of the benefit of fundamental research to human health.

Current Topics in Developmental Biology, Volume 126
ISSN 0070-2153
https://doi.org/10.1016/bs.ctdb.2017.09.001
© 2018 Elsevier Inc.
All rights reserved.

generation of fundamental knowledge stemming from pure academic research, which underpins short- and long-term insight and advances, has been stalled or at least slowed. The field also faces challenges common to all healthcare interventions, while there are also some unique barriers to address, in both developmental translation of the therapeutic and obtaining reimbursement approval. This is independent of the number of people globally who suffer directly and indirectly from skeletal muscle degeneration or degradation. This chapter covers key issues facing skeletal muscle intervention translation, problems that seem to be routinely occurring, followed by suggestions on what can and should be done differently.

1. INTRODUCTION

It is extremely difficult to reach the healthcare marketplace for skeletal muscle disease-related medical interventions, whether they are diagnostic, therapeutic, a medical device, or a care-related procedure or process, and to then have it reimbursed. From the start of clinical development through to market authorization, the healthcare success rate averages only 10% (DiMasi, Grabowski, & Hansen, 2016), without accounting for convincing healthcare providers to purchase the intervention. The underlying issue is that very few people die from the majority of skeletal muscle diseases (Arango-Lopera, Arroyo, Gutierrez-Robledo, Perez-Zepeda, & Cesari, 2013; Kiadaliri, Woolf, & Englund, 2017); primary morbidities such as sarcopenia (Marcell, 2003) or rare neuromuscular disorders infrequently result in mortality because of the muscle disease itself but rather the body suffers an accumulation of strain as muscle tissue deteriorates. Conversely, a large number of common diseases, such as many forms of cancer, chronic obstructive pulmonary disease (COPD), chronic kidney disease (CKD), osteoporosis, and almost all cardiovascular disease, results in skeletal muscle degeneration as comorbidity thereby increasing the healthcare cost (Androga, Sharma, Amodu, & Abramowitz, 2017; Byun, Cho, Chang, Ahn, & Kim, 2017; Fernández et al., 2016; Harada et al., 2017; Von Haehling & Anker, 2014). Therefore, there is a clear and significant health economic impact of damage to the tissue, but in the scheme of decision-making on reimbursing novel interventions, and the present unsustainable nature of health expenditure economics, unless the issue is life-threatening, reimbursement is difficult to obtain.

Obtaining reimbursement is the driving factor for all forms of health care intervention; someone or some entity has to pay for it, and if the costs of the intervention are not covered, then it is not implemented. As a result, the private sector finds it difficult to focus exclusively on skeletal muscle diseases. This has resulted in an ad hoc patchwork between academia, industry, public funding agencies, and charities/foundations, which at present is only partly succeeding. It is to be expected that if research is financed by private entities, the bottom line will be placed on economic gains rather than patient gains. This focus has generated some poor decision-making and ignores many of the existing and significant knowledge gaps that require a system that promotes fundamental research. It should be noted that these issues are not exclusive to the field of skeletal muscle research; nonetheless, this chapter seeks to outline the present situation and make recommendations regarding where changes can be made.

2. THE INNOVATION SYSTEM PIPELINE LEADING TO CLINICAL INTERVENTION CREATION

Progressive muscle diseases do not lead to rapid death, and during their progression, the accompanying loss of muscle function leads to a decrease in quality of life, and the resultant healthcare costs are exorbitant (Beaudart, Rizzoli, Bruyère, Reginster, & Biver, 2014; Lo et al., 2017). Simultaneously, healthcare bodies, due to financial constraints, are more inclined to address diseases that result in mortality or very high long-term health care costs, which have directly measurable and very large healthcare burdens. The calculations made that underlie reimbursement decisions, based on quality adjusted life years (QALYs for Europe) or disability adjusted life years (DALYs for the United States), are clearer and more conclusive for high direct burden diseases, less so for slow nonmortal degenerative diseases of the skeletal muscle.

Investigators seeking funding who are in academic or research institutes are confronted with the task of proposing a cohesive and timely program coupled with addressing the societal impact of their proposal. For the young investigator, this is as much a challenge as for established investigators; this exercise can result in rather imaginative policy statements. If Gregor Mendel (Miko, 2008) sought funding today to address how sweet pea traits were determined, he would have to justify the work in the context of agriculture and sustainable farming. In the relevance/impact section, he would not be in a position to know that within a 150 years, his data would lead to a revolution in genetics and medicine, and if he were to state such a claim, it is unlikely the proposal would be funded. It is impossible to predict impact, yet since the Bayh-Dole Act of 1980 (Bayh-Dole Act, n.d.; Rhines, 2005) which has been interpreted to mean that academic innovation has enormous value, such economic and socioeconomic valuations have become a staple of many funding schemes. The Bayh-Dole Act was a US legislation that permitted any entity receiving federal government funds for research to be able to own and commercially exploit the outcomes. Prior to this, any inventions belonged to the federal government if they funded the research generating them. The global impact of this one policy and its nationally mutated versions around the world has been huge. To justify the tax expense of funding public research to the general population, most national-based grants require some level of justification based on a measurable socioeconomic impact. However, the generation of knowledge, whose future value, strategic business case for commercialization, and date of impact are totally unknown, cannot be measured using short-term metrics, because the impact of the generated knowledge may be felt in several generations. The reason we know that the possibility of generating a short-term high impact from fundamental research is low is because we know how much it costs and the probability of successfully translating it into a therapeutic (DiMasi et al., 2016; Sertkaya, Birkenbach, Berlind, & Eyraud, 2014). Such pressure encourages applicants to claim that the proposed research will treat the primary morbidity and might be repositioned for other diseases as a successful healthcare intervention, when in reality this is impossible. As Bianco and colleagues stated in 2013, "A model of 'translational medicine' has been subliminally accepted by many scientists. The scheme is driven by the pressure to effect the rapid translation of data from the bench. Translation cannot be aimed for a priori; not everything can, or needs to, be translated" (Bianco et al., 2013).

The value of fundamental research was understood as a result of its impact during the Second World War, which led globally to the creation of tax revenue-based publicly funded research agencies throughout the world (Bianco, 2012): the United States-based NIH and the French-based CNRS are two known examples. The intention was clear, and the generation of knowledge for the purpose of generating knowledge was necessary in itself; it is an indication of a society with a long-term vision when it chooses to investigate the unknown, on the premise that at some indeterminate point, the knowledge will be beneficial. This worked for the best part of 35 years, with the assumption that when knowledge was identified to have a societal value, it would be translated. A very different model to that operating now insists government-funded research must have an identified societal impact with a route or pathway to it being achieved to justify being funded. However, the impact of knowledge on society is random, which reflects the total heterogeneity and randomness of global society. The instance that predetermination of the application of knowledge is performed, the impact of that knowledge becomes limited because restrictions are assigned to its usage—this arises "when knowledge for knowledge's sake" is replaced by knowledge for measurable benefit. Knowledge-for-benefit, enforced upon academia as an approach, does not help healthcare simply because the time needed for development is so long as opposed to IT and to a certain extent engineering. In the context of skeletal muscle research, this has created a significant problem. Basically there are a limited number of researchers performing fundamental skeletal muscle research so the knowledge foundation is already limited. There is increasing pressure for funding in the field to perform semitranslational work, irrespective of the fact that there is insufficient information to sustain this approach. It should not be forgotten that fundamental research can and does inform some fairly spectacular advances in life sciences research. One example of this type of advance is the identification of Taq polymerase leading to the development of PCR, which is now used with healthcare research within most molecular diagnostic laboratories. It was not researched with the aim of generating a healthcare benefit but rather stemmed from an initial curiosity-driven project to explore how bacteria grow in hot springs (Brock, 1997). By forcing academic skeletal muscle research to serve societal benefit, it has generated an opposite effect such that the knowledge coffers are not being maintained, expanded, and replenished as quickly as they could. The long-term result is a decreased healthcare benefit.

Only a very specific segment of the professionals on the medical intervention development process fully understand what it means to have a medical intervention launched successfully and correctly within a healthcare marketplace including most importantly that the invention is reimbursed or paid for the potential benefit of patients. To illustrate this point, Fig. 1 represents an industrial standard drug development pathway, ending at generic release. Most life science professionals, irrespective of sector and discipline, are familiar with this image, but will likely not understand it nor the pivotal rules within it. For academic research, the generation of knowledge is not on this chart but rather it is somewhere to the left of it. Without some level of reconfiguration of this innovation cycle, including academic collaboration (Dando & Weiss, 2013), skeletal muscle clinical transition remains problematic. There are several reasons why these problems remain: First,

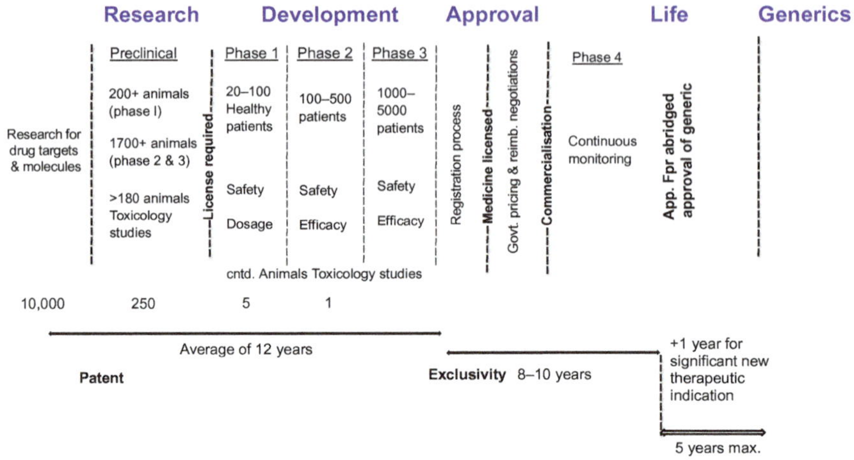

Fig. 1 Typical drug development numbers.

there is limited flexibility in the animal and human numbers indicated; there are disease specificities, but the numbers correspond to what needs to be performed or engaged to move from one phase to another, with the aim of reaching the final phase adhering to the correct regulations while simultaneously generating statistically relevant data. Statistical relevance in the context of generalizability will be addressed later as the actual physical dimensions of skeletal muscle means that obtaining this one critical point is difficult. The costs of this critical path of translation are actually fairly well fixed (Sertkaya et al., 2014). The issues therefore lie elsewhere. Second, most academic institutes do not have the sufficient expertise and/or infrastructure in-house to do any of this; they are very expensive, have annual maintenance costs, and need to be used significantly to justify their expense. It is not uncommon to witness conversations have held in academic centres with educated specialists in which clinical translation units have been proposed with a suggested composition of the centre consisting of a clinical director and an assistant and the rest is far more poorly thought out but makes for publicity. This naïveté is worrying for the future as it suggests that educated specialists now think that translating their work into the marketplace is straightforward and simple. After 30 years of being told that their research must be demonstrably pivotal with short-term measurable metrics, and after a lifetime barrage of "innovation success" statistics and easy solutions or platforms that facilitate this process, it is likely time to recalibrate. Indeed the popular press has raised concerns regarding the level of nonrelevant academic staff in universities that have nourished a corporate culture in place of an academic one and the potential negative impact (Spicer, 2017). It is clear that other than a small handful of academic institutions, almost all of the rest do not have the knowledge, expertise, financial, or logistic resources required to perform preclinical and clinical translation work correctly. The development of a single intervention might require 1700 mice; however, if this number were proposed in a research grant application, it would unlikely meet with success. Moreover, obtaining the actual license to perform this level of animal work would generate overwhelming paperwork, and many institutes with high animal costs, large numbers of mice would exhaust a standard grant budget leaving not residual funds to pay personnel nor other required reagents.

There are several other knowledge gaps: one key area in the preclinical-to-clinical transition relates to quality control of the intervention being tested in the preclinical phase. Few entities know that regulatory bodies require that preclinical work be performed using interventions manufactured under regulatory and market relevant conditions (current good manufacturing practice or cGMP), to ensure that the manufacturing process does not introduce changes to the intervention profiles, so that the preclinical outcomes can be extrapolated to humans (Plaford, 2015). The cost of these requirements carries a cost between 0.5 million and 2.5 million for the cGMP, ideally the latter because you want enough cGMP product to get you to the end of phase IIa. In the context of biomaterial-based regenerative medicine approaches, this is one of the major reasons experimental concepts do not reach the clinic as scaled-up manufacture results in the loss of the key characteristics of efficacy. In the context of academic research, these points should preclude such work from most publicly funded entities as the resources are simply not available.

In contrast, there are some very innovative models that have been implemented, albeit with varying degrees of success, but with the correct concept, by private charities for some time now, such as Cancer Research UK and their in-house Research and Innovation division (Blackburn, 2017) that successfully bridge this gap and continually strive to find better models to make it work. This will be revisited later, in the recommendations sector with regard to what can be done within muscle research.

The third and final point is that the diagram glosses over some critical issues, specifically late-stage development and the approval phase. What happens during the clinical trial design and implementation, when the trials have finished, how the national health agencies and health insurers perceive the outcome of the intervention, and how they make the decision on whether to pay for the intervention fundamentally defines the success of the intervention? These issues are relevant equally to all pertinent stakeholders in healthcare intervention design, implementation, and financing.

3. THE RELEVANCE OF HEALTH ECONOMICS AND POPULATION HEALTH IN DECISION-MAKING AND THE COMPLEXITY OF A DEGENERATING MUSCLE

Some of the biggest issues to be contended with for all healthcare interventions are economic based, but unless one has access to a Health Economics studies or Population Health department (Abrams, 2014; EUPATI resources, 2016; Saltman, 2013), has relevant specialists or a dedicated unit, then they will be ignorant of this critical point in any detail. Large pharma and hospitals have these resources, some, but not all, universities have them; it is rare to find them in research institutes and even rarer in small companies. The impact of the economic importance is best understood if the issue is reverse engineered from the process of an intervention being paid for and where in the healthcare system it is being used, back through the precise decision gates that lead to its application, into early-stage academic research. Once this complete path has been identified, to then forward engineer and project manage the process to define who the stakeholders and regulators are, and precisely identify what information and data they require, to then see where optimization, change, and redesign can be implemented to move toward a defined target.

There is a risk during this planning phase if erroneous economic valuations are used to justify the plan: economic valuations are often used during commercially focused grant applications and in the start-up phase of new companies. These valuations rely on an evaluation model call risk-adjusted net present value (rNPV). NPV calculations are standard in business and have been adapted for the high risk long development times of healthcare by introducing a "risk" adjustment; the model aim is to look at the estimated market value of the final product, reverse calculate through the different phases of development, and identify the products present value. If the NPV is positive, then the investment "can" be considered worthwhile. In contrast if the NPV is negative, the investment is considered detrimental. To achieve NPV one has to know the market estimated valuations at each time point, and this works for many businesses with short to medium life-cycles. However, in health, the usage and interpretation of rNPV differ significantly from large industry to early-stage innovators: in large industry the perception is always long term, using the precise patient population that the anticipated primary endpoint defines, while for early-stage innovators, an abbreviated segment of the plan from above is used. Abbreviated in that estimated final and often total market size of a disease is interpreted to mean the value of the intervention at the end of phase III, and therefore risk calculations mean that phase II has a percentage value of phase III, phase I has a percentage value of phase II, and so on. The error is that final market estimates and phase III data are not synonymous.

In the Fig. 1 scheme, probably one of the largest misunderstandings within many stakeholders is what data is collected during a phase III clinical trial, and how it is used to appraise reimbursement. Common perspectives are that phase III data value is predominantly clinical metric based and that once the data is approved, then its job is done apart from postmarketing control, and just a matter of sales. What actually happens is that prior to a sales force convincing hospitals and trusts to purchase the intervention, it actually has to be approved for reimbursement, and clinical metric data is a small component of that decision approval.

Within the healthcare agencies responsible for deciding whether a therapeutic should be approved for reimbursement, the data is evaluated via a Medical Technology Evaluation Programme, and for the purposes of convenience we will use the National Institute for Health and Care Excellence (NICE) as the model process (Sprange & Clift, 2012); there are some national differences but we will not review them here (another good alternative for educational purposes is the Australian Pharmaceutical Benefits Advisory Committee guidelines) (PBAC, 2015). For review, this typically requires the submission of a reimbursement/Health Technology Assessment (HTA) dossier (Goodman, 2014), which the body reviews as well as performing its own research. The reimbursement bodies look at the scope of the intervention, review the dossier itself which contains information on clinical effectiveness and an estimate of cost-effectiveness based on QALY or DALY calculations, and then generate an appraisal including their decision on if and under what conditions the intervention could be reimbursed.

There are several sections in the HTA dossier and the Scope section has significant relevance for trying to obtain reimbursement for a nonmortal disease-based intervention such as those of the skeletal muscle.

The first component of this section is the target population, and specifically which part of the pathology in this target population the intervention aims to correct; this has direct relevance to the clinical trial design as by phase III the intervention is looking to target a primary outcome (FDA draft guidance for industry, 2017) of the disease which therefore defines its usage; the intervention will not cure everything, and therefore by definition the agency is only going to reimburse for that defined usage, which then becomes the real market.

The second section is a comparator (Lathyris, Patsopoulos, Salanti, & Ioannidis, 2010; Medina & Alvarez-Nunez, 2011); for muscle diseases this becomes problematic since there are few if any and most target pain and inflammation as opposed to muscle repair. Furthermore the reimbursement agency typically performs the evaluation of an intervention based on it being better and the same price, the same and cheaper or ideally better and cheaper than the comparator, with relevant supporting clinical information. While having a unique therapeutic may seem attractive at the start of an innovation cycle, without a comparator, a lot more clinical data need to be collected to convince the agency, which of course increases the cost.

The third section is effectiveness and quality of life (Piantadosi, 2013): quality of life measurements are made based on a standardized questionnaire or patient reported outcome measure called the Euroqol EQ-5D (Euroqol, 2017) which are completed by the person responsible for interviewing patients during the clinical trial. A sarcopenia disease-specific version has recently been created, SarQoL®, but the authors suggest further validation is still needed (Beaudart, Reginster, Geerinck, Locquet, & Bruyère, 2017). EQ-5D examples can be easily found on the internet, and it is interesting to look at these documents in the context of the level of expense and innovation that has been engaged to get the intervention that far and how this is reflected in a human response. The questionnaire is performed via direct interaction with a patient, and many of the questions are understandably based on the patient communicating how they feel.

For skeletal muscle diseases this creates a significant problem; if this is the primary focus of the intervention, it is likely that the patient has had a slow progressive development of this issue, to which they will have adapted over time and for which there will be day-to-day changes. If this could be supported by additional clinical measurements and/or validated biomarkers, we could envisage generating a comprehensive evidence package of success. However, effective functional and biological diagnostics for muscle integrity are lacking; the standard strength or functional tests such as hand grip strength or walking time and distance are subject to changes based on diet and comorbidites. Biopsies are inconvenient, costly and biased by location of extraction. The absence of precise biomarkers, more effective measurement techniques, as well as management products and processes for muscle disease is inhibiting progress.

The large problem with these assessments are the complications of the comorbidities which are going to significantly influence QALY/DALY assessments, as well as prioritizing which intervention to make. If the patient has heart disease or cancer, and skeletal muscle disease, an economic choice needs to be made: this means the generation of a reimbursable skeletal muscle disease targeted intervention is rather difficult.

The fourth component relates to the components of the HTA dossier: the major component of this is the systematic review (Drummond, 2013; Hart, Lundh, & Bero, 2012) and meta-analysis of the clinical effectiveness. Here, agencies are not only looking at data generated by the manufacturer of the intervention but examine all potential associated clinical trial data, and a confidence interval (CI) analysis (Stevens, 2001) is performed to generate an assessment against the null hypothesis. The null hypothesis is typically that the existing market intervention is better than the proposed new intervention, and it is the design of the clinical trials (patient recruitment, inclusion/exclusion, drop out, and trial structure), which can have a significant impact on defining if the null hypothesis is true, or that there is statistically relevant evidence that the new intervention works.

Note clinical trials, plural, as a CI analysis normally require a significant number of independently performed clinical studies with the proposed intervention (Stevens, 2001): the more trials that have been performed, the greater the possibility that the intervention looks worthwhile.

For most medical interventions, the aim is *generalizability* (Bonell, Oakley, Hargreaves, Strange, & Rees, 2006; Sculpher et al., 2004), in that any patient can show up with that specific disease at any time point in their life and be prescribed the intervention and benefit from the desired effect. Without generalizability the restrictions on prescription and reimbursement increase, thereby further reducing the market potential. However, the actual physiology and scale of human skeletal muscle tissue itself, how it differs from person to person, age to age, lifestyle to lifestyle, matched with all forms of the muscle degenerative process, makes generalizability difficult to achieve if this is the primary tissue target. If the disease is widespread and a lot of tissue needs to be reached by the intervention, but the intervention only reaches a low amount of tissue then the statistical and physiological relevance of the clinical data generated is consequently much lower. It therefore becomes much more difficult to demonstrate the effect of an intervention in a single human without considering the next step of larger scale usage and generalizability. When routes of administration and dosage requirements are also considered there is a high capacity for adverse events such as nausea and gastrointestinal-related events, again accounting for height, weight, levels of activity, and lifestyle choices, without integrating in complications arising from the comorbidities.

All of the above points need to be factored in to all stages of the clinical trial design and also back into the early-stage innovation design process. This may read as dissuasive; that is not the aim. The aim is to communicate how truly complicated clinical development and real medical intervention is, and that academia may need to redefine its role away from these routes.

One solution to addressing the complexity would be to integrate Population Health specialists with biological research, which focuses exclusively on knowledge generation could partially alleviate this problem. Population health looks, among many things, at how health outcomes are impacted by societal factors: the environment, lifestyle choices, the economic status of the individual, and the status of the healthcare itself. Surprisingly, the latter has been estimated to be only 20% of the resultant costs of healthcare.

Exploratory and fundamental research generates hypotheses that almost always include a biological component with a human homologue. Whether it is a biomarker or a protein target in many cases there are in many instances human analytical studies published. If population health and earlier stage biological studies are combined, this should reveal human characteristics that could possibly be measured in animal models to generate an expanded insight on muscle degeneration outside of traditional parameters.

If the above work is further stratified by known comorbidities, and then mapped across other muscle pathologies matched with the progressive comorbidities, this should generate significant fundamental knowledge that will permit other specialists to conceptualize and design incremental, progressive, and radical interventions. In concrete terms, from the cardiology, oncology diabetes, rare disease and kidney disease clinic if there are patients with muscle degeneration and another disease, then there will be both differentiators and shared clinical characteristics. There will be potential factors that initiate, amplify, or accelerate disease manifestation; if this information is then replicated in preclinical or exploratory studies using studies of animal models, in which similar morbidity/comorbidity models are used, this will create pivotal fundamental knowledge that maps with the human being. Through this significant amounts of key knowledge will be generated that better informs intervention design and development, now and in 100 years.

The rationale is that to be able to identify avenues that can launch an intervention for skeletal muscle, we have to demonstrate an impact on the comorbidity, which typically means giving the patient more than one drug, a phenomenon known as polypharmacy. Resolving polypharmacy-related issues (Masoudi & Krumholz, 2003; Stawicki et al., 2015) or considering them as part of the impact on the tissue itself will open new avenues. This work should also be performed with Population Health specialists; if overlaps do appear, it should be determined whether patterns can be identified within a larger animal population, that will provide critical information to correctly choose the animal model for modeling the pathology, and when comorbidities start or accelerate inducing rapid decline define which key mechanisms change when this occurs.

Beyond this point in the path of healthcare design, direct academic researcher involvement in intervention development per se should probably stop. If concepts are going to be transitioned into development, it needs to be done properly, and significantly more fundamental knowledge is necessary for this to proceed. It is not cost-effective for academic institutions to perform this, therefore mimicking other initiatives, it may be time to generate key accelerator laboratories, cofunded and coordinated by charities, with the support of regional funders, national funders, and industry. If the selection process is correct and the work coordinated among the different accelerators to obtain independent reproducibility, this should result in the creation of evidence that permits a potential intervention to be tested and developed. This stated, the process must also be based on a reverse engineering from reimbursement with a clearly defined critical path and clinical study plan.

4. THE PRICE OF FAILURE

The average actual total capitalized cost for drug development is now estimated at $2.6 billion, and insufficient returns are making such expenditures questionable (DiMasi et al., 2016). Since 2010, Deloitte have reported the industry average of returns on investment (ROI) on R&D in the large pharma industry: 10.1%, 7.6%, 7.3%, 4.8%, 5.5%, 4.2%, and 3.7%, each year from 2010 to 2016 (Terry & Remnant, 2016). Stern publishes the different industry costs of capital each year: the 2016 figures, the reported weight adjusted cost of capital for biotechnology drugs and pharmaceutical drugs are 9.25% and 7.58%, respectively (Stern Communication, 2017). The cost of capital is what any provider of funds (bank loan, investor) expects to make from their investment and is calculated to include risk-free returns (bonds), plus a financial risk, plus a business risk. If the ROI is lower than the industry-specific cost of capital it means that by performing this investment, one is destroying the value of the investment itself making it better to invest elsewhere.

The critical path for clinical development is actually surprisingly cheap: noncapitalized costs of a phase I–phase III trial have been calculated to be 114.2 million, while the direct clinical trial costs of phase I through to phase IV range from $42 million to $118 million depending on the disease (DiMasi et al., 2016; Sertkaya et al., 2014). However, only around 10% of those interventions starting phase I actually make it to phase IV (where on average $6000 per patient per year is required for the marketing control) (Mcquire, 2011). The 90% failure rate is what gives rise to the $2.6 billion cost. DiMasi calculated that the clinical development phase, when accounting for failures, is around $1.4 billion which reflects direct costs added to around $1.3 billion worth of failure, or alternatively when considering why failures occur, this is $1.3 billion worth of knowledge.

Plaford listed the major areas during the clinical development phase where most failure points occur (Plaford, 2015), which were principally linked to misrepresenting the interventions safety profile, insufficient proof of concept data and trial designs inconsistent with clinical endpoints. Most, if not all, of these failure points are due to an absence of knowledge and/or poor decision making. Avoiding these failure points requires a good quality partnership between academia and industry to ensure transparency and foster trust. Ultimately, few want to be responsible for communicating high cost failures. Therefore, publicly funded bodies have an enormous role to play during the clinical development. Failures, therefore, directly contribute to all forms of fundamental knowledge, but they are rarely communicated.

Medical intervention can be schizophrenic: in other sectors, once an idea has been transferred from the public sector to the private sector, all further actions are performed exclusively by the private sector and regulation of that product is based on industrial standards. The purchase of the product becomes a choice of the consumer. For academic innovations based on engineering and IT that do not touch healthcare and in which the innovation has been transferred entirely into the private sector, this results in some world changing innovation. In contrast, within healthcare, the public and private sector perform a perpetual dance, while the stakeholders try to identify key outputs that demonstrate that the field making progress targeting a patient population with a poorly defined market size. There are already many sources providing information on clinical translation that are unnecessary to reproduce. However, Table 1 indicates those that are freely available

Table 1 Online Resources for Busy Professionals Who Want to Start Learning About Key Aspects of the Intervention Development Process

Course Title	Institute	Online Provider
Drug discovery	UCSD	Coursera
Drug development	UCSD	Coursera
Drug commercialization	UCSD	Coursera
Health technology assessment	University of Sheffield	Future learn
Measuring and valuing health	University of Sheffield	Future learn
Healthcare marketplace specialization	University of Minnesota	Coursera
Entrepreneurship in emerging economies	HarvardX	EdX
Meta-analysis and systematic review	John Hopkins	Coursera
Design and interpretation of clinical trials	John Hopkins	Coursera
Data management for clinical research	Vanderbilt	Coursera
Epidemiology	University of North Carolina	Coursera
Population Health	University of Manchester	Coursera

addressing this very complex field, for those that are interested in learning more about the whole process, we also suggest the many reviews and books, which an interested party can easily find in their library. For example, the Oxford handbooks series are rather good on Health Economics and the Economics of the Biopharmaceutical Industry.

It is important to understand in detail the key components that can be obtained from these resources, and tie this information into strategic decision-making and the socioeconomic impact of muscle diseases. It is necessary to consider the socioeconomic relevance of muscle disease because there are few therapies being reimbursed even though the disease range is broad, while the potential to overlap innovation is huge. The future relevance of muscle disease is significantly larger than most people realize, because muscle disease has enormous impact on the changing pertinence of primary morbidity and comorbidity in all diseases and for all ages This means that interventions need to be designed that target primary muscle diseases as well as diseases that have a significant secondary effect on muscle tissue (e.g., new cardiac drugs will need to demonstrate a beneficial effect on an associated tissue such as skeletal muscle, whereas a cardiac-directed treatment will alleviate the secondary skeletal muscle comorbidity).

Most people suffering from muscular dystrophy die from cardiomyopathy (Van Ruiten et al., 2016). People with cardiac disease have associated muscle wasting. People with cancer, COPD, and diabetes have muscle wasting, as do people with osteoporosis and CKD (Androga et al., 2017; Byun et al., 2017; Fernández et al., 2016; Harada et al., 2017; Von Haehling & Anker, 2014). Aging people with sarcopenia present cardiac comorbidities. These strong associations of diseases are not difficult to understand as the whole body is being strained. As muscle not only serves as a tissue for the body to move but is also one of the largest metabolic "sinks" in the body, muscle will work overtime to maintain body homeostasis. This is important within the contexts of modern health economics such that polypharmacy is a growing problem in healthcare (Beloosesky, Nenaydenko, Gross Nevo, Adunsky, & Weiss, 2013); it induces liver disease, toxic events, and drug failure, and reimbursement agencies now want drugs that impact the primary morbidity and the comorbidity, otherwise it is just another drug to be paid for with questionable efficacy. At any given time, an elderly patient takes, on average, four or five prescription drugs and two over-the-counter medications, and the number of drugs taken increases each decade of life, starting from 50 years of age onward (Flesch & Erdmann, 2006; Lewis, 2017). There is a very real possibility that the efficacy of one therapeutic cancels out the effect of another, with the additional real need to reduce prescription drug costs. This is further complicated by unproven dietary supplements that are routinely used without prescription nor the knowledge of the primary physician that induce metabolic changes subsequently altering the efficacy of the therapeutic, potentially inducing both drug failure and liver injury (Table 2).

Snake oil salesman pitch herbal and dietary supplement remedies that have not been clinically tested (FDA 101: Dietary Supplements, 2017; Roberts, 2017), while other therapies are being pitched and semiapproved for application despite their being little relevant clinical evidence that they work, especially when metrics of therapeutic efficacy are compared to healthy individuals as opposed to placebo controls. Entities whose underlying ethics are honorable can be distorted when shareholder value and publicity risk to outweigh the needs of the patients. These same entities then tie in academic specialists who have been encouraged to seek industry funding in order to demonstrate to the national agencies to justify a model of questionable sustainability. This problem is further exacerbated by an increasing

Table 2 Estimated Global Hospital-Related Healthcare Cost for Caring with Patients With a Skeletal Muscle Disease[a]

Primary Morbidity	Worldwide Frequency (%)	Number (Millions)	Number in Millions Who Also Present Muscle Masting (%)	Healthcare Cost ($ Billions)
COPD	11	825	206 (25)	412
CKD	11	825	74 (9)	148
CVD	31	2325	674 (29)	1348
Osteoporosis	2.6	200	82 (41)	164
Cancer	0.2	15	7.5 (50)	15
Rare diseases	0.02	1.5	1.5 (100)	3

[a]Note that these values are not "market sizes," which are defined by primary outcomes but correspond to the healthcare-related cost of caring for patients with muscle disease: to arrive at a market size primary healthcare and patient support would need to be included, and then a detailed cost analysis performed to understand what proportion of those costs are based on medical interventions, which for prescription drugs are only around 10% of costs. Following this one would need to understand what proportion of the population could be targeted and would benefit from the intervention. We have excluded Sarcopenia alone as without doubt there will be comorbidities in the aged, so there will be overlap with the other morbidities. When looking at primary morbidity plus muscle wasting, note that for many of the primary morbidities there are other tissues affected, e.g., in COPD; however, if you were to develop an intervention that targeted the primary morbidity and muscle wasting, this would already be complicated (but worthwhile), however, to generate an intervention to target multiple morbidities would pose unique problems in clinical trial design which would likely prevent any statistically relevant data being generated for the reimbursement agencies to effectively appraise. We are also using western culture healthcare costs, it is very likely that in many other countries the costs are lower, therefore the values are undoubtedly overestimated. For rare muscle diseases we are using figures for all the rare neuromuscular disorders with significant skeletal muscle involvement.
Facts: In 2017 the world population is 7.5 billion people.
Assumptions: The average minimal hospital-related cost for caring for patient just for their muscle disease is $2000/year, on top of any other medical care.

intimacy between academics and the private that present-day conflict of interest statements do not solve.

Where do we go from here? There are several ways that the problems outlined above can be resolved. There is a growing trend animal work needs to be performed as if it was a clinical trial during the transition phase of exploratory to preclinical work. This means performing animal studies via multisite independent and anonymized or blinded studies to prevent exploratory work that is premature from moving into "development" without being effectively and stringently validated. Additionally, while xenografts and transgenics are good for experimental work, the harsh reality is that they do not always reflect the pathology in humans. One example of this is the mouse model for Duchenne dystrophy that carries the same mutation as found in humans (Partridge, 2013). Yet while the mice are not perfectly healthy, they do not die much sooner than unaffected mice and certainly make it to the equivalent of "middle age" in sharp contrast to afflicted boys. If by working with Population Health specialists reveal focus points of interest, which animal models should be used, as well as when and how to better mimic the natural progression of the disease, this will better inform the transition into clinical development.

The prices of failure, and more importantly the knowledge it generates, need to be integrated into the decision-making plan. In and out of skeletal muscle research we are not doing this very well. If clinical development work generates $1.3 billion worth of negative data for each successful drug, then the value of this negative data is significantly greater that its monetary value. The closer we can move clinical development to its critical path value the better it is for all.

Making this happen will be complicated; however, the data is generated for the benefit of private entities within public entities and if pseudoanonymized should be shareable. If we can trade options on negative data like a junk bond, in that be having access to it enables one entity to lower its development cost then revenue is shared or resultant cost reductions result in a financial reward being paid to the provider of the negative data. The more negative data that is shared the more obvious what we should not, but also more importantly what we should be doing with humans with specific diseases.

It should also be possible to use negative data outcomes and map them back into the preclinical and exploratory stages. It is constantly baffling to watch technicians, Ph.D. students, and Postdoctoral fellows work in isolation, present their best data, and get a publication. Yet no one shares avenues or techniques that did not work or were not worthy of a high impact publication. Put in a different context, if all the muscle charities insisted that the work they support also share negative outcomes on a centralized database, new avenues of research would be much better informed this is critical when one considers that 95% of fundamental research results in disproving a given hypothesis, and these results are subsequently not communicated. Significant duplicated efforts would be avoided is such information was shared.

5. CONCLUDING COMMENTS

The reality within all of healthcare is that it is caught in a cycle of increasing costs and needs and that budgets to address this situation are decreasing. The large pharmaceutical model is only viable providing healthcare systems reimburse their interventions and healthcare systems are viable providing they reduce costs and maintain health. Caught in-between are academic entities generating knowledge along with charities trying their hardest to support patients and their families. From a strategic perspective, the field needs to redesign the risk assessment and management controls, optimize the portfolio designs, and tie in all stages of intervention development with the clinical need to fundamentally reset the clock. Roles and responsibilities of all the actors need to be redefined as a function of the knowledge gaps. This needs to be implemented transparently such that all failures are identified as early as possible and communicated whether their source is private or public. In parallel, national governments need to rededicate their commitment to "generating" knowledge and reconsider the viewpoint that research investment is somehow directly linked to curing a specific disease as well as stimulating the economy, at least in a manner that is under their control. In the case of skeletal muscle disease, there needs to be a significant increase in knowledge creation and availability, for the exclusive purpose of creating that knowledge and providing insight, prior to any further translational steps being taken. This will require integrating our knowledge of the tissue, its biology with population information, and major decision-making structures to accelerate new therapies.

Development of medical interventions for skeletal muscle diseases is necessary; skeletal muscle degeneration plays a key role, from rare diseases in children through to octogenarians who are major sufferers of muscle function decline. As few people actually die from skeletal muscle degeneration directly, it poses a conundrum for clinical development of therapeutics, especially in the context of how healthcare reimbursement decisions are made. Because loss of muscle function has a clear negative impact upon the quality and length of life when copresenting with other morbidities, and children with rare diseases die from those morbidities, it is without question, worthwhile to pursue this line of research. With correct planning and a collective fundamental shift in how all forms of data are perceived as well as a recognition of the value it generates, in light of the unsustainability of existing healthcare infrastructures, this change may result in alleviating costs and offering solutions that so far have been difficult to identify or envision.

REFERENCES

Abrams, M. N. (2014). Toward population health management. Retrieved from http://www.hhnmag.com/articles/4078-toward-population-health-management.

Androga, L., Sharma, D., Amodu, A., & Abramowitz, M. K. (2017). Sarcopenia, obesity, and mortality in US adults with and without chronic kidney disease. *Kidney International Reports*, *2*, 201–211.

Arango-Lopera, V. E., Arroyo, P., Gutierrez-Robledo, L. M., Perez-Zepeda, M. U., & Cesari, M. (2013). Mortality as an outcome of sarcopenia. *The Journal of Nutrition, Health & Aging*, *17*, 259–262.
Bayh-Dole Act: Landmark law helped Universities lead the way (n.d.). Retrieved from https://www.autm.net/advocacy-topics/government-issues/bayh-dole-act/
Beaudart, C., Reginster, J., Geerinck, A., Locquet, M., & Bruyère, O. (2017). Current review of the SarQoL®: A health-related quality of life questionnaire specific to sarcopenia. *Expert Review of Pharmacoeconomics & Outcomes Research*, *17*, 335–341.
Beaudart, C., Rizzoli, R., Bruyère, O., Reginster, J.-Y., & Biver, E. (2014). Sarcopenia: Burden and challenges for public health. *Archives of Public Health*, *72*, 45.
Beloosesky, Y., Nenaydenko, O., Gross Nevo, R. F., Adunsky, A., & Weiss, A. (2013). Rates, variability, and associated factors of polypharmacy in nursing home patients. *Clinical Interventions in Aging*, *8*, 1585–1590.
Bianco, P. (2012). *A bitter pill. Longitude. Vol. 13* (pp. 25–30). http://users.unimi.it/unistem/wp/wp-content/uploads/Longitude-Paolo-Bianco-2012.pdf.
Bianco, P., Barker, R., Brüstle, O., Cattaneo, E., Clevers, H., Daley, G. Q., et al. (2013). Regulation of stem cell therapies under attack in Europe: For whom the bell tolls. *The EMBO Journal*, *32*(11), 1489–1495.
Blackburn, N. (2017). An audience with. *Nature Reviews Drug Discovery*, *16*, 596.
Bonell, C., Oakley, A., Hargreaves, J., Strange, V., & Rees, R. (2006). Assessment of generalisability in trials of health interventions: Suggested framework and systematic review. *BMJ: British Medical Journal*, *333*(7563), 346–349.
Brock, T. D. (1997). The value of basic research: Discovery of *Thermus aquaticus* and other extreme thermophiles. *Genetics*, *146*, 1207–1210.
Byun, M. K., Cho, E. N., Chang, J., Ahn, C. M., & Kim, H. J. (2017). Sarcopenia correlates with systematic inflammation in COPD. *International Journal of COPD*, *12*, 669–675.
Dando, J., & Weiss, I. (2013). *Maximising outputs from early stage research collaborations. Global virtual conference-business management: Vol. 1* (pp. 34–40). http://www.gv-conference.com/archive/?vid=1&aid=31&kid=30101-125&q=f1&s=azaid62.
DiMasi, J. A., Grabowski, H. G., & Hansen, R. W. (2016). Innovation in the pharmaceutical industry: New estimates of R&D costs. *Journal of Health Economics*, *47*, 20–33.
Drummond, M. (2013). Twenty years of using economic evaluations for drug reimbursement decisions: What has been achieved? *Journal of Health Politics, Policy and Law*, *38*, 1081–1102.
EUPATI resources. (2016). *HTA systems in Europe*. Retrieved from https://www.eupati.eu/health-technology-assessment/hta-systems-in-europe/.
Euroqol. (2017). *EQ-5D instruments*. Retrieved from https://euroqol.org/eq-5d-instruments/.
FDA 101: Dietary Supplements. (2017). Retrieved from https://www.fda.gov/ForConsumers/ConsumerUpdates/ucm050803.htm.
FDA draft guidance for industry on multiple endpoints in clinical trials. (2017). Retrieved from https://www.fda.gov/downloads/drugs/guidancecomplianceregulatoryinformation/guidances/ucm536750.pdf.
Fernández, C., Oliveri, B., Bagur, A., Glorioso, D. G., González, D., Mastaglia, S., et al. (2016). High prevalence of sarcopenia in women with osteoporotic fractures. *Journal of Osteoporosis and Physical Activity*, *4*, 1–4.
Flesch, M., & Erdmann, E. (2006). The problem of polypharmacy in heart failure. *Current Cardiology Reports*, *8*, 217–225.
Goodman, C. S. (2014). *HTA 101: Introduction to health technology assessment*. Bethesda, MD: National Library of Medicine.
Harada, H., Kai, H., Shibata, R., Niiyama, H., Nishiyama, Y., Murohara, T., et al. (2017). New diagnostic index for sarcopenia in patients with cardiovascular diseases. *PLoS One*, *12*(5), e0178123.

Hart, B., Lundh, A., & Bero, L. (2012). Effect of reporting bias on meta-analyses of drug trials: Reanalysis of meta-analyses. *BMJ: British Medical Journal*, *344*, 1–11.

Kiadaliri, A. A., Woolf, A. D., & Englund, M. (2017). Musculoskeletal disorders as underlying cause of death in 58 countries, 1986-2011: Trend analysis of WHO mortality database. *BMC Musculoskeletal Disorders*, *18*, 1–12.

Lathyris, D. N., Patsopoulos, N. A., Salanti, G., & Ioannidis, J. P. A. (2010). Industry sponsorship and selection of comparators in randomized clinical trials. *European Journal of Clinical Investigation*, *40*, 172–182.

Lewis, N. (2017). *Too many drugs: Solutions to polypharmacy issues.* Retrieved from http://medicaleconomics.modernmedicine.com/medical-economics/news/too-many-drugs-solutions-polypharmacy-problems.

Lo, Y.-T. C., Wahlqvist, M. L., Huang, Y.-C., Chuang, S.-Y., Wang, C.-F., & Lee, M.-S. (2017). Medical costs of a low skeletal muscle mass are modulated by dietary diversity and physical activity in community-dwelling older Taiwanese: A longitudinal study. *The International Journal of Behavioral Nutrition and Physical Activity*, *14*, 31.

Marcell, T. J. (2003). Sarcopenia: Causes, consequences and preventions. *Journal of Gerontology*, *58A*, 911–916.

Masoudi, F. A., & Krumholz, H. M. (2003). Polypharmacy and comorbidity in heart failure: Most patients have comorbidities that need to be addressed. *BMJ: British Medical Journal*, *327*(7414), 513–514.

Mcquire, R. (2011). *US phase IV budgets top $12, 000 per patient.* Retrieved from https://www.cuttingedgeinfo.com/2011/us-phase-iv-budgets/.

Medina, C., & Alvarez-Nunez, F. (2011). *Evaluating regulatory risk for comparator drug products.* Retrieved from www.raps.org/WorkArea/DownloadAsset.aspx?id=3478.

Miko, I. (2008). Gregor Mendel and the principles of inheritance. *Nature Education*, *1*, 134.

Partridge, T. A. (2013). The mdx mouse model as a surrogate for Duchenne muscular dystrophy. *The FEBS Journal*, *280*(17), 4177–4186. https://doi.org/10.1111/febs.12267.

PBAC—Pharmaceutical Benefits Advisory Committee guidelines. (2015). *Guidelines for preparing submissions to the pharmaceutical benefits Advisory Committee.* (Version 4.5). Retrieved from https://pbac.pbs.gov.au/content/information/archived-versions/pbac-guidelines-v4-5.pdf.

Piantadosi, S. (2013). *Clinical trials: A methodologic perspective.* United Kingdom: John Wiley & Sons.

Plaford, C. (2015). *Why do most clinical trials fail?* Retrieved from https://www.clinicalleader.com/doc/why-do-most-clinical-trials-fail-0001.

Rhines, R. (2005). *Consequences of the Bayh-Dole act.* Retrieved from http://web.mit.edu/lawclub/www/Bayh-Dole%20Act.pdf.

Roberts, R. (2017). *Homeopathic remedies are 'nonsense and risk significant harm'. The Independent.* Retrieved from http://www.independent.co.uk/news/uk/home-news/homeopathy-nonsense-risk-harm-29-european-academies-science-advisory-council-remedies-a7963786.html.

Saltman, D. (2013). *An introduction to European market Access.* Retrieved from https://www1.imperial.ac.uk/resources/7F5B8EAC-E83A-4C71-9192-7BACDFCAFBA7/europeanmafeb27.pdf.

Sculpher, M., Pang, F., Manca, A., Drummond, M., Golder, S., & Urdahl, H. (2004). Generalisability in economic evaluation studies in healthcare: A review and case studies. *Health Technology Assessment*, *8*, 1–192.

Sertkaya, A., Birkenbach, A., Berlind, A., & Eyraud, J. (2014). *Examination of clinical trial costs and barriers for drug development.* Retrieved from https://aspe.hhs.gov/report/examination-clinical-trial-costs-and-barriers-drug-development.

Spicer, A. (2017). *Universities are broke. So let's cut the pointless admin and get back to teaching. The Guardian.* Retrieved from https://www.theguardian.com/commentisfree/2017/aug/21/universities-broke-cut-pointless-admin-teaching.

Sprange, K., & Clift, M. (2012). The NICE medical technologies evaluation programme (MTEP): Manufacturer submission challenges. *Journal of the Royal Society of Medicine, 105*(Suppl. 1), S4–S11.

Stawicki, S. P., Kalra, S., Jones, C., Justiniano, C. F., Papadimos, T. J., Galwankar, S. C., et al. (2015). Comorbidity polypharmacy score and its clinical utility: A pragmatic practitioner's perspective. *Journal of Emergencies, Trauma, and Shock, 8*(4), 224–231. https://doi.org/10.4103/0974-2700.161658.

Stern Communication. (2017). *Costs of capital by sector (2017)*. Retrieved from http://people.stern.nyu.edu/adamodar/New_Home_Page/datafile/wacc.htm.

Stevens, A. (2001). *The advanced handbook of methods in evidence based healthcare*. United Kingdom: Sage Publishing.

Terry, C., & Remnant, J. (2016). *Deloitte insights—Measuring the return from pharmaceutical innovation 2016: Balancing the R&D equation*. Retrieved from https://www2.deloitte.com/uk/en/pages/life-sciences-and-healthcare/articles/measuring-return-from-pharmaceutical-innovation.html.

Van Ruiten, H. J., Marini Bettolo, C., Cheetham, T., Eagle, M., Lochmuller, H., Straub, V., et al. (2016). Why are some patients with Duchenne muscular dystrophy dying young: An analysis of causes of death in North East England. *European Journal of Paediatric Neurology, 6*, 904–909.

Von Haehling, S., & Anker, S. D. (2014). Prevalence, incidence and clinical impact of cachexia: Facts and numbers—Update 2014. *Journal of Cachexia, Sarcopenia and Muscle, 5*, 261–263.

FURTHER READING

Azoulay, P., Fons-Rosen, C., & Zivin, J. S. G. (2015). *Does science advance one funeral at a time*. NBER Working Paper 21788.

Denison, H. J., Cooper, C., Sayer, A. A., & Robinson, S. M. (2015). Prevention and optimal management of sarcopenia: A review of combined exercise and nutrition interventions to improve muscle outcomes in older people. *Clinical Interventions in Aging, 10*, 859–869.

Robinder, J. S., & Hasni, S. (2017). Pathogenesis and management of sarcopenia. *Clinics in Geriatric Medicine, 33*, 17–26.

Solomon, C. Y. Y., Kareeann, S. F. K., Agathe, D. J., & Visvanathan, R. (2016). Clinical screening tools for sarcopenia and its management. *Current Gerontology and Geriatrics Research, 2016*, 1–10.

The drug development and approval process. (2016). Retrieved from http://www.fdareview.org/03_drug_development.php.

Yu, J. (2015). The etiology and exercise implications of sarcopenia in the elderly. *International Journal of Nursing Sciences, 2*, 199–203.

CHAPTER TWO

The Muscle Stem Cell Niche in Health and Disease

Omid Mashinchian*,†,2, Addolorata Pisconti‡,2, Emmeran Le Moal§,2, C. Florian Bentzinger§,1

*Nestlé Institute of Health Sciences, Lausanne, Switzerland
†École Polytechnique Fédérale de Lausanne, Doctoral Program in Biotechnology and Bioengineering, Lausanne, Switzerland
‡Institute of Integrative Biology, University of Liverpool, Liverpool, United Kingdom
§Faculté de médecine et des sciences de la santé, Université de Sherbrooke, Sherbrooke, QC, Canada
[1]Corresponding author: e-mail address: cf.bentzinger@usherbrooke.ca

Contents

Abstract

The regulation of stem cells that maintain and regenerate postnatal tissues depends on extrinsic signals originating from their microenvironment, commonly referred to as the stem cell niche. Complex higher-order regulatory interrelationships with the tissue and factors in the systemic circulation are integrated and propagated to the stem cells through the niche. The stem cell niche in skeletal muscle tissue is both a paradigm for a structurally and functionally relatively static niche that maintains stem cell quiescence during tissue homeostasis, and a highly dynamic regenerative niche that is subject to extensive structural remodeling and a flux of different support cell populations. Conditions ranging from aging to chronically degenerative skeletal muscle diseases affect the composition of the niche and thereby impair the regenerative potential of muscle stem cells. A holistic and integrative understanding of the extrinsic mechanisms

[2] Equal contribution.

Current Topics in Developmental Biology, Volume 126
ISSN 0070-2153
https://doi.org/10.1016/bs.ctdb.2017.08.003
© 2018 Elsevier Inc.
All rights reserved.

regulating muscle stem cells in health and disease in a broad systemic context will be imperative for the identification of regulatory hubs in the niche interactome that can be targeted to maintain, restore, or enhance the regenerative capacity of muscle tissue. Here, we review the microenvironmental regulation of muscle stem cells, summarize how niche dysfunction can contribute to disease, and discuss emerging therapeutic implications.

1. INTRODUCTION

Skeletal muscle is the most abundant tissue of the human body (Hoppeler & Fluck, 2002; Waterlow, 1984). Its contractile properties are essential for vital functions such as locomotion, postural support, and breathing. In addition, skeletal muscle has important endocrine and paracrine functions, and regulates thermogenesis and systemic metabolism (Schnyder & Handschin, 2015). Apart from plastic adaptations to exercise or disuse, skeletal muscle function and mass remain relatively stable until the third or fourth decade of life (Janssen, Heymsfield, Wang, & Ross, 2000; Silva et al., 2010). In aged individuals, skeletal muscle becomes naturally smaller and weaker, and its metabolic activity decreases (Demontis, Piccirillo, Goldberg, & Perrimon, 2013). However, an active healthy lifestyle and balanced nutritional intake can help to preserve muscle mass and quality to some extent into old age (Montero-Fernandez & Serra-Rexach, 2013).

As a consequence of contractile activity and stretch, the skeletal muscle fiber plasma membrane and t-tubule system can be affected by microlesions. Skeletal muscle injuries at a larger scale, such as strains, contusions, and lacerations, can occur due to trauma or surgery. While microlesions in the muscle fiber membrane are immediately sealed through a repair-patch containing specialized lipids and proteins, muscle fibers undergoing necrosis are rebuilt through cell–cell fusion of muscle precursors called myoblasts (Cooper & McNeil, 2015; Wang & Rudnicki, 2011). In addition, myoblasts might contribute to tissue repair by fusing to nonlethally damaged muscle fibers (Rochlin, Yu, Roy, & Baylies, 2010).

Myoblasts are generated by tissue-resident stem cells termed "satellite cells" or muscle stem cells (MuSCs) (Mauro, 1961; Scharner & Zammit, 2011). Following injury, the MuSC pool expands and gives rise to a subpopulation of transiently expanding cells that are committed to differentiation (Wang & Rudnicki, 2011). Committed MuSCs subsequently become myoblasts. Myoblasts can still proliferate but will ultimately differentiate into postmitotic myocytes that fuse to existing fibers or to each other. Due to self-renewal mechanisms, such as asymmetric division, stochastic expansion, and differentiation, the MuSC pool remains at constant size following bouts of regeneration (Kuang, Gillespie, & Rudnicki, 2008).

Owing to the MuSCs, skeletal muscle tissue has an outstanding ability to recover from damage and can undergo multiple cycles of de- and regeneration without any loss of functionality or major consequences on tissue architecture (Bentzinger, Wang, & Rudnicki, 2012). However, a range of conditions such as aging, muscular dystrophy, cancer cachexia, and diabetes affect the composition of muscle tissue and can impair the function of MuSCs (Ali & Garcia, 2014). These pathologies can lead to a vicious cycle of impaired muscle functionality and regenerative failure. For instance, in aging, hip arthritis or osteoporotic fractures can require invasive surgery that causes significant muscle damage (Petis, Howard, Lanting, & Vasarhelyi, 2015). Due to the age-associated reduction in MuSC function, recovery from such muscle damage is slow. Prolonged immobilization during the healing process may in turn accelerate disuse atrophy, weaken the patients further, and increase the risk for falls that will necessitate additional surgical interventions. Muscular dystrophy is another example of a condition in which regenerative failure is a contributor to pathology. This group of diseases is largely caused by mutations that lead to an instability of muscle fibers inducing chronic de- and regeneration of the muscle tissue. At a certain point, MuSCs in dystrophic muscles become incapable to compensate for fiber degeneration and the tissue deteriorates progressively (Serrano & Munoz-Canoves, 2017). Thus, strategies to improve the function of MuSCs hold great therapeutic promise and may help to maintain or restore functional skeletal muscle mass in aging and disease.

MuSC function is regulated at two levels. First, through intrinsic preprogrammed mechanisms and, second, through extrinsic regulation imposed by the stem cell microenvironment, the so-called stem cell niche (Fig. 1). The stem cell niche concept was first proposed by Schofield based on the observation that hematopoietic stem cells require the association with support cells to maintain their stem cell character (Schofield, 1978). The stem cell niche has subsequently been defined as a specific anatomic location that regulates how stem cells participate in tissue generation, maintenance, and repair (Scadden, 2006). The niche maintains adult stem cells throughout life and instructs fundamental stem cell behaviors such as quiescence, self-renewal, lineage progression, and differentiation. In skeletal muscle, the

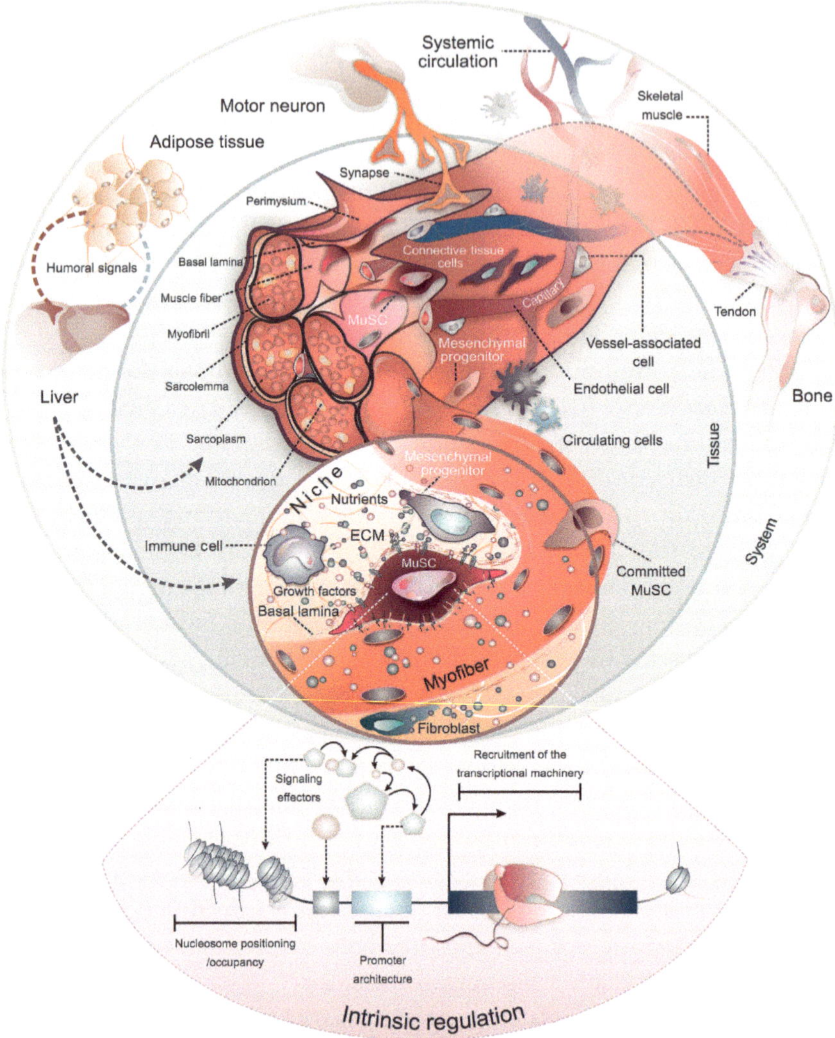

Fig. 1 Intrinsic and extrinsic regulation of MuSCs. The niche microenvironment is composed of structural elements, locally bound and secreted signaling molecules, and cell–cell interactions. Systemic signals such as nutrients, hormones, and circulating growth factors can affect MuSCs directly, or influence them through effects mediated on the tissue level or imposed by support cells. Intrinsic mechanisms such as epigenetic adaptations, telomerase activity, and constitutively activated or repressed signaling loops act in concert with extrinsic mechanisms to regulate MuSC function.

niche is relatively static under homeostatic conditions but becomes dynamically remodeled following injury. During the regenerative response, MuSC function is controlled by a spatiotemporally tightly coordinated flux of different cell types. Regulatory signals originating from these cells involve the remodeling and deposition of extracellular matrix (ECM), the release of growth factors, and cell–cell interactions.

In this chapter, we will review the composition and the regulatory function of the MuSC niche. In addition, we summarize how aging and disease affect the niche and discuss how these processes can be targeted for the development of stem cell-based therapeutic approaches.

2. THE PERINATAL MuSC NICHE

Perinatally, skeletal muscles go through a phase of intense growth (Gokhin, Ward, Bremner, & Lieber, 2008). At this stage MuSCs are highly proliferative and account for about 30% of sublaminal nuclei (Allbrook, Han, & Hellmuth, 1971; Schultz, 1974). Thus, active MuSCs comprise a major cell population in growing postnatal muscle. Autoregulatory deposition of ECM components by MuSCs has emerged as an important mechanism shaping the niche in this period (Tierney et al., 2016). Fetal and postnatal MuSCs express high levels of certain ECM molecules some of which are differentially regulated when compared to quiescent or active adult MuSCs. For instance, both fetal and quiescent adult MuSCs synthesize high levels of the ECM protein collagen VI, while activated adult MuSCs do not deposit this factor into their niche (Tierney et al., 2016; Urciuolo et al., 2013). Collagen VI is involved in bridging other ECM components and likely serves as a modulator of biomechanical properties of the perinatal and quiescent MuSC niche. As opposed to collagen VI, the high-molecular-weight ECM glycoprotein fibronectin is expressed in the fetal stage as well as in activated adult MuSCs but displays very low expression in quiescent adult cells (Bentzinger et al., 2013; Tierney et al., 2016). Fibronectin is a ligand for integrin and syndecan receptors that are expressed at high levels by MuSCs and has been shown to be a critical factor mediating the adhesion of proliferating cells to the niche ECM (Burkin & Kaufman, 1999; Cornelison, Filla, Stanley, Rapraeger, & Olwin, 2001; Gnocchi, White, Ono, Ellis, & Zammit, 2009; Lukjanenko et al., 2016; Rozo, Li, & Fan, 2016). Interestingly, only fetal MuSCs appear to secrete tenascin-C into their niche (Tierney et al., 2016). Loss of this factor reduces the proliferation and differentiation potential of fetal but not adult MuSCs. While its molecular function in the perinatal MuSC niche remains to be further investigated, in other tissue niches, tenascin-C is required for modulation of growth factor signals and controls stem cell proliferation and differentiation (Garcion, Halilagic, Faissner, & ffrench-Constant, 2004; Nakamura-Ishizu et al., 2012). Globally, the

ECM in developing muscle tissue progresses from an embryonic state that is rich in large chondroitin sulfate proteoglycans implicated in cell adhesion, migration, and proliferation, to a mixture of growth factor-regulating dermatan sulfate, chondroitin sulfate, and heparan sulfate glycosaminoglycans in the perinatal period (Young, Carrino, & Caplan, 1990).

Next to regulatory structural elements, the perinatal MuSC niche contains several different cell populations that modulate MuSC function. For instance, a subpopulation of PW1+ interstitial cells (PICs) expressing the marker stem cell antigen 1 at intermediate levels is present in early postnatal muscle (Pannerec, Formicola, Besson, Marazzi, & Sassoon, 2013). PICs secrete follistatin and insulin-like growth factor-1 (IGF1), which promote proliferation and differentiation of MuSCs (Formicola, Marazzi, & Sassoon, 2014). Vice versa, Pax7 knockout mice that are devoid of MuSCs after puberty display a marked increase in PIC numbers (Mitchell et al., 2010). Thus, MuSCs appear to restrict the number of PICs in postnatal muscle through yet to be identified signals. Apart from PICs, connective tissue fibroblasts are highly abundant in perinatal muscle where they promote the differentiation of MuSCs and the maturation of muscle fiber types (Mathew et al., 2011). Cells constituting blood vessels, such as endothelial cells, smooth muscle cells, and pericytes, represent another important component of the perinatal MuSC niche that is differentially regulated when compared to the adult situation. The microcirculation in perinatal muscle is characterized by small intercapillary distances, short capillary lengths, and a tortuous network geometry (Sarelius, Damon, & Duling, 1981). Endothelial cells have been shown to secrete several different growth factors that promote MuSC proliferation (Christov et al., 2007; Kostallari et al., 2015). Postnatally, MuSCs become increasingly associated with pericytes lining the vessel walls. These cells secrete IGF1 and angiopoietin 1, which promotes the progressive withdrawal of MuSCs from the cell cycle and their transition into the quiescent state (Abou-Khalil et al., 2009; Kostallari et al., 2015). Another notable difference between the perinatal and adult niche may originate from neuromuscular signals. Recent work has revealed that neuromuscular junction (NMJ) remodeling can induce regionalized fusion of MuSCs (Liu, Wei-LaPierre, Klose, Dirksen, & Chakkalakal, 2015). In contrast to adult muscle fibers that generally contain only one NMJ, muscle fibers in newborn animals are polyinnervated (Redfern, 1970). Therefore, augmented synaptic signals may contribute to the differential regulation of perinatal MuSCs.

Ephrin receptor (Eph) tyrosine kinases, which are involved in cell contact-dependent signaling of MuSCs with muscle fibers, show a highly dynamic expression during development (Alonso-Martin et al., 2016; Stark, Karvas, Siegel, & Cornelison, 2011). During the fetal to postnatal transition, MuSCs induce the expression of EphA2 and EphB1. Moreover, the ephrin ligands EfnA1 and EfnB2 are upregulated at this stage. Next to Eph/ ephrin signaling, notch receptors and their ligands, for instance delta-like 1, delta-like 4, jagged 1, and jagged 2, have emerged as signals regulating MuSC function perinatally (Geffers et al., 2007; Ladi et al., 2005). During embryonic development, a major source of delta-like 1 appears to originate from proliferating committed myogenic progenitors (Delfini, Hirsinger, Pourquie, & Duprez, 2000; Hirsinger et al., 2001; Schuster-Gossler, Cordes, & Gossler, 2007). Reduced levels of delta-like 1 or loss of the notch downstream effector RBP-J in fetal muscle progenitors lead to muscle hypotrophy and MuSC exhaustion (Schuster-Gossler et al., 2007; Vasyutina et al., 2007). Conversely, constitutive notch signaling blocks differentiation and arrests lineage progression (Mourikis, Gopalakrishnan, Sambasivan, & Tajbakhsh, 2012). Thus, notch signaling is required for the temporal specification of MuSCs during embryonic development and in the perinatal period. Interestingly, notch signaling in perinatal MuSCs also appears to stimulate them to adhere to developing myofibers and to secrete basal lamina ECM components (Brohl et al., 2012).

Taken together, autoregulatory ECM components and intralineage signals from committed myogenic descendants are critical components of the perinatal MuSC niche. Given the high abundance of MuSCs at all stages of lineage progression in perinatal skeletal muscle tissue, contributions of heterologous supportive cell types appear to play a less dominant role when compared to the adult regenerative niche.

3. THE QUIESCENT MuSC NICHE

During puberty, postnatal muscle growth decreases and the MuSC pool completes the transition into quiescence (Tajbakhsh, 2009; White, Bierinx, Gnocchi, & Zammit, 2010). In 1961, two pioneering studies have employed electron microscopy to provide the first description of adult quiescent MuSCs adopting a "satellite cell position" in the periphery of muscle fibers in rat and frog muscles (Katz, 1961; Mauro, 1961). The quiescent state is characterized by a very low cytoplasmic to nuclear volume ratio, a low metabolic activity and mitotic inactivity (Cheung & Rando, 2013).

Quiescent MuSCs are wedged in-between the plasma membrane of their associated host fiber and the ECM of the basal lamina (Fig. 2). Thus, their niche is highly polarized and characterized by ECM interactions on their apical pole and cell–cell interactions with the muscle fiber on the basal pole.

The basal lamina ECM is rich in members of the collagen and laminin family, in particular in laminin containing $\alpha2$, $\beta1$, and $\gamma1$ subunits (also called laminin-2) and collagen IV (Holmberg & Durbeej, 2013). The polymerized collagen and laminin networks in the basal lamina are structurally linked through the glycoprotein nidogen (Fox et al., 1991; Kohfeldt, Sasaki,

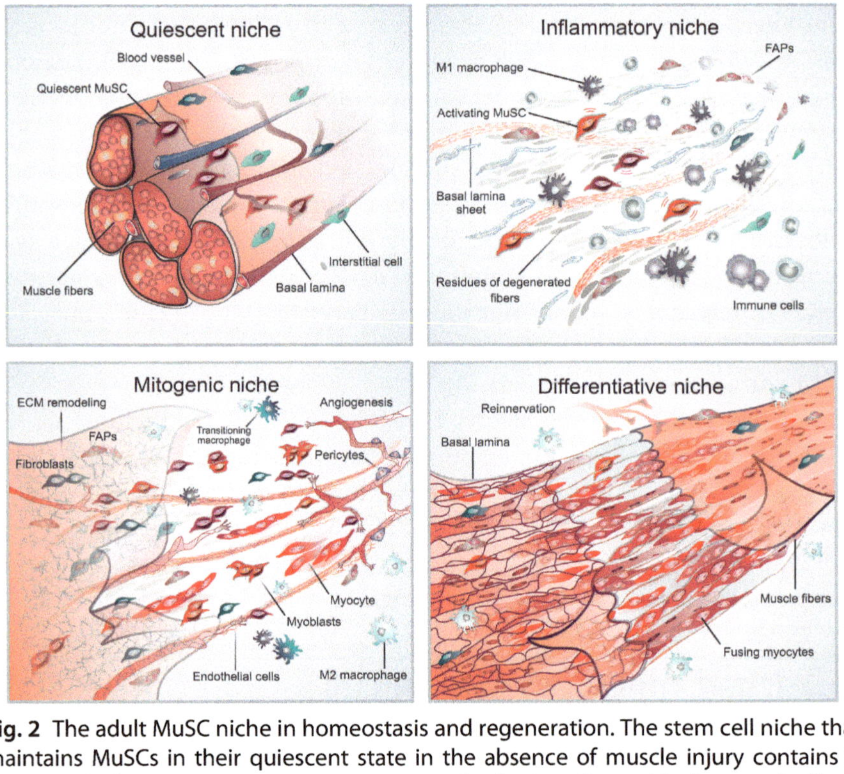

Fig. 2 The adult MuSC niche in homeostasis and regeneration. The stem cell niche that maintains MuSCs in their quiescent state in the absence of muscle injury contains is composed of two major compartments, namely, the interface with the muscle fibers and the basement membrane. In the immediate phase following injury, the niche contains debris of degenerated muscle fibers and a high abundance of proinflammatory immune cells. Subsequently, the niche changes into a milieu that promotes the proliferation of MuSCs and that is characterized by extensive ECM synthesis by fibroblastic cells and angiogenesis. In the differentiative phase, antiinflammatory macrophage subsets become dominant and MuSC-derived myoblasts fuse into young muscle fibers that are reinnervated, and basement membranes mature.

Gohring, & Timpl, 1998). Heparan sulfate is the glycosaminoglycan component of several heparan sulfate proteoglycans present in the basal lamina and on the surface of cells, including MuSCs and muscle fibers (Ghadiali, Guimond, Turnbull, & Pisconti, 2017). Heparan sulfates are linear polysaccharides composed of repeated disaccharide units of *N*-acetylglucosamine and uronic acid which can be variably sulfated in several different positions. Perlecan, which can also bind to laminin and collagen, is an example of an abundant heparan sulfate proteoglycan of the basal lamina that plays important roles in the local sequestration of growth factors (Gohring, Sasaki, Heldin, & Timpl, 1998).

The linkage of quiescent MuSCs to the basal lamina is established through the apically localized membrane receptors α7β1 integrin and dystroglycan (Blanco-Bose, Yao, Kramer, & Blau, 2001; Cohn et al., 2002; Dumont et al., 2015; Rozo et al., 2016). MuSC-specific loss of integrin β1 in adult mice leads to a break in quiescence and aberrant cell cycle entry (Rozo et al., 2016). On the other hand, loss of laminin α2 in mice reduces the number of myogenic progenitors generated during development and the MuSC pool in perinatal muscle fails to undergo the normal progressive reduction in cell number relative to fetal muscles (Nunes et al., 2017). Together with the observation that postnatal laminin α2 deficiency also leads to increased expression of the differentiation marker myogenin, this supports the idea that impaired basal lamina interactions prevent or disrupt MuSC quiescence. Interestingly, laminin α2 deficiency induces a secondary loss of the α7β1 integrin and, vice versa, integrin α7 knockout mice display reduced levels of laminin α2 (Rooney et al., 2006; Vachon et al., 1997).

Gene expression studies revealed that quiescent MuSCs contribute to the ECM in their own microenvironment (Bentzinger et al., 2013; Fukada et al., 2007). Candidate autoregulatory ECM components include vitronectin, laminins, perlecan, decorin, nidogen, biglycan, and collagen VI. Quiescent MuSCs also express the transmembrane proteoglycans syndecan-3 and -4, which carry extracellular heparan sulfate and chondroitin sulfate chains allowing for binding of several growth factors such as fibroblast growth factors (FGFs), hepatocyte growth factor (HGF), vascular endothelial growth factor (VEGF), and transforming growth factor beta 1 (TGFβ1) (Carey, 1997; Cornelison et al., 2001; Xian, Gopal, & Couchman, 2010). In addition, syndecans can serve as coreceptors for integrins. Loss of syndecan-3 leads to spontaneous MuSC activation in adult muscles, while deletion of syndecan-4 has no effect on the stem cell pool (Cornelison et al., 2004;

Pisconti et al., 2016). Since syndecan-3 knockout cells exhibit increased extracellular signal-regulated kinase-1 (ERK) map kinase phosphorylation following stimulation with FGF2 or HGF, it has been suggested that they are missing an inhibitory signal rendering them overly sensitive to growth factors. Apart from syndecan-3, MuSCs employ multiple other strategies to limit the impact of growth factor signals in their niche. These include the attenuation of FGF-mediated ERK signaling by the receptor tyrosine kinase inhibitor sprouty 1, as well as the expression of the insulin-like growth factor-2 (IGF2) inhibitor insulin-like growth factor-binding protein 6 (Pallafacchina et al., 2010; Shea et al., 2010). Notably, the importance of restricted growth factor signaling in maintaining the dormant stem cell state is underlined by the observation that a culture medium containing FGF and HGF receptor inhibitors promote MuSC in quiescence in vitro (Quarta et al., 2016).

Apart from the basal lamina, the muscle fiber membrane represents another major compartment of the quiescent MuSC niche. Quiescent MuSCs and muscle fibers interact through the sialomucin CD34 and m-cadherin (Beauchamp et al., 2000; Bornemann & Schmalbruch, 1994; Irintchev, Zeschnigk, Starzinski-Powitz, & Wernig, 1994). Interestingly, m-cadherin knockout mice display no overt muscle phenotype and it has been postulated that this could be due to compensation by other cadherins, in particular n-cadherin (Hollnagel, Grund, Franke, & Arnold, 2002; Krauss, 2010). Quiescent MuSCs also express the calcitonin receptor, suggesting that electrical signals from innervated myofibers are involved in the regulation of the dormant state (Fukada et al., 2007). Loss of the calcitonin receptor specifically from MuSCs leads to a break in quiescence, cell cycle entry, and extravasation into the interstitial space (Yamaguchi et al., 2015). Furthermore, notch ligands present at the surface of myofibers can likely bind to receptors at the surface of MuSCs to promote the maintenance of quiescence. Genetic ablation of the notch effector RBP-J in MuSCs results in spontaneous activation and terminal differentiation (Bjornson et al., 2012; Mourikis et al., 2012). Conversely, constitutive expression of the notch intracellular domain in myoblasts inhibits S-phase entry and Ki67 expression, and stimulates expression of the self-renewal marker Pax7 (Wen et al., 2012). In addition, syndecan-3 interacts with notch receptors and is required for notch processing and signal transduction (Pisconti, Cornelison, Olguin, Antwine, & Olwin, 2010). Recently, it has been shown that myofibers express an E3 ubiquitin ligase family member called mind bomb 1 that allows the activation of notch signaling to prime MuSC toward quiescence (Kim et al., 2016). Notably, mind bomb 1 is induced by sex hormones, which connects alterations in the quiescent stem cell niche to the systemic circulation.

In addition to the myofiber, blood vessels have been implicated in the maintenance of the quiescent MuSC state. Histological observations have revealed a close proximity between MuSCs and endothelial cells, and a linear relationship between MuSC numbers and capillarization has been demonstrated (Christov et al., 2007). Vessel-associated smooth muscle cells and pericytes have been shown to secrete angiopoietin 1 that binds to Tie2 receptors on the surface of MuSCs and thereby promotes their quiescence (Abou-Khalil et al., 2009; Kostallari et al., 2015).

Traditionally, the study of quiescence has been hindered by the fact that MuSCs become activated within a very short time window after isolation, making it difficult to examine them in a true dormant state in vitro. Notably, the Rando group has recently described a culture model for the maintenance of isolated quiescent MuSCs (Quarta et al., 2016). The authors developed collagen-based artificial myofibers with an elasticity around 1.3 kPa that were functionalized with α4β1 integrin and coated with a layer of laminin. In combination with a specialized medium, this engineered microenvironment allowed for the prolonged in vitro maintenance of mouse and human MuSCs that display key characteristics of quiescent cells.

Overall, a wide range of niche-mediated mechanisms regulate MuSC quiescence. In particular, attachment sites in the basal lamina, cell–cell receptors presented by muscle fibers, and ECM that sequesters growth factors are critical characteristics of the quiescent MuSC niche.

4. THE REGENERATIVE MuSC NICHE

Much of our knowledge about muscle regeneration and MuSC function originates from experimental injury models in rodents. Multiple protocols, including freeze injury, intramuscular injection of barium chloride, glycerol, or the snake venoms notexin and cardiotoxin, have been well established in the research community. Depending on the injury paradigm variation in the kinetics of the MuSC response, the engagement of support cells and revascularization dynamics needs to be taken into consideration (Hardy et al., 2016; Lukjanenko, Brachat, Pierrel, Lach-Trifilieff, & Feige, 2013). The regenerative response that follows all types of muscle injury can be broadly divided into three distinct phases, an initial inflammatory phase during which the tissue is cleared of necrotic muscle fibers, a mitogenic phase characterized by extensive muscle progenitor and support cell proliferation, and a differentiation phase during which new muscle fibers are generated through myoblast fusion (Fig. 2). In the following paragraphs, we discuss how the cellular and acellular microenvironment regulates MuSCs during these different phases of muscle regeneration.

4.1 The Inflammatory Niche

At early time points following muscle injury, necrotic muscle fibers hypercontract within their basal lamina sheet (Bischoff, 1990). The remaining basal laminae, that have been termed "ghost fibers," serve as a template for the subsequent de novo formation of muscle fibers during regenerative myogenesis and guide motor neuron growth cones for reinnervation at original synaptic sites (Caldwell, Mattey, & Weller, 1990; Koskinen et al., 2002; Sanes, Marshall, & McMahan, 1978; Schmalbruch, 1976; Vracko & Benditt, 1972; Webster, Manor, Lippincott-Schwartz, & Fan, 2016). Destruction of the basal lamina tubes during muscle injury using trypsin leads to a disorientation of myoblasts and transiently irregular myofiber pattern (Caldwell et al., 1990). Moreover, activated MuSCs that happen to migrate away from the basal lamina ghost and start to divide in the interstitial space form branched disorganized myotubes (Webster et al., 2016). Migration of MuSCs on the basal lamina has been shown to depend on the laminin receptor integrin $\alpha7\beta1$ (Siegel, Atchison, Fisher, Davis, & Cornelison, 2009). Transcriptional profiling of regenerating muscles revealed that early after muscle injury remnant basal lamina is modified by matrix remodeling enzymes that contribute to the liberation of growth factors and cytokines from it, while in later stages of regeneration extensive de novo synthesis of ECM components is induced (Goetsch, Hawke, Gallardo, Richardson, & Garry, 2003; Kherif et al., 1999).

Immediately after muscle injury, circulating fibrin is deposited at the injury site where it stabilizes the tissue and provides a scaffold for the engagement of infiltrating immune cells (Mann et al., 2011). Necrosis of damaged muscle fibers leads to leakage of normally muscle-compartmentalized factors into the surrounding tissue and the circulation (Tidball, Dorshkind, & Wehling-Henricks, 2014; Tidball & Villalta, 2010). These damage-associated molecular patterns (DAMPs), which include DNA, heat shock proteins, and the redox-sensitive high-mobility group box 1, collectively contribute to immune cell recruitment mediated by toll-like receptors (Hindi & Kumar, 2016). In a zebrafish model, a tissue-scale gradient of hydrogen peroxide observed in the first minutes after muscle injury has been linked to leukocyte recruitment at the damaged site (Niethammer, Grabher, Look, & Mitchison, 2009). This observation suggests that a cascade of both chemical and biological molecules orchestrates the earliest events of the immune response.

Muscle-resident mastocytes and macrophages sense DAMPs and recruit neutrophils by releasing proinflammatory cytokines, including the chemoattractants CXC-chemokine ligand 1 and CC-chemokine ligand 2 (CCL2) (Brigitte et al., 2010). Cytotoxic lymphocytes have recently also been described to contribute to CCL2 production early after muscle damage (Zhang et al., 2014). CCL2 signaling is critical for the early immune response as genetic ablation of its receptor CC-chemokine receptor 2 reduces subsequent macrophage recruitment and impairs regeneration (Sun et al., 2009). Neutrophils peak their number early after the induction of injury and initiate the phagocytic clearance of muscle fiber debris (Stoick-Cooper, Moon, & Weidinger, 2007; Tidball & Villalta, 2010). Eosinophils become abundant in the injured area after about 1 day and secrete interleukin (IL)-4, which promotes the proliferation of fibro–adipogenic progenitors (FAPs) and stimulates them to contribute to tissue clearance (Heredia et al., 2013). Following the onset of neutrophil infiltration, circulating blood monocytes, also referred to as patrolling monocytes, extravasate and enter the regenerating muscles (Arnold et al., 2007). These cells are then primed toward a proinflammatory M1 macrophage phenotype by T helper 1 cytokines (Varga, Mounier, Horvath, et al., 2016). M1 macrophages further support phagocytosis of cellular debris and release cytokines and nitric oxide-related molecules that generate a proinflammatory and nitrosative environment (Saclier et al., 2013; Tidball, 2005). Both Neutrophils and macrophages release the cytokine tumor necrosis factor (TNF) α which curbs the high expression of Pax7 and notch in quiescent cells to levels permissive for myogenesis (Acharyya et al., 2010; Palacios et al., 2010).

Early studies have reported that HGF expression is steeply upregulated in injured muscles (Jennische, Ekberg, & Matejka, 1993). Subsequently, it was shown that quiescent MuSCs express the HGF receptor c-Met and enter the cell cycle upon HGF stimulation (Allen, Sheehan, Taylor, Kendall, & Rice, 1995; Tatsumi, Anderson, Nevoret, Halevy, & Allen, 1998). HGF is deposited in the basal lamina and liberated during injury-induced regeneration due to matrix metalloproteinase (MMP) 2 activation, which is mediated by nitric oxide derived from infiltrating cells (Filippin et al., 2011; Yamada et al., 2008). The earliest stage of MuSC activation has been shown to involve a HGF-induced transition from the quiescent state to a poised alert state (Rodgers et al., 2014). Interestingly, a transition into the alert state also occurs in muscles distant from the injury site due to proteolytic activation of HGF by systemic hepatocyte growth factor activator (Rodgers, Schroeder, Ma, & Rando, 2017). Next to HGF, FGF2 has also been shown to be involved in MuSC activation. The ability of both HGF and FGF2 to induce MAP kinase signaling in MuSCs depends on syndecan-4 (Cornelison et al., 2004). Syndecan-4 knockout MuSCs display delayed entry into the cell cycle and an impaired upregulation of the myogenic commitment marker MyoD. An additional mechanism involved in the activation of MuSCs involves the quiescence marker CD34 that, under homeostatic conditions, interfaces with the muscle fiber (Beauchamp et al., 2000). MuSCs in CD34 knockout mice show impaired entry into proliferation and delayed myogenic progression (Alfaro et al., 2011). Thus, CD34 is required for the activation of MuSCs in early stages after injury, possibly by sensing the altered membrane properties of necrotic fibers.

4.2 The Mitogenic Niche

Following the inflammatory response that characterizes the early stages of muscle regeneration, the MuSC niche becomes permissive to proliferation. During this period the MuSC pool expands and a large number of transiently amplifying myoblasts committed to differentiation are generated. The fibrinolytic proteases uPA and plasmin are critical for resolving the transient proinflammatory fibrin-rich matrix that formed at the site of muscle injury (Lluis et al., 2001; Suelves et al., 2002). The importance of this process is illustrated by the persistent accumulation of fibrin, and the occurrence of chronic inflammation and fibrosis in transgenic mice lacking uPA and plasmin. MMPs are an additional class of proteolytic enzymes involved in remodeling of the ECM during the expansion of the MuSC pool (Alameddine & Morgan, 2016). For instance, MMP10 released from vascular cells and MuSCs is induced in early stages of muscle regeneration (Bobadilla et al., 2014). This protease degrades a variety of ECM components and can activate other MMPs (Nakamura, Fujii, Ohuchi, Yamamoto, & Okada, 1998; Nicholson, Murphy, & Breathnach, 1989). MMP10 knockout mice display impaired MuSC proliferation and differentiation, and contain fewer arterioles. Paradoxically, regenerating muscle tissue of MMP10 knockout mice contains lower levels of collagen, laminin, and fibronectin, which is likely due to the compensatory upregulation and activation of other MMPs (Bobadilla et al., 2014).

Apart from proteolytic modifications, the structural composition of the niche during the proliferative response of MuSCs and their support cells is dramatically modified through the de novo deposition of several ECM components. One of the most strongly induced components of this regenerative matrix is the glycoprotein fibronectin (Lukjanenko et al., 2016). In activated MuSCs, fibronectin binds to integrins containing β1 subunits and to the Frizzled-7/Syndecan-4 (Fzd7/Sdc4) coreceptor complex (Bentzinger et al., 2013; Rozo et al., 2016). Fibronectin is produced by a range of cells in muscle tissue including fibroblasts, FAPs, cells of the hematopoietic fibrogenic lineage, as well as MuSCs themselves (Lukjanenko et al., 2016; Singh, Carraher, & Schwarzbauer, 2010). Adhesion to fibronectin is required to prevent anchorage-dependent cell death and regulates asymmetric division and strand segregation of MuSCs (Bentzinger et al., 2013; Yennek, Burute, Thery, & Tajbakhsh, 2014). Linking niche adhesion of MuSCs to growth factor signaling, integrin β1 has been shown to cooperate with FGF2, while the Fzd7/Sdc4 coreceptor complex potentiates Wnt7a signaling (Bentzinger et al., 2013; Rozo et al., 2016). Both, the FGF receptor 1 and the Fzd7/Sdc4 complex have been demonstrated to be critical for self-renewal mechanisms regulating the MuSC pool during regenerative myogenesis (Bentzinger et al., 2013; Bernet et al., 2014; Le Grand, Jones, Seale, Scime, & Rudnicki, 2009). Collagen VI, released from fibroblasts, represents another ECM protein that is upregulated during the peak of MuSC proliferation (Urciuolo et al., 2013). Deposition of collagen VI in the MuSC niche is critical for the regulation of its mechanical properties. Collagen VI knockout mice show impaired MuSC function and muscle regeneration that is due to tissue elasticity diverging from the optimal 12 kPa of normal muscle to around 7 kPa (Gilbert et al., 2010).

Several growth factors, released from the niche ECM or originating directly from cellular sources, have been implicated in the regulation of the MuSC pool during the regenerative response (Yin, Price, & Rudnicki, 2013). For instance, basal lamina-bound FGF2 that is derived from MuSCs, myofibers, and fibroblasts promotes expansion and maintenance of MuSCs by repressing terminal myogenic differentiation (Anderson, Mitchell, McGeachie, & Grounds, 1995; Bernet et al., 2014; Chakkalakal, Jones, Basson, & Brack, 2012; DiMario, Buffinger, Yamada, & Strohman, 1989; Flanagan-Steet, Hannon, McAvoy, Hullinger, & Olwin, 2000; Hannon, Kudla, McAvoy, Clase, & Olwin, 1996; Rao et al., 2013; Rapraeger, Krufka, & Olwin, 1991). IGF1 originating from the systemic circulation or locally produced in muscle tissue represents another factor that has been shown to promote an expansion of the MuSC pool (Allen & Boxhorn, 1989; Machida, Spangenburg, & Booth, 2003; Musaro et al., 2001). IGF1 downregulates expression of the cell cycle inhibitor p27kip and thereby increases MuSC proliferation (Chakravarthy, Abraha, Schwartz, Fiorotto, & Booth, 2000; Machida et al., 2003).

The mitogenic response of MuSCs is coordinated by a range of resident and infiltrating cells. In particular, immune cells have emerged as pivotal regulators of MuSCs in this phase. Activated MuSCs attract monocytes that eventually differentiate into macrophages and mediate cell–cell contact-induced inhibition of apoptosis (Chazaud et al., 2003). M1 macrophages that are still present in the niche during the proliferative MuSC response, release TNFα, IL-6, IL-1β, and VEGF to stimulate MuSC proliferation and limit early differentiation (Arnold et al., 2007; Saclier et al., 2013). Secretion of interferon (IFN)γ by M1 macrophages, lymphocytes, and natural killer cells correlates with MuSC expansion (Cheng, Nguyen, Fantuzzi, & Koh, 2008). IFNγ induces the myogenic differentiation repressor MHC Class II transactivator that prevents myogenic differentiation (Londhe & Davie, 2013). In recent years, a role for muscle regulatory T (Treg) cells within the mitogenic MuSC niche has emerged (Burzyn et al., 2013). These resident cells with high clonogenic potential accumulate in injured skeletal muscle around the time when MuSC proliferation is maximal, and their genetic ablation impairs muscle regeneration and increases collagen deposition. This might be due to altered transition of M1 to M2 macrophages, as well as to a loss of direct supportive signals for MuSCs. Illustrating the complexity of cellular cross talk in the niche, IL-33 expressed from FAPs-like cells has been shown to be required for the recruitment of Treg cells to injured muscle (Kuswanto et al., 2016).

At more advanced time points during the proliferative expansion of the MuSC pool, when the first myoblasts begin to differentiate, M2 macrophages become dominant over M1 macrophages in the regenerating tissue (Arnold et al., 2007; Tidball, 2017). M1 deactivation is required for sequential steps of macrophage activation (Perdiguero et al., 2011). IGF1 is essential for this process, and its genetic loss from myeloid cells impairs M1 to M2 skewing, dysregulates tissue cytokine levels, and impairs muscle repair (Tonkin et al., 2015). Similarly, a null mutation of IL-10 amplifies M1 at the expense of M2 macrophages due to blunted skewing and leads to premature differentiation of MuSCs (Deng, Wehling-Henricks, Villalta, Wang, & Tidball, 2012). Supporting these observations, diphtheria toxin-mediated ablation of intramuscular macrophages during the transitional period leads to defective regeneration (Arnold et al., 2007). Thus, efficient skewing of macrophage phenotypes is critical for the regulation of the MuSC pool. Transcriptomic profiling of muscle macrophage subsets from regenerating muscles revealed that M2 macrophages express a large set of ECM-related genes such as thrombospondin 4, microfibrillar-associated protein 5, lumicam, and dermatopontin, and thereby appear to contribute to structural remodeling of the niche (Varga, Mounier, Horvath, et al., 2016).

Next to immune cells, the expansion of the MuSC pool is supported by fibroblasts that display comparable proliferation kinetics (Murphy, Lawson, Mathew, Hutcheson, & Kardon, 2011). Partial ablation of fibroblasts during muscle regeneration reduces the number of MuSCs, leads to premature differentiation, and reduces the size of myofibers formed in late myogenic stages. Since fibroblasts are major producers of ECM, these effects are likely due to a deregulation of the niche architecture. FAPs represent another rapidly expanding niche cell type that supports MuSCs in response to injury (Joe et al., 2010; Uezumi, Fukada, Yamamoto, Takeda, & Tsuchida, 2010). FAPs regulate MuSCs through secretion of promyogenic cytokines and ECM. Inhibition of FAP-induced transient collagen deposition using the pharmacological tyrosine kinase inhibitor Nilotinib impairs MuSC expansion and results in defective muscle regeneration (Fiore et al., 2016).

Angiogenesis and myogenesis proceed in parallel after skeletal muscle injury (Luque, Pena, Martin, Jimena, & Vaamonde, 1995; Scholz, Thomas, Sass, & Podzuweit, 2003). Proangiogenic factors such as VEGF, angiopoietin 1 and 2 increase markedly in regenerating muscle tissue (Wagatsuma, 2007). Cycling MuSCs are largely colocalized with endothelial cells and release VEGF to stimulate microvascular sprout frequency and enlargement (Christov et al., 2007; Rhoads et al., 2009). Reciprocally, endothelial cells appear to promote MuSC proliferation through the release of several growth factors including IGF1 and FGF2.

4.3 The Differentiative Niche

After the proliferative phase, myoblasts extensively fuse to generate multinucleated muscle fibers and MuSCs that resisted myogenic commitment will begin to transition back into the quiescent state. During this period the vasculature enlacing the basal lamina tubes of the newly generated muscle fibers becomes denser and more organized, and pericytes and smooth muscle cells are recruited to stabilize its structure (Birbrair et al., 2014;

Hardy et al., 2016; Luque et al., 1995). Analogous to the mechanism that promotes the acquisition of quiescence during postnatal development, the activation of the Tie2 receptor in MuSCs by pericyte-derived angiotensin 1 progressively promotes their withdrawal from the cell cycle (Abou-Khalil et al., 2009; Abou-Khalil, Mounier, & Chazaud, 2010; Kostallari et al., 2015). This process is essential to guarantee the availability of sufficient quiescent MuSCs that can be activated for subsequent rounds of regeneration.

During myoblast fusion, changes in the heparan sulfate content of the niche ECM lead to an increased retention of proproliferative and anti-differentiative growth factors (Ghadiali et al., 2017). For instance, differentiating myoblasts sequester FGF2 through the lipid raft-associated proteoglycan glypican-1 and, at the same time, downregulate the expression of syndecans, FGF and HGF receptors (Brandan & Gutierrez, 2013; Gutierrez & Brandan, 2010). Certain growth factors such as IGF2 also promote differentiation directly. IGF2 induces the expression of the cyclin-dependent kinase inhibitor p21 to promote cell cycle exit and cell fusion (Duan, Ren, & Gao, 2010; Florini et al., 1991; Stewart, James, Fant, & Rotwein, 1996).

During the differentiation phase, immune cells are mainly primed to limit inflammatory responses and to promote tissue repair through the release of antiinflammatory cytokines such as IL-10 (Deng et al., 2012). In macrophages, the lipid-activated peroxisome proliferator-activated receptor gamma (PPARγ) has been shown to be required for the transcription of growth differentiation factor 3, which decreases myoblast proliferation and promotes fusion (Varga, Mounier, Patsalos, et al., 2016). FAPs have also been suggested to promote myogenic differentiation through the secretion of IL-6 (Joe et al., 2010). As myogenesis progresses and muscle fibers become more mature FAP numbers are reduced through apoptosis induced by macrophage-derived TNFα (Lemos et al., 2015).

Interestingly, pericytes themselves as well as PDGF receptor negative PICs have been shown to participate in muscle fiber formation during the differentiative stage of regeneration (Birbrair et al., 2013; Dellavalle et al., 2011, 2007; Mitchell et al., 2010; Pannerec et al., 2013). The niche signals required for fate decisions and recruitment of these interstitial cell types to myogenesis remain to be elucidated and appear to involve a complex interplay with other regenerative processes such as reinnervation, angiogenesis, fibrosis, and adipogenesis (Birbrair et al., 2014; Pannerec, Marazzi, & Sassoon, 2012).

Altogether, the MuSC niche instructs all stages of regenerative myogenesis through the timed availability of signals originating from supportive cell types and from the ECM. The different cell types involved in these processes communicate extensively and coordinate their function to create the transitory environmental conditions that control MuSC activation, proliferation, self-renewal, differentiation, and return to quiescence. As outlined in the subsequent paragraphs, these processes are highly susceptive to perturbation and their deregulation in aging and disease can have profound consequences on muscle regeneration.

5. THE MuSC NICHE IN AGING

Virtually every tissue of the body undergoes a decline as an organism ages. The molecular mechanisms of aging are complex and for many tissues have been shown to involve a failure of stem cells to maintain tissue integrity and function over time (Goodell & Rando, 2015). In case of human skeletal muscle, one of the most prominent features of aging is the loss of muscle mass and strength, also known as sarcopenia, which is accompanied by an impaired regenerative capacity and a pathological increase in baseline ECM levels (Lightfoot, McCormick, Nye, & McArdle, 2014).

The implications of MuSCs in muscle aging are subject to a long-lasting debate. One hypothesis, the numerical stem cell aging theory, is that MuSCs become less abundant with aging thus can contribute less to muscle maintenance and regeneration. Indeed, several studies have provided evidence that a decline in MuSC numbers occurs with aging in both humans and rodents (Brack, Bildsoe, & Hughes, 2005; Day, Shefer, Shearer, & Yablonka-Reuveni, 2010; Renault, Thornell, Eriksson, Butler-Browne, & Mouly, 2002; Sajko et al., 2004; Shefer, Rauner, Yablonka-Reuveni, & Benayahu, 2010; Shefer, Van de Mark, Richardson, & Yablonka-Reuveni, 2006; Verdijk et al., 2007). However, recent results suggest that ablation of a large fraction of MuSCs does not accelerate nor exacerbate sarcopenia (Fry et al., 2015). It is important to consider that some MuSCs were surviving the ablation procedure employed in this study, and it is not yet known "how few is too few MuSCs" for maintenance of muscle tissue throughout life (Collins, Zammit, Ruiz, Morgan, & Partridge, 2007). Notably, transplantation of just a few myofibers carrying no more than a few dozens of associated MuSCs into regenerating host muscles has been shown to contribute to large portions of the recipient tissue (Collins et al., 2005; Hall, Banks, Chamberlain, & Olwin, 2010). It has also been described that next to a complete loss of regenerative capacity following injury, fibrosis was increased in old mice that had had their MuSC pool depleted earlier in life (Fry et al., 2015). This phenomenon has been recently mechanistically explained by the same group with a loss of MuSC-mediated inhibition of fibroblast activity through exosomes (Fry, Kirby, Kosmac, McCarthy, & Peterson, 2017).

A second hypothesis on muscle aging, the functional stem cell aging theory, argues that old MuSCs become exhausted and fail to maintain the tissue due to environmental changes and cell-intrinsic mechanisms. Strong evidence has been provided in the last two decades in support of this theory, and important links with the numeric stem cell aging hypothesis have emerged (Sousa-Victor & Munoz-Canoves, 2016). Compared to organs such as the gut or the skin, most skeletal muscles have a relatively low cellular turnover and MuSCs persist in a quiescent state with low metabolic activity over prolonged periods of life (Keefe et al., 2015; Pawlikowski, Pulliam, Betta, Kardon, & Olwin, 2015; Rando, 2006). Therefore, an intrinsic mechanism that limits number of divisions or the function of the cells through a combination of telomere attrition, DNA and oxidative damage, misfolded proteins, or mitochondrial dysfunction would appear to be less dominant in contributing to MuSC aging than altered environmental signals in the surrounding tissue. Interestingly, recent work has shown that even aspects of aging that have traditionally been considered intrinsic can be influenced by environmental stimuli. For instance, telomere length is conserved in MuSCs from healthy and physically active individuals (Ponsot, Lexell, & Kadi, 2008). This observation suggests that the control of telomere length is a dynamic process that can be affected by environmental stimuli and therefore may not be exclusively intrinsically regulated. Similarly, epigenetic adaptations occurring with age can be modulated by environmental signals. A recent study by the Conboy group has shown that the loci of cyclin-dependent kinase inhibitors p21 and p16 are less epigenetically silenced in MuSCs from old mice than in MuSCs from young mice (Li, Han, Cousin, & Conboy, 2015). FGF2 signaling can silence the p21 locus in old MuSCs and thereby restores cell proliferation. Lastly, several signaling pathways have been shown to be rewired in a sustained fashion in aged MuSCs contributing to impaired proliferation or function (Bernet et al., 2014; Chakkalakal et al., 2012; Cosgrove et al., 2014; Price et al., 2014; Tierney et al., 2014). Depending on definitions, this phenomenon may be considered an intrinsic or "cell-autonomous" property since it persists to some degree in isolated cells. However, recent studies indicate that disrupted environmental signals originating from the stem cell niche ECM may be upstream mechanisms leading to some of these long-lasting changes (Lukjanenko et al., 2016; Rozo et al., 2016).

First evidence that the aged environment plays a critical role in regulating muscle regenerative capacity was provided by transplantation experiments in which extensor digitorum longus muscles were grafted between young and old donor rats (Carlson & Faulkner, 1989). The results showed that when the recipient was young, no difference in the regenerative capacity of either young or old donor grafts was observed. In contrast, when the recipient was old, both young and old donor grafts failed to regenerate efficiently. These results clearly suggested that the old environment prevents regeneration. Subsequently, Conboy et al. showed that replacing the systemic environment of old mice with that of young mice via heterochronic parabiosis was able to improve the regenerative potential of old mice (Conboy et al., 2005). This observation supports the idea that age-affected systemic signals can either influence MuSCs directly or promote the function of supportive cells in the niche. Furthermore, several groups have shown that MuSCs explanted from both old and young mice show only minor changes in their replicative and differentiative potential in culture (Conboy, Conboy, Smythe, & Rando, 2003). In contrast, if MuSCs are explanted along with their accompanying myofiber and basement membrane ECM, which constitute a significant portion of the stem cell niche, old cells show a strongly impaired proliferative capacity (Bernet et al., 2014; Conboy et al., 2003). Thus, under conditions where the native aged environment is partially preserved, the regenerative potential of old MuSCs is impaired, while in isolation, the regenerative potential of old and young MuSCs becomes largely equalized. However, this aspect remains controversial for human cells for which some groups observed no difference in culture when MuSCs were obtained from either young or old donors, while others reported a dramatic reduction in proliferative potential and accelerated senescence of aged cells (Alsharidah et al., 2013; Corbu et al., 2010).

In the course of the heterochronic parabiosis experiments outlined earlier, notch signaling was identified to be involved in mediating the positive effects of a youthful systemic environment on MuSCs. Inhibition of notch impairs regeneration in young muscle while forced expression of notch restores regeneration of old muscle (Conboy et al., 2003). The notch pathway has also been studied in the context of cross talk with other aging mechanisms, especially with TGFβ signals. In aged MuSCs, forced activation of notch blocks TGFβ-induced expression of several cyclin-dependent kinase inhibitors, namely, p15, p15, p21, and p27 via Smad3 (Carlson, Hsu, & Conboy, 2008). Notably, the effect of TGFβ on myoblast proliferation is dose dependent with small amounts promoting MuSC proliferation and higher concentrations having inhibitory effects (Carlson et al., 2009). Blood levels of TGFβ1 increase with age, both in humans and in mice, possibly reaching concentrations that exceed the threshold for negative effects on MuSCs. Similarly, growth differentiation factor 11, another member of the TGFβ superfamily, is found in increased levels in the aged systemic circulation and inhibits MuSC function (Egerman et al., 2015).

Heparan sulfate chains of proteoglycans are a major binder of growth factors in the ECM (Turnbull, Powell, & Guimond, 2001). In aged mouse muscles, the heparan sulfate composition changes and its binding capacity for growth factors becomes altered. Heparan sulfate extracted from young muscle can, for instance, inhibit the mitotic activity of FGF2, while extracts from old muscle have inverse effects (Ghadiali et al., 2017). Thus, heparan sulfate from aged muscle fails to efficiently sequester growth factors, which may lead to a break in MuSC quiescence (Chakkalakal et al., 2012). Interestingly, the higher availability of FGF2 in the aged niche is paralleled by decreased sensitivity of the FGF receptor 1 (Bernet et al., 2014; Rozo et al., 2016). Next to FGF2, heparan sulfate also binds signaling molecules of the Wnt family, which display increased activity in aged muscle (Brack et al., 2007). These age-associated increases in Wnt signaling have been connected to an activation of β-catenin-mediated canonical signaling in myoblasts that trigger an increased conversion into fibroblasts. Fibroblasts are able to secrete copious amounts of certain ECM components and further exacerbate fibrosis of old muscles. Since MuSCs seeded on decellularized ECM derived from old muscles express fibrogenic markers and are less myogenic than cells seeded on ECM from young muscle, the altered aged structural environment appears to also directly contribute to fibrogenic conversion (Stearns-Reider et al., 2017). FAPs are another fibrogenic cell type closely related to fibroblasts that is resident in skeletal muscles and that appears to contribute to tissue fibrosis. Recent work has revealed that limiting the proliferation of FAPs through facilitating the expression of a truncated intronic variant of PDGFRα that acts as a decoy that limits PDGF signaling can reduce fibrosis in aged muscles (Mueller, van Velthoven, Fukumoto, Cheung, & Rando, 2016). In contrast to the age-associated increase in collagen-rich ECM in homeostatic muscles, the transient regenerative matrix that instructs MuSCs following injury is reduced in aged muscles (Lukjanenko et al., 2016). For instance, reduced upregulation of fibronectin leads to insufficient activation of integrin receptors, reduced sensitivity to FGF2 signaling, and to increased anchorage-dependent cell death of MuSCs (Lukjanenko et al., 2016; Rozo et al., 2016). Reactivation of integrin or restoration of youthful levels fibronectin in old regenerating muscles rescues MuSC function and improves the regenerative potential of aged muscles.

Next to changes that manifest at the level of the ECM, aged skeletal muscle tissue is also characterized by altered microvascular structure and function, mild chronic inflammation, and increased production of reactive oxygen species (Hearon & Dinenno, 2016; Le Moal et al., 2017). These phenomena appear to be closely interconnected. For instance, higher numbers of immune cells in the aged niche increase reactive oxygen species levels, which in turn might affect nitric oxide signaling and thereby lead to altered vascular function (Brune et al., 2013; Hearon & Dinenno, 2016). The chronic inflammatory state that characterizes aged muscle is also present in other tissues and has been termed as "inflammaging" (Franceschi & Campisi, 2014). In skeletal muscle this phenomenon is associated with altered levels of several cytokines (Coletti, Moresi, Adamo, Molinaro, & Sassoon, 2005; Coletti, Yang, Marazzi, & Sassoon, 2002; Degens, 2010; Guttridge, Mayo, Madrid, Wang, & Baldwin, 2000; Miller, Ito, Blau, & Torti, 1988). TNFα appears to decrease myoblast differentiation in aged muscles and thereby impairs regeneration. IL-6, which is similarly increased with aging, might also perturb MuSC function (McKay et al., 2013; Tierney et al., 2014). Jack/STAT signaling, which is downstream of IL-6, is increased in aged MuSCs, and its inhibition restores self-renewal and myogenic potential (Price et al., 2014; Tierney et al., 2014). Yet another inflammation-related factor that increases with aging and might directly affect MuSC function is osteopontin secreted by aged CD11+ macrophages. Neutralization of osteopontin promotes MuSC differentiation and improves the regenerative capacity of old muscles (Paliwal, Pishesha, Wijaya, & Conboy, 2012).

In conclusion, extensive evidence supports the notion that the MuSC niche is significantly altered with aging and that such changes inevitably affect the maintenance and regenerative capacity of skeletal muscle tissue. The development of interventions that holistically stall the aging process or even rejuvenate muscle will be facilitated by the identification of the physiological upstream triggers leading to the deterioration of the stem cell niche with old age.

6. THE PATHOLOGIC MuSC NICHE

Several diseases, including primary myopathies, neuromuscular, metabolic, and inflammatory disorders, are associated with changes in the MuSC niche. The best-studied group of primary myopathies in the context of pathologic changes in the MuSC niche are the muscular dystrophies. In the majority of muscular dystrophies described so far, muscle fibers are structurally more fragile and tend to rupture under the mechanical stress of contraction (Sahenk & Mendell, 2011). This causes protein leakage which attracts inflammatory cells such as neutrophils, macrophages, natural killer cells, and lymphocytes (Tidball & Wehling-Henricks, 2005). The presence of several foci of injury in dystrophic muscle, which develop asynchronously and continuously, triggers an inflammatory response that is different from the one caused by acute trauma-induced injury (Dadgar et al., 2014). In muscular dystrophy, inflammation becomes chronic and the continuous presence of immune cells dramatically alters the MuSC niche through thickening and stiffening of the ECM and the accumulation of ectopic adipose tissue (Serrano & Munoz-Canoves, 2017). In the long run, these sustained changes in the extracellular milieu impair MuSC function up to the point where they can no longer compensate for the dystrophic muscle fiber degeneration. This state of regenerative failure leads to a deterioration of the tissue and progressively impairs muscle functionality.

Early studies using traditional biochemical and histological approaches to characterize the extracellular environment have identified an accumulation of collagen type I, III, IV, and V in human dystrophic muscles (Myllyla, Myllyla, Tolonen, & Kivirikko, 1982; Peltonen, Myllyla, Tolonen, & Myllyla, 1982). In vitro studies carried out with isolated myogenic cells obtained from dystrophic patients suggested that the increase in collagen deposition in these conditions was due to augmented cell-autonomous secretion (Ionasescu & Ionasescu, 1982). However, this simplified view has been revised and it is now clear that dystrophic muscle fibrosis is the result of highly complex processes involving several cell types that act in concert with myogenic cells, namely, FAPs, macrophages, pericytes, and other cells of mesenchymal or hematopoietic origin (Serrano & Munoz-Canoves, 2017).

Next to collagens, various heparan sulfate proteoglycans are increased in the ECM in dystrophic muscle compared to healthy muscle (Alvarez, Fadic, & Brandan, 2002; Caceres et al., 2000). These molecules include perlecan, syndecan-3, and glypican-1 in muscle of Duchenne muscular dystrophy patients and decorin in a mouse model of this disease. Moreover, the expression and levels of ECM modifiers, such as MMPs and their endogenous inhibitors, the tissue inhibitors of metalloproteinases (TIMPs), are also affected in muscular dystrophy (Alameddine & Morgan, 2016; Fukushima et al., 2007; Sun, Zhao, Li, & Jiang, 2006; von Moers et al., 2005). These observations suggest that not only the abundance of ECM molecules but also their structural arrangement becomes altered as a consequence of the disease.

In addition to MMPs and TIMPs, several serine proteases and their endogenous serine protease inhibitors (Serpins) are differentially regulated in muscular dystrophy. These include members of the coagulation cascade, such as the PAI-1/uPA system and those produced by neutrophils, such as neutrophil elastase and its inhibitor serpinb1a and the latent TGFβ-binding protein 4 (Ardite et al., 2012; Arecco et al., 2016; Heydemann et al., 2009; Suelves et al., 2007). The imbalance described for several proteolytic systems in the dystrophic MuSC niche can affect MuSCs in diverse ways. First, it affects the structure of the ECM and therefore adhesion signaling, which is critical for MuSC function (Lukjanenko et al., 2016; Rozo et al., 2016). Second, it affects the bioavailability and activation status of growth factors, cytokines, and other signaling molecules such as TGFβ, CTGF, IL-1, IL-6, and Ang2 that can simultaneously promote fibrosis and inhibit myogenesis (Clancy, Henry, Sullivan, & Martin, 2017; Meyer-Hoffert & Wiedow, 2011; Painemal, Acuna, Riquelme, Brandan, & Cabello-Verrugio, 2013; Serrano & Munoz-Canoves, 2017; Vial et al., 2008; Yoshida et al., 2013). Third, it might affect MuSCs directly through protease-activated receptor (PAR) 1 and 2 (Chevessier, Hantai, & Verdiere-Sahuque, 2001; Chinni et al., 2000).

Matricellular proteins are ECM bound nonstructural proteins that interact with various integrins, growth factor receptors, and growth factors to modulate their function and activity (Viloria & Hill, 2016). Proteomics-based profiling of dystrophic muscle has revealed an upregulation of several matricellular proteins such as dermatopontin, decorin, asporin, prolargin, periostin, lumican, and fibrinogen, while levels of nidogen and fibrillin were found to be decreased (Arecco et al., 2016; Holland et al., 2015; Holland, Murphy, Dowling, & Ohlendieck, 2016; Thakur & Mishra, 2016). The TGFβ-inducible protein periostin that can bind to collagen, fibronectin, tenascin-C, bone morphogenetic protein (BMP)1, and notch 1 was found to play a critical role in the pathogenesis of muscular dystrophy (Lorts, Schwanekamp, Baudino, McNally, & Molkentin, 2012). Genetic deletion of periostin in δ-sarcoglycan-null mice, which display a severe form of muscular dystrophy phenotypically similar to Duchenne muscular dystrophy, promotes regeneration and reduces fibrogenesis through inhibition of TGFβ signaling. Likewise, decorin that is upregulated in dystrophic muscles modulates members of the TGFβ superfamily, especially TGFβ1, TGFβ2, myostatin, BMP2, and connective tissue growth factor signaling, and may thereby lead to a deregulation of MuSCs (Cabello-Verrugio & Brandan, 2007; Goetsch & Niesler, 2016; Gutierrez, Osses, & Brandan, 2006; Miura et al., 2006; Riquelme et al., 2001; Vial, Gutierrez, Santander, Cabrera, & Brandan, 2011). Lastly, the chronic matricellular deposition of fibrinogen in dystrophic muscles favors inflammation through sustained leukocyte activation (Vidal et al., 2012). Inhibition of fibrinogen binding to the αMβ2 integrin receptor limits macrophage activation and ameliorates disease progression in a mouse model of Duchenne muscular dystrophy.

As a consequence of chronic inflammation and the infiltration of CD4+ and CD8+ T cells, macrophages, eosinophils, and natural killer T cells, the MuSC niche in dystrophic muscle is highly enriched in signaling molecules such as prostaglandins, cytokines, and chemokines (Villalta, Rosenberg, & Bluestone, 2015). These molecules can influence the proliferation, differentiation, migration, fusion, and survival of myoblasts, and their pathologic deregulation will ultimately affect normal MuSC function and thereby contribute to regenerative failure of dystrophic muscle (Tidball & Villalta, 2010). Increased prostaglandin levels have been described in muscles of mice and humans with muscular dystrophy (McArdle, Foxley, Edwards, & Jackson, 1991; Nakagawa et al., 2013; Okinaga et al., 2002). Similarly, several cytokines and chemokines, including TNFα, IL-1β, IL-4, IL-6, and IFNγ, are present in higher levels in dystrophic muscles (Kumar & Boriek, 2003; Kuru et al., 2003; Messina et al., 2011; Villalta, Nguyen, Deng, Gotoh, & Tidball, 2009; Villalta, Rinaldi, et al., 2011). Pharmacologic normalization of IKK/NF-κB-mediated cytokine signaling, neutralization of TNFα, or genetic ablation of IFNγ, all decrease inflammation and promote regeneration in mouse models of muscular dystrophy (Acharyya et al., 2007; Radley, Davies, & Grounds, 2008; Villalta, Deng, Rinaldi, Wehling-Henricks, & Tidball, 2011). Supporting the assumption that these molecules are largely derived from immune cells, experimental depletion of either myeloid or lymphocyte populations greatly reduces dystrophic pathology (Muller & Kapr, 1975; Wehling, Spencer, & Tidball, 2001; Wehling-Henricks et al., 2008). Moreover, immunosuppressive therapy with glucocorticoids can improve muscle strength in patients and in animal models of muscular dystrophy (Hodgetts, Radley, Davies, & Grounds, 2006; Hussein et al., 2006; Manzur, Kuntzer, Pike, & Swan, 2008; Wehling-Henricks, Lee, & Tidball, 2004).

Diseases other than muscular dystrophy can affect the niche and thereby the regenerative capacity of MuSCs. Diabetes mellitus (DM) is defined as a group of metabolic disorders characterized by hyperglycemia due to lack of insulin production or function. As the largest site for glucose uptake in the body and a main regulator of glycemia, skeletal muscle contributes significantly to the pathogenesis of DM. Although much is known about DM-associated changes in skeletal muscle composition and health, it remains subject to ongoing research how the diabetic environment affects MuSCs (D'Souza, Al-Sajee, & Hawke, 2013). Correlating with the general impairment in wound-healing capacity in diabetic individuals, it has been shown that muscle regeneration is affected in type 1 and 2 DM (Aragno et al., 2004; Jeong, Conboy, & Conboy, 2013; Krause et al., 2013; Krause, Moradi, Nissar, Riddell, & Hawke, 2011; Nguyen, Cheng, & Koh, 2011; Nunan, Harding, & Martin, 2014; Peterson, Bryner, & Alway, 2008). Hyperglycemia has been shown to increase free radical species and reducing antioxidant levels in diabetic rats (Aragno et al., 2004; Henriksen, Diamond-Stanic, & Marchionne, 2011). This phenomenon may involve nitric oxide synthase in endothelial cells that produces more reactive oxygen and nitrogen species when exposed to increased glucose levels (Zou, Shi, & Cohen, 2002). Administration of antioxidant compounds in these animals counteracts muscle fiber proteolysis and improves markers of muscle regeneration (Aragno et al., 2004; Henriksen et al., 2011). Muscles in obese or diabetic mice and humans also present an increased infiltration of M1 macrophages leading to fibrotic deposition of collagen and abnormal levels of TGFβ, ILs, and TNFα, which may collectively impair normal MuSC function (Berria et al., 2006; Hong et al., 2009; Richardson et al., 2005; Watts, McAinch, Dixon, O'Brien, & Cameron-Smith, 2013). Lastly, the accumulation of proteins or lipids that become glycated as a result of continuous exposure to sugars exerts negative effects on both human and mouse myoblasts (Chiu et al., 2016).

Cachexia is a dramatic loss of skeletal muscle mass that can occur as a consequence of several conditions, including cancer, acquired immunodeficiency syndrome (AIDS), chronic obstructive pulmonary disease (COPD), and heart failure (Morley, Thomas, & Wilson, 2006). It is plausible that such dramatic and rapid loss of muscle mass also alters the MuSC niche and affects MuSC function. Indeed, the onset of muscle atrophy usually triggers MuSC activation, while later stages appear to be characterized by reduced MuSC numbers (Ferreira et al., 2006; Kuschel, Yablonka-Reuveni, & Bornemann, 1999; Mitchell & Pavlath, 2004; Verdijk et al., 2012). In contrast to dystrophic or diabetic skeletal muscles, muscles from cachectic patients or tumor-bearing mice show no signs of infiltrating immune cells but exhibit abnormalities in the structure of the basal lamina and the muscle fiber membrane (Acharyya et al., 2005; He et al., 2013). This altered niche, in combination with changes in circulating serum factors, leads to hyperactivation of both MuSCs and other muscle-resident cell populations such as PICs and pericytes. Interestingly, under these conditions MuSCs display sustained expression of the self-renewal factor Pax7 leading to a suppression of the differentiation program and compromised muscle repair. Other adaptations in the cachectic muscles include altered levels of (I) TNFα, which was originally called "cachectin" because of its strong association with cachexia, (II) angiotensin 2, (III) myostatin, (IV) oxidative stress, and (V) corticosteroids (Anker et al., 2003; Beutler, Mahoney, Le Trang, Pekala, & Cerami, 1985; Brzeszczynska et al., 2016; Murphy, Chee, et al., 2011; Sanders, Russell, & Tisdale, 2005; Schakman, Kalista, Barbe, Loumaye, & Thissen, 2013; Zhou et al., 2010).

Taken together, changes in structural and soluble signals in the MuSC niche contribute to the pathology of several types of muscle disease. Altered dynamics of inflammatory cells appear to be an important upstream trigger involved in most of these adaptations. Thus, further investigation of the intricate interplay of immune cells, MuSCs, and other supportive niche cells will provide us with a much-needed improved understanding of the pathological adaptations in diseased muscle, and will help to device novel therapeutic strategies to maintain, normalize, or restore the endogenous repair potential of MuSCs.

7. TARGETING THE NICHE FOR THERAPY

Several strategies that promote youthful MuSC function and that maintain the regenerative capacity of skeletal muscle into old age have been trialed in both humans and animal models. Arguably the most successful antiaging intervention, among those tested so far, is a moderate but constant level of exercise, which supports MuSCs through a spectrum of adaptations in the niche. Exercise has been shown to promote vascular function and angiogenesis, which are systemically compromised in aging, leaving the MuSC niche depleted of nutrients, oxygen, and endothelial cell-mediated support (Di Francescomarino, Sciartilli, Di Valerio, Di Baldassarre, & Gallina, 2009; Groen et al., 2014; Ryan et al., 2006). An important mechanism by which physical activity exerts positive effects on the vasculature appears to involve the release of nitric oxide, a regulator of vessel cell proliferation, vascular tone, and leukocyte–endothelial cell adhesion

(Franzoni et al., 2004; Green, Maiorana, O'Driscoll, & Taylor, 2004; Nyberg et al., 2012; Sessa, 2009). Exercise promotes the release of proangiogenic factors such as VEGF that are also known to modulate MuSC function directly (Christov et al., 2007; Gustafsson & Kraus, 2001). Chronic inflammation and oxidative stress have emerged as additional age-associated processes that can be limited through exercise. Physical activity has been shown to correlate with reduced levels of proinflammatory markers in elderly individuals and stabilizes the muscle redox system (Hollander et al., 2001; McArdle & Jackson, 2000; Reuben, Judd-Hamilton, Harris, Seeman, & MacArthur Studies of Successful Aging, 2003). The positive effects of exercise on the MuSC niche are not limited to better vascular health, reduced inflammation, and oxidative stress but appear to also involve ECM remodeling. Although the response to exercise is globally blunted in older individuals, it still leads to rapid expression of ECM proteins and certain matrix remodeling enzymes, especially various collagens, MMP2 and MMP9 (Dennis et al., 2008; Garg & Boppart, 2016; Han et al., 1999; Kjaer et al., 2006; Rullman et al., 2007). Niche elasticity critically influences MuSC function and aging is generally associated with stiffer muscles (Gilbert et al., 2010; Kovanen & Suominen, 1988; Lacraz et al., 2016). Notably, muscles of aged trained individuals are characterized by reduced collagen crosslinking and therefore display an improved elasticity that may promote MuSC function (Gosselin, Adams, Cotter, McCormick, & Thomas, 1998). Taken together, a moderate but constant training regime is associated with the reversal of a wide spectrum of age-associated changes in the MuSC niche. An important additional benefit of physical activity is that it contributes to reducing adipose mass and thus counteracts the development or progression of age-related diabetes, which can also affect MuSC function.

As outlined earlier, the excessive fibrotic deposition of certain ECM molecules that characterizes aging and degenerative muscle diseases can impair MuSC function through a broad spectrum of mechanisms. Moreover, secreted profibrotic signaling factors often also inhibit MuSCs directly. The TGFβ pathway represents an obvious target for the therapeutic reduction of profibrotic signaling in the pathologic niche. Inhibition of TGFβ using neutralizing antibodies or pharmacologic inhibition of its downstream effectors reduces fibrosis and improves MuSC function (Biressi, Miyabara, Gopinath, Carlig, & Rando, 2014; Cabrera et al., 2014; Gosselin, Williams, Personius, & Farkas, 2007; Li et al., 2016; Morales et al., 2013). The peptide hormone angiotensin 2 promotes TGFβ signaling at several levels in the TGFβ signaling pathway (Murphy, Wong, & Bezuhly, 2015). Losartan is an angiotensin 2 receptor antagonist drug approved for the treatment of high blood pressure, which has been shown to reduce skeletal muscle fibrosis in mouse models of both Duchenne and congenital muscular dystrophy (Bish et al., 2011; Cohn et al., 2007; Elbaz et al., 2012; Meinen, Lin, & Ruegg, 2012; Spurney et al., 2011). Similarly, the angiotensin-converting enzyme inhibitor Lisinopril has been shown to efficiently reduce fibrosis in dystrophic muscle (Rafael-Fortney et al., 2011).

Chronic inflammation is a major contributor to pathologic changes in the MuSC niche in aging and muscle disease, and can both promote fibrosis and inhibit myogenesis. Antiinflammatory treatment with glucocorticoids represents the current standard of care for Duchenne muscular dystrophy. Muscles of prednisone-treated Duchenne muscular dystrophy patients become stronger and contain lower numbers of mononuclear inflammatory cells (Hussein et al., 2006; Manzur et al., 2008). Unfortunately, antiinflammatory treatments can only delay loss of ambulation by a few years and side effects, such as weight gain and reduced bone mineral density, are often quite significant (Angelini, 2007). In support of the idea that reducing inflammation ameliorates pathology in muscular dystrophy by rendering the niche more permissive to MuSC function, the glucocorticoid Deflazacort increases muscle strength and stimulates myogenic differentiation in dystrophic mice (Anderson, Weber, & Vargas, 2000). Alternative antiinflammatory strategies under investigation for the treatment of muscle diseases involve biologics targeting cytokines, chemokines, or other proinflammatory signals. Examples include biologics with anti-TNFα activity, which have proven to ameliorate fibrosis, but may negatively impact cardiac function (Ermolova et al., 2014; Hodgetts et al., 2006). A more direct way to reduce chronic inflammation and to reduce peripheral side effects is to directly target-specific proinflammatory cell types. For instance, neutrophil depletion via a cytotoxic antimouse granulocyte antibody has been shown to reduce muscle necrosis in dystrophic mice (Hodgetts et al., 2006).

Deregulation of redox control is another hallmark of several muscle disorders and muscle aging. A number of natural and synthetic antioxidant compounds have been trialed for improving muscle health and regeneration. In the context of aging, the green tea component epigallocatechin gallate, vitamin C, vitamin E, beta-carotene, retinol, and melatonin have been described to be beneficial (Cesari et al., 2004; Coto-Montes, Boga, Tan, & Reiter, 2016; Meador et al., 2015). Epigallocatechin gallate, vitamin C, and melatonin have also been demonstrated to improve pathologic features in mouse models of Duchenne muscular dystrophy, while several antioxidative vitamins, zinc, and selenium supplementation have shown efficacy in patients with facioscapulohumeral muscular dystrophy (Dorchies et al., 2006; Hibaoui, Reutenauer-Patte, Patthey-Vuadens, Ruegg, & Dorchies, 2011; Nakae et al., 2012; Passerieux et al., 2015; Tonon et al., 2012). Mitochondrial dysfunction appears to be a critical downstream effect of oxidative stress in muscle pathology (Zuo & Pannell, 2015). Importantly, damaged mitochondria themselves contribute to the overproduction of reactive oxygen and nitrogen species and thereby exacerbate oxidative damage. Idebenone is a synthetic analogue of coenzyme Q10 and as such has a protective effect on mitochondria. Clinical trials on patients affected by Duchenne muscular dystrophy have shown efficacy of Idebenone based on respiratory function outcomes (Buyse et al., 2015).

In conclusion, aging and degenerative muscle diseases are highly complex processes leading to multifactorial defects in the MuSC niche. Tackling specific aspects of the pathology has proven to ameliorate MuSC dysfunction and future integrative strategies combining treatments that keep immune cells, fibrosis, and oxidative stress in check hold great promise. Moreover, the raise of exercise mimetics may open entirely new avenues for the reversal or prevention of pathologic changes in the MuSC niche (Handschin, 2016).

8. CONCLUDING REMARKS

Experimental models allowing to genetically tag, modify, or ablate cellular components of the MuSC niche have provided important insights into the enormous complexity and interconnectivity of niche elements, as well as their susceptibility to signals arising on the tissue or systemic level. With the emergence of transformative new technologies such as mass cytometry, single cell sequencing, and superresolution imaging, the field is poised to leap forward toward a comprehensive road map of the niche regulation of MuSCs in quiescence, growth, and during regenerative myogenesis. In the long run, these advancements will provide us with the toolkit and the molecular targets to develop strategies to efficiently boost stem cell function and thereby promote the endogenous regenerative capacity of healthy and diseased muscle. The clinical development of personalized therapies is very slow and faces significant challenges, in particular in the case of heterogeneous rare muscle diseases (Aartsma-Rus, Ginjaar, & Bushby, 2016; Bello & Pegoraro, 2016; Niks & Aartsma-Rus, 2017). Thus, therapeutic approaches such as the stimulation of endogenous repair, which may be applicable over a spectrum of muscle diseases and conditions, represent an important complementary path.

REFERENCES

Aartsma-Rus, A., Ginjaar, I. B., & Bushby, K. (2016). *Journal of Medical Genetics, 53*, 145–151.

Abou-Khalil, R., Le Grand, F., Pallafacchina, G., Valable, S., Authier, F. J., Rudnicki, M. A., et al. (2009). *Cell Stem Cell, 5*, 298–309.

Abou-Khalil, R., Mounier, R., & Chazaud, B. (2010). *Cell Cycle, 9*, 892–896.

Acharyya, S., Butchbach, M. E., Sahenk, Z., Wang, H., Saji, M., Carathers, M., et al. (2005). *Cancer Cell, 8*, 421–432.

Acharyya, S., Sharma, S. M., Cheng, A. S., Ladner, K. J., He, W., Kline, W., et al. (2010). *PLoS One, 5*, e12479.

Acharyya, S., Villalta, S. A., Bakkar, N., Bupha-Intr, T., Janssen, P. M., Carathers, M., et al. (2007). *The Journal of Clinical Investigation, 117*, 889–901.

Alameddine, H. S., & Morgan, J. E. (2016). *Journal of Neuromuscular Diseases, 3*, 455–473.

Alfaro, L. A., Dick, S. A., Siegel, A. L., Anonuevo, A. S., McNagny, K. M., Megeney, L. A., et al. (2011). *Stem Cells, 29*, 2030–2041.

Ali, S., & Garcia, J. M. (2014). *Gerontology, 60*, 294–305.

Allbrook, D. B., Han, M. F., & Hellmuth, A. E. (1971). *Pathology, 3*, 223–243.

Allen, R. E., & Boxhorn, L. K. (1989). *Journal of Cellular Physiology, 138*, 311–315.

Allen, R. E., Sheehan, S. M., Taylor, R. G., Kendall, T. L., & Rice, G. M. (1995). *Journal of Cellular Physiology, 165*, 307–312.

Alonso-Martin, S., Rochat, A., Mademtzoglou, D., Morais, J., de Reynies, A., Aurade, F., et al. (2016). *Frontiers in Cell and Development Biology, 4*, 58.

Alsharidah, M., Lazarus, N. R., George, T. E., Agley, C. C., Velloso, C. P., & Harridge, S. D. (2013). *Aging Cell, 12*, 333–344.

Alvarez, K., Fadic, R., & Brandan, E. (2002). *Journal of Cellular Biochemistry, 85*, 703–713.

Anderson, J. E., Mitchell, C. M., McGeachie, J. K., & Grounds, M. D. (1995). *Experimental Cell Research, 216*, 325–334.

Anderson, J. E., Weber, M., & Vargas, C. (2000). *Cell Transplantation, 9*, 551–564.

Angelini, C. (2007). *Muscle & Nerve, 36*, 424–435.

Anker, S. D., Negassa, A., Coats, A. J., Afzal, R., Poole-Wilson, P. A., Cohn, J. N., et al. (2003). *Lancet, 361*, 1077–1083.

Aragno, M., Mastrocola, R., Catalano, M. G., Brignardello, E., Danni, O., & Boccuzzi, G. (2004). *Diabetes, 53*, 1082–1088.

Ardite, E., Perdiguero, E., Vidal, B., Gutarra, S., Serrano, A. L., & Munoz-Canoves, P. (2012). *The Journal of Cell Biology, 196*, 163–175.

Arecco, N., Clarke, C. J., Jones, F. K., Simpson, D. M., Mason, D., Beynon, R. J., et al. (2016). *Scientific Reports, 6*, 24708.

Arnold, L., Henry, A., Poron, F., Baba-Amer, Y., van Rooijen, N., Plonquet, A., et al. (2007). *The Journal of Experimental Medicine, 204*, 1057–1069.

Beauchamp, J. R., Heslop, L., Yu, D. S., Tajbakhsh, S., Kelly, R. G., Wernig, A., et al. (2000). *The Journal of Cell Biology, 151*, 1221–1234.

Bello, L., & Pegoraro, E. (2016). *Acta Myologica, 35*, 122–127.

Bentzinger, C. F., Wang, Y. X., & Rudnicki, M. A. (2012). *Cold Spring Harbor Perspectives in Biology, 4*, 1–16.

Bentzinger, C. F., Wang, Y. X., von Maltzahn, J., Soleimani, V. D., Yin, H., & Rudnicki, M. A. (2013). *Cell Stem Cell, 12*, 75–87.

Bernet, J. D., Doles, J. D., Hall, J. K., Kelly Tanaka, K., Carter, T. A., & Olwin, B. B. (2014). *Nature Medicine, 20*, 265–271.

Berria, R., Wang, L., Richardson, D. K., Finlayson, J., Belfort, R., Pratipanawatr, T., et al. (2006). *American Journal of Physiology. Endocrinology and Metabolism, 290*, E560–5.

Beutler, B., Mahoney, J., Le Trang, N., Pekala, P., & Cerami, A. (1985). *The Journal of Experimental Medicine, 161*, 984–995.

Birbrair, A., Zhang, T., Wang, Z. M., Messi, M. L., Enikolopov, G. N., Mintz, A., et al. (2013). *Stem Cells and Development*, *22*, 2298–2314.

Birbrair, A., Zhang, T., Wang, Z. M., Messi, M. L., Mintz, A., & Delbono, O. (2014). *Frontiers in Aging Neuroscience*, *6*, 245.

Biressi, S., Miyabara, E. H., Gopinath, S. D., Carlig, P. M., & Rando, T. A. (2014). *Science Translational Medicine*, *6*, 267ra176.

Bischoff, R. (1990). *Development*, *109*, 943–952.

Bish, L. T., Yarchoan, M., Sleeper, M. M., Gazzara, J. A., Morine, K. J., Acosta, P., et al. (2011). *PLoS One*, *6*, e20856.

Bjornson, C. R., Cheung, T. H., Liu, L., Tripathi, P. V., Steeper, K. M., & Rando, T. A. (2012). *Stem Cells*, *30*, 232–242.

Blanco-Bose, W. E., Yao, C. C., Kramer, R. H., & Blau, H. M. (2001). *Experimental Cell Research*, *265*, 212–220.

Bobadilla, M., Sainz, N., Rodriguez, J. A., Abizanda, G., Orbe, J., de Martino, A., et al. (2014). *Stem Cells*, *32*, 447–461.

Bornemann, A., & Schmalbruch, H. (1994). *The Anatomical Record*, *239*, 119–125.

Brack, A. S., Bildsoe, H., & Hughes, S. M. (2005). *Journal of Cell Science*, *118*, 4813–4821.

Brack, A. S., Conboy, M. J., Roy, S., Lee, M., Kuo, C. J., Keller, C., et al. (2007). *Science*, *317*, 807–810.

Brandan, E., & Gutierrez, J. (2013). *Matrix Biology*, *32*, 289–297.

Brigitte, M., Schilte, C., Plonquet, A., Baba-Amer, Y., Henri, A., Charlier, C., et al. (2010). *Arthritis and Rheumatism*, *62*, 268–279.

Brohl, D., Vasyutina, E., Czajkowski, M. T., Griger, J., Rassek, C., Rahn, H. P., et al. (2012). *Developmental Cell*, *23*, 469–481.

Brune, B., Dehne, N., Grossmann, N., Jung, M., Namgaladze, D., Schmid, T., et al. (2013). *Antioxidants & Redox Signaling*, *19*, 595–637.

Brzeszczynska, J., Johns, N., Schilb, A., Degen, S., Degen, M., Langen, R., et al. (2016). *Aging (Albany NY)*, *8*, 1690–1702.

Burkin, D. J., & Kaufman, S. J. (1999). *Cell and Tissue Research*, *296*, 183–190.

Burzyn, D., Kuswanto, W., Kolodin, D., Shadrach, J. L., Cerletti, M., Jang, Y., et al. (2013). *Cell*, *155*, 1282–1295.

Buyse, G. M., Voit, T., Schara, U., Straathof, C. S., D'Angelo, M. G., Bernert, G., et al. (2015). *Lancet*, *385*, 1748–1757.

Cabello-Verrugio, C., & Brandan, E. (2007). *The Journal of Biological Chemistry*, *282*, 18842–18850.

Cabrera, D., Gutierrez, J., Cabello-Verrugio, C., Morales, M. G., Mezzano, S., Fadic, R., et al. (2014). *Skeletal Muscle*, *4*, 6.

Caceres, S., Cuellar, C., Casar, J. C., Garrido, J., Schaefer, L., Kresse, H., et al. (2000). *European Journal of Cell Biology*, *79*, 173–181.

Caldwell, C. J., Mattey, D. L., & Weller, R. O. (1990). *Neuropathology and Applied Neurobiology*, *16*, 225–238.

Carey, D. J. (1997). *The Biochemical Journal*, *327*(Pt. 1), 1–16.

Carlson, M. E., Conboy, M. J., Hsu, M., Barchas, L., Jeong, J., Agrawal, A., et al. (2009). *Aging Cell*, *8*, 676–689.

Carlson, B. M., & Faulkner, J. A. (1989). *The American Journal of Physiology*, *256*, C1262–6.

Carlson, M. E., Hsu, M., & Conboy, I. M. (2008). *Nature*, *454*, 528–532.

Cesari, M., Pahor, M., Bartali, B., Cherubini, A., Penninx, B. W., Williams, G. R., et al. (2004). *The American Journal of Clinical Nutrition*, *79*, 289–294.

Chakkalakal, J. V., Jones, K. M., Basson, M. A., & Brack, A. S. (2012). *Nature*, *490*, 355–360.

Chakravarthy, M. V., Abraha, T. W., Schwartz, R. J., Fiorotto, M. L., & Booth, F. W. (2000). *The Journal of Biological Chemistry*, *275*, 35942–35952.

Chazaud, B., Sonnet, C., Lafuste, P., Bassez, G., Rimaniol, A. C., Poron, F., et al. (2003). *The Journal of Cell Biology*, *163*, 1133–1143.
Cheng, M., Nguyen, M. H., Fantuzzi, G., & Koh, T. J. (2008). *American Journal of Physiology. Cell Physiology*, *294*, C1183–91.
Cheung, T. H., & Rando, T. A. (2013). *Nature Reviews. Molecular Cell Biology*, *14*, 329–340.
Chevessier, F., Hantai, D., & Verdiere-Sahuque, M. (2001). *Journal of Cellular Physiology*, *189*, 152–161.
Chinni, C., de Niese, M. R., Jenkins, A. L., Pike, R. N., Bottomley, S. P., & Mackie, E. J. (2000). *Journal of Cell Science*, *113*(Pt. 24), 4427–4433.
Chiu, C. Y., Yang, R. S., Sheu, M. L., Chan, D. C., Yang, T. H., Tsai, K. S., et al. (2016). *The Journal of Pathology*, *238*, 470–482.
Christov, C., Chretien, F., Abou-Khalil, R., Bassez, G., Vallet, G., Authier, F. J., et al. (2007). *Molecular Biology of the Cell*, *18*, 1397–1409.
Clancy, D. M., Henry, C. M., Sullivan, G. P., & Martin, S. J. (2017). *The FEBS Journal*, *284*, 1712–1725.
Cohn, R. D., Henry, M. D., Michele, D. E., Barresi, R., Saito, F., Moore, S. A., et al. (2002). *Cell*, *110*, 639–648.
Cohn, R. D., van Erp, C., Habashi, J. P., Soleimani, A. A., Klein, E. C., Lisi, M. T., et al. (2007). *Nature Medicine*, *13*, 204–210.
Coletti, D., Moresi, V., Adamo, S., Molinaro, M., & Sassoon, D. (2005). *Genesis*, *43*, 120–128.
Coletti, D., Yang, E., Marazzi, G., & Sassoon, D. (2002). *The EMBO Journal*, *21*, 631–642.
Collins, C. A., Olsen, I., Zammit, P. S., Heslop, L., Petrie, A., Partridge, T. A., et al. (2005). *Cell*, *122*, 289–301.
Collins, C. A., Zammit, P. S., Ruiz, A. P., Morgan, J. E., & Partridge, T. A. (2007). *Stem Cells*, *25*, 885–894.
Conboy, I. M., Conboy, M. J., Smythe, G. M., & Rando, T. A. (2003). *Science*, *302*, 1575–1577.
Conboy, I. M., Conboy, M. J., Wagers, A. J., Girma, E. R., Weissman, I. L., & Rando, T. A. (2005). *Nature*, *433*, 760–764.
Cooper, S. T., & McNeil, P. L. (2015). *Physiological Reviews*, *95*, 1205–1240.
Corbu, A., Scaramozza, A., Badiali-DeGiorgi, L., Tarantino, L., Papa, V., Rinaldi, R., et al. (2010). *Neurological Research*, *32*, 63–72.
Cornelison, D. D., Filla, M. S., Stanley, H. M., Rapraeger, A. C., & Olwin, B. B. (2001). *Developmental Biology*, *239*, 79–94.
Cornelison, D. D., Wilcox-Adelman, S. A., Goetinck, P. F., Rauvala, H., Rapraeger, A. C., & Olwin, B. B. (2004). *Genes & Development*, *18*, 2231–2236.
Cosgrove, B. D., Gilbert, P. M., Porpiglia, E., Mourkioti, F., Lee, S. P., Corbel, S. Y., et al. (2014). *Nature Medicine*, *20*, 255–264.
Coto-Montes, A., Boga, J. A., Tan, D. X., & Reiter, R. J. (2016). *International Journal of Molecular Sciences*, *17*, 1–15.
Dadgar, S., Wang, Z., Johnston, H., Kesari, A., Nagaraju, K., Chen, Y. W., et al. (2014). *The Journal of Cell Biology*, *207*, 139–158.
Day, K., Shefer, G., Shearer, A., & Yablonka-Reuveni, Z. (2010). *Developmental Biology*, *340*, 330–343.
Degens, H. (2010). *Scandinavian Journal of Medicine & Science in Sports*, *20*, 28–38.
Delfini, M. C., Hirsinger, E., Pourquie, O., & Duprez, D. (2000). *Development*, *127*, 5213–5224.
Dellavalle, A., Maroli, G., Covarello, D., Azzoni, E., Innocenzi, A., Perani, L., et al. (2011). *Nature Communications*, *2*, 499.
Dellavalle, A., Sampaolesi, M., Tonlorenzi, R., Tagliafico, E., Sacchetti, B., Perani, L., et al. (2007). *Nature Cell Biology*, *9*, 255–267.

Demontis, F., Piccirillo, R., Goldberg, A. L., & Perrimon, N. (2013). *Disease Models & Mechanisms, 6*, 1339–1352.
Deng, B., Wehling-Henricks, M., Villalta, S. A., Wang, Y., & Tidball, J. G. (2012). *Journal of Immunology, 189*, 3669–3680.
Dennis, R. A., Przybyla, B., Gurley, C., Kortebein, P. M., Simpson, P., Sullivan, D. H., et al. (2008). *Physiological Genomics, 32*, 393–400.
Di Francescomarino, S., Sciartilli, A., Di Valerio, V., Di Baldassarre, A., & Gallina, S. (2009). *Sports Medicine, 39*, 797–812.
DiMario, J., Buffinger, N., Yamada, S., & Strohman, R. C. (1989). *Science, 244*, 688–690.
Dorchies, O. M., Wagner, S., Vuadens, O., Waldhauser, K., Buetler, T. M., Kucera, P., et al. (2006). *American Journal of Physiology. Cell Physiology, 290*, C616–25.
D'Souza, D. M., Al-Sajee, D., & Hawke, T. J. (2013). *Frontiers in Physiology, 4*, 379.
Duan, C., Ren, H., & Gao, S. (2010). *General and Comparative Endocrinology, 167*, 344–351.
Dumont, N. A., Wang, Y. X., von Maltzahn, J., Pasut, A., Bentzinger, C. F., Brun, C. E., et al. (2015). *Nature Medicine, 21*, 1455–1463.
Egerman, M. A., Cadena, S. M., Gilbert, J. A., Meyer, A., Nelson, H. N., Swalley, S. E., et al. (2015). *Cell Metabolism, 22*, 164–174.
Elbaz, M., Yanay, N., Aga-Mizrachi, S., Brunschwig, Z., Kassis, I., Ettinger, K., et al. (2012). *Annals of Neurology, 71*, 699–708.
Ermolova, N. V., Martinez, L., Vetrone, S. A., Jordan, M. C., Roos, K. P., Sweeney, H. L., et al. (2014). *Neuromuscular Disorders, 24*, 583–595.
Ferreira, R., Neuparth, M. J., Ascensao, A., Magalhaes, J., Vitorino, R., Duarte, J. A., et al. (2006). *European Journal of Applied Physiology, 97*, 340–346.
Filippin, L. I., Cuevas, M. J., Lima, E., Marroni, N. P., Gonzalez-Gallego, J., & Xavier, R. M. (2011). *Nitric Oxide, 24*, 43–49.
Fiore, D., Judson, R. N., Low, M., Lee, S., Zhang, E., Hopkins, C., et al. (2016). *Stem Cell Research, 17*, 161–169.
Flanagan-Steet, H., Hannon, K., McAvoy, M. J., Hullinger, R., & Olwin, B. B. (2000). *Developmental Biology, 218*, 21–37.
Florini, J. R., Magri, K. A., Ewton, D. Z., James, P. L., Grindstaff, K., & Rotwein, P. S. (1991). *The Journal of Biological Chemistry, 266*, 15917–15923.
Formicola, L., Marazzi, G., & Sassoon, D. A. (2014). *Frontiers in Aging Neuroscience, 6*, 328.
Fox, J. W., Mayer, U., Nischt, R., Aumailley, M., Reinhardt, D., Wiedemann, H., et al. (1991). *The EMBO Journal, 10*, 3137–3146.
Franceschi, C., & Campisi, J. (2014). *The Journals of Gerontology. Series A, Biological Sciences and Medical Sciences, 69*(Suppl. 1), S4–9.
Franzoni, F., Galetta, F., Morizzo, C., Lubrano, V., Palombo, C., Santoro, G., et al. (2004). *Clinical Science (London, England), 106*, 329–335.
Fry, C. S., Kirby, T. J., Kosmac, K., McCarthy, J. J., & Peterson, C. A. (2017). *Cell Stem Cell, 20*, 56–69.
Fry, C. S., Lee, J. D., Mula, J., Kirby, T. J., Jackson, J. R., Liu, F., et al. (2015). *Nature Medicine, 21*, 76–80.
Fukada, S., Uezumi, A., Ikemoto, M., Masuda, S., Segawa, M., Tanimura, N., et al. (2007). *Stem Cells, 25*, 2448–2459.
Fukushima, K., Nakamura, A., Ueda, H., Yuasa, K., Yoshida, K., Takeda, S., et al. (2007). *BMC Musculoskeletal Disorders, 8*, 54.
Garcion, E., Halilagic, A., Faissner, A., & ffrench-Constant, C. (2004). *Development, 131*, 3423–3432.
Garg, K., & Boppart, M. D. (2016). *Journal of Applied Physiology, 121*, 1053–1058.
Geffers, I., Serth, K., Chapman, G., Jaekel, R., Schuster-Gossler, K., Cordes, R., et al. (2007). *The Journal of Cell Biology, 178*, 465–476.

Ghadiali, R. S., Guimond, S. E., Turnbull, J. E., & Pisconti, A. (2017). *Matrix Biology, 59*, 54–68.
Gilbert, P. M., Havenstrite, K. L., Magnusson, K. E., Sacco, A., Leonardi, N. A., Kraft, P., et al. (2010). *Science, 329*, 1078–1081.
Gnocchi, V. F., White, R. B., Ono, Y., Ellis, J. A., & Zammit, P. S. (2009). *PLoS One, 4*, e5205.
Goetsch, S. C., Hawke, T. J., Gallardo, T. D., Richardson, J. A., & Garry, D. J. (2003). *Physiological Genomics, 14*, 261–271.
Goetsch, K. P., & Niesler, C. U. (2016). *Biochemical and Biophysical Research Communications, 479*, 351–357.
Gohring, W., Sasaki, T., Heldin, C. H., & Timpl, R. (1998). *European Journal of Biochemistry, 255*, 60–66.
Gokhin, D. S., Ward, S. R., Bremner, S. N., & Lieber, R. L. (2008). *The Journal of Experimental Biology, 211*, 837–843.
Goodell, M. A., & Rando, T. A. (2015). *Science, 350*, 1199–1204.
Gosselin, L. E., Adams, C., Cotter, T. A., McCormick, R. J., & Thomas, D. P. (1998). *Journal of Applied Physiology, 85*, 1011–1016.
Gosselin, L. E., Williams, J. E., Personius, K., & Farkas, G. A. (2007). *Muscle & Nerve, 35*, 208–216.
Green, D. J., Maiorana, A., O'Driscoll, G., & Taylor, R. (2004). *The Journal of Physiology, 561*, 1–25.
Groen, B. B., Hamer, H. M., Snijders, T., van Kranenburg, J., Frijns, D., Vink, H., et al. (2014). *Journal of Applied Physiology, 116*, 998–1005.
Gustafsson, T., & Kraus, W. E. (2001). *Frontiers in Bioscience, 6*, D75–89.
Gutierrez, J., & Brandan, E. (2010). *Molecular and Cellular Biology, 30*, 1634–1649.
Gutierrez, J., Osses, N., & Brandan, E. (2006). *Journal of Cellular Physiology, 206*, 58–67.
Guttridge, D. C., Mayo, M. W., Madrid, L. V., Wang, C. Y., & Baldwin, A. S., Jr. (2000). *Science, 289*, 2363–2366.
Hall, J. K., Banks, G. B., Chamberlain, J. S., & Olwin, B. B. (2010). *Science Translational Medicine, 2*, 57ra83.
Han, X. Y., Wang, W., Komulainen, J., Koskinen, S. O., Kovanen, V., Vihko, V., et al. (1999). *Pflügers Archiv, 437*, 857–864.
Handschin, C. (2016). *Pharmacological Research, 103*, 158–166.
Hannon, K., Kudla, A. J., McAvoy, M. J., Clase, K. L., & Olwin, B. B. (1996). *The Journal of Cell Biology, 132*, 1151–1159.
Hardy, D., Besnard, A., Latil, M., Jouvion, G., Briand, D., Thepenier, C., et al. (2016). *PLoS One, 11*, e0147198.
He, W. A., Berardi, E., Cardillo, V. M., Acharyya, S., Aulino, P., Thomas-Ahner, J., et al. (2013). *The Journal of Clinical Investigation, 123*, 4821–4835.
Hearon, C. M., Jr., & Dinenno, F. A. (2016). *The Journal of Physiology, 594*, 2261–2273.
Henriksen, E. J., Diamond-Stanic, M. K., & Marchionne, E. M. (2011). *Free Radical Biology & Medicine, 51*, 993–999.
Heredia, J. E., Mukundan, L., Chen, F. M., Mueller, A. A., Deo, R. C., Locksley, R. M., et al. (2013). *Cell, 153*, 376–388.
Heydemann, A., Ceco, E., Lim, J. E., Hadhazy, M., Ryder, P., Moran, J. L., et al. (2009). *The Journal of Clinical Investigation, 119*, 3703–3712.
Hibaoui, Y., Reutenauer-Patte, J., Patthey-Vuadens, O., Ruegg, U. T., & Dorchies, O. M. (2011). *Journal of Pineal Research, 51*, 163–171.
Hindi, S. M., & Kumar, A. (2016). *The Journal of Pathology, 239*, 125–128.
Hirsinger, E., Malapert, P., Dubrulle, J., Delfini, M. C., Duprez, D., Henrique, D., et al. (2001). *Development, 128*, 107–116.

Hodgetts, S., Radley, H., Davies, M., & Grounds, M. D. (2006). *Neuromuscular Disorders, 16*, 591–602.

Holland, A., Dowling, P., Meleady, P., Henry, M., Zweyer, M., Mundegar, R. R., et al. (2015). *Proteomics, 15*, 2318–2331.

Holland, A., Murphy, S., Dowling, P., & Ohlendieck, K. (2016). *Proteomics, 16*, 345–366.

Hollander, J., Fiebig, R., Gore, M., Ookawara, T., Ohno, H., & Ji, L. L. (2001). *Pflügers Archiv, 442*, 426–434.

Hollnagel, A., Grund, C., Franke, W. W., & Arnold, H. H. (2002). *Molecular and Cellular Biology, 22*, 4760–4770.

Holmberg, J., & Durbeej, M. (2013). *Cell Adhesion & Migration, 7*, 111–121.

Hong, E. G., Ko, H. J., Cho, Y. R., Kim, H. J., Ma, Z., Yu, T. Y., et al. (2009). *Diabetes, 58*, 2525–2535.

Hoppeler, H., & Fluck, M. (2002). *The Journal of Experimental Biology, 205*, 2143–2152.

Hussein, M. R., Hamed, S. A., Mostafa, M. G., Abu-Dief, E. E., Kamel, N. F., & Kandil, M. R. (2006). *International Journal of Experimental Pathology, 87*, 451–461.

Ionasescu, V., & Ionasescu, R. (1982). *Journal of the Neurological Sciences, 54*, 79–87.

Irintchev, A., Zeschnigk, M., Starzinski-Powitz, A., & Wernig, A. (1994). *Developmental Dynamics, 199*, 326–337.

Janssen, I., Heymsfield, S. B., Wang, Z. M., & Ross, R. (2000). *Journal of Applied Physiology, 89*, 81–88.

Jennische, E., Ekberg, S., & Matejka, G. L. (1993). *The American Journal of Physiology, 265*, C122–8.

Jeong, J., Conboy, M. J., & Conboy, I. M. (2013). *Acta Pharmacologica Sinica, 34*, 1052–1060.

Joe, A. W., Yi, L., Natarajan, A., Le Grand, F., So, L., Wang, J., et al. (2010). *Nature Cell Biology, 12*, 153–163.

Katz, B. (1961). *Philosophical Transactions of the Royal Society of London. Series B, Biological Sciences, 243*, 221–240.

Keefe, A. C., Lawson, J. A., Flygare, S. D., Fox, Z. D., Colasanto, M. P., Mathew, S. J., et al. (2015). *Nature Communications, 6*, 7087.

Kherif, S., Lafuma, C., Dehaupas, M., Lachkar, S., Fournier, J. G., Verdiere-Sahuque, M., et al. (1999). *Developmental Biology, 205*, 158–170.

Kim, J. H., Han, G. C., Seo, J. Y., Park, I., Park, W., Jeong, H. W., et al. (2016). *Nature Cell Biology, 18*, 930–940.

Kjaer, M., Magnusson, P., Krogsgaard, M., Boysen Moller, J., Olesen, J., Heinemeier, K., et al. (2006). *Journal of Anatomy, 208*, 445–450.

Kohfeldt, E., Sasaki, T., Gohring, W., & Timpl, R. (1998). *Journal of Molecular Biology, 282*, 99–109.

Koskinen, S. O., Ahtikoski, A. M., Komulainen, J., Hesselink, M. K., Drost, M. R., & Takala, T. E. (2002). *Pflügers Archiv, 444*, 59–72.

Kostallari, E., Baba-Amer, Y., Alonso-Martin, S., Ngoh, P., Relaix, F., Lafuste, P., et al. (2015). *Development, 142*, 1242–1253.

Kovanen, V., & Suominen, H. (1988). *Comprehensive Gerontology Section A, 2*, 18–23.

Krause, M. P., Al-Sajee, D., D'Souza, D. M., Rebalka, I. A., Moradi, J., Riddell, M. C., et al. (2013). *PLoS One, 8*, e70971.

Krause, M. P., Moradi, J., Nissar, A. A., Riddell, M. C., & Hawke, T. J. (2011). *Diabetes, 60*, 1964–1972.

Krauss, R. S. (2010). *Experimental Cell Research, 316*, 3042–3049.

Kuang, S., Gillespie, M. A., & Rudnicki, M. A. (2008). *Cell Stem Cell, 2*, 22–31.

Kumar, A., & Boriek, A. M. (2003). *The FASEB Journal, 17*, 386–396.

Kuru, S., Inukai, A., Kato, T., Liang, Y., Kimura, S., & Sobue, G. (2003). *Acta Neuropathologica, 105*, 217–224.

Kuschel, R., Yablonka-Reuveni, Z., & Bornemann, A. (1999). *The Journal of Histochemistry and Cytochemistry, 47*, 1375–1384.

Kuswanto, W., Burzyn, D., Panduro, M., Wang, K. K., Jang, Y. C., Wagers, A. J., et al. (2016). *Immunity, 44*, 355–367.

Lacraz, G., Rouleau, A. J., Couture, V., Sollradl, T., Drouin, G., Veillette, N., et al. (2016). *PLoS One, 11*, e0167661.

Ladi, E., Nichols, J. T., Ge, W., Miyamoto, A., Yao, C., Yang, L. T., et al. (2005). *The Journal of Cell Biology, 170*, 983–992.

Le Grand, F., Jones, A. E., Seale, V., Scime, A., & Rudnicki, M. A. (2009). *Cell Stem Cell, 4*, 535–547.

Le Moal, E., Pialoux, V., Juban, G., Groussard, C., Zouhal, H., Chazaud, B., et al. (2017). *Antioxidants & Redox, 27*, 276–310.

Lemos, D. R., Babaeijandaghi, F., Low, M., Chang, C. K., Lee, S. T., Fiore, D., et al. (2015). *Nature Medicine, 21*, 786–794.

Li, J., Han, S., Cousin, W., & Conboy, I. M. (2015). *Stem Cells, 33*, 951–961.

Li, H., Hicks, J. J., Wang, L., Oyster, N., Philippon, M. J., Hurwitz, S., et al. (2016). *Biomaterials, 87*, 147–156.

Lightfoot, A. P., McCormick, R., Nye, G. A., & McArdle, A. (2014). *Current Opinion in Pharmacology, 16*, 116–121.

Liu, W., Wei-LaPierre, L., Klose, A., Dirksen, R. T., & Chakkalakal, J. V. (2015). *eLife, 4*, 1–15.

Lluis, F., Roma, J., Suelves, M., Parra, M., Aniorte, G., Gallardo, E., et al. (2001). *Blood, 97*, 1703–1711.

Londhe, P., & Davie, J. K. (2013). *Science Signaling, 6*, ra107.

Lorts, A., Schwanekamp, J. A., Baudino, T. A., McNally, E. M., & Molkentin, J. D. (2012). *Proceedings of the National Academy of Sciences of the United States of America, 109*, 10978–10983.

Lukjanenko, L., Brachat, S., Pierrel, E., Lach-Trifilieff, E., & Feige, J. N. (2013). *PLoS One, 8*, e71084.

Lukjanenko, L., Jung, M. J., Hegde, N., Perruisseau-Carrier, C., Migliavacca, E., Rozo, M., et al. (2016). *Nature Medicine, 22*, 897–905.

Luque, E., Pena, J., Martin, P., Jimena, I., & Vaamonde, R. (1995). *Anatomia, Histologia, Embryologia, 24*, 87–89.

Machida, S., Spangenburg, E. E., & Booth, F. W. (2003). *Journal of Cellular Physiology, 196*, 523–531.

Mann, C. J., Perdiguero, E., Kharraz, Y., Aguilar, S., Pessina, P., Serrano, A. L., et al. (2011). *Skeletal Muscle, 1*, 21.

Manzur, A. Y., Kuntzer, T., Pike, M., & Swan, A. (2008). *Cochrane Database of Systematic Reviews*, CD003725.

Mathew, S. J., Hansen, J. M., Merrell, A. J., Murphy, M. M., Lawson, J. A., Hutcheson, D. A., et al. (2011). *Development, 138*, 371–384.

Mauro, A. (1961). *The Journal of Biophysical and Biochemical Cytology, 9*, 493–495.

McArdle, A., Foxley, A., Edwards, R. H., & Jackson, M. J. (1991). *Biochemical Society Transactions, 19*, 177S.

McArdle, A., & Jackson, M. J. (2000). *Journal of Anatomy, 197*(Pt. 4), 539–541.

McKay, B. R., Ogborn, D. I., Baker, J. M., Toth, K. G., Tarnopolsky, M. A., & Parise, G. (2013). *American Journal of Physiology. Cell Physiology, 304*, C717–28.

Meador, B. M., Mirza, K. A., Tian, M., Skelding, M. B., Reaves, L. A., Edens, N. K., et al. (2015). *The Journal of Frailty & Aging, 4*, 209–215.

Meinen, S., Lin, S., & Ruegg, M. A. (2012). *Skeletal Muscle, 2*, 18.

Messina, S., Vita, G. L., Aguennouz, M., Sframeli, M., Romeo, S., Rodolico, C., et al. (2011). *Acta Myologica, 30*, 16–23.

Meyer-Hoffert, U., & Wiedow, O. (2011). *Current Opinion in Hematology, 18*, 19–24.
Miller, S. C., Ito, H., Blau, H. M., & Torti, F. M. (1988). *Molecular and Cellular Biology, 8*, 2295–2301.
Mitchell, K. J., Pannerec, A., Cadot, B., Parlakian, A., Besson, V., Gomes, E. R., et al. (2010). *Nature Cell Biology, 12*, 257–266.
Mitchell, P. O., & Pavlath, G. K. (2004). *American Journal of Physiology. Cell Physiology, 287*, C1753–62.
Miura, T., Kishioka, Y., Wakamatsu, J., Hattori, A., Hennebry, A., Berry, C. J., et al. (2006). *Biochemical and Biophysical Research Communications, 340*, 675–680.
Montero-Fernandez, N., & Serra-Rexach, J. A. (2013). *European Journal of Physical and Rehabilitation Medicine, 49*, 131–143.
Morales, M. G., Gutierrez, J., Cabello-Verrugio, C., Cabrera, D., Lipson, K. E., Goldschmeding, R., et al. (2013). *Human Molecular Genetics, 22*, 4938–4951.
Morley, J. E., Thomas, D. R., & Wilson, M. M. (2006). *The American Journal of Clinical Nutrition, 83*, 735–743.
Mourikis, P., Gopalakrishnan, S., Sambasivan, R., & Tajbakhsh, S. (2012). *Development, 139*, 4536–4548.
Mueller, A. A., van Velthoven, C. T., Fukumoto, K. D., Cheung, T. H., & Rando, T. A. (2016). *Nature, 540*, 276–279.
Muller, C., & Kapr, J. (1975). *Ceskoslovenské Zdravotnictví, 23*, 68–71.
Murphy, K. T., Chee, A., Gleeson, B. G., Naim, T., Swiderski, K., Koopman, R., et al. (2011). *American Journal of Physiology. Regulatory, Integrative and Comparative Physiology, 301*, R716–26.
Murphy, M. M., Lawson, J. A., Mathew, S. J., Hutcheson, D. A., & Kardon, G. (2011). *Development, 138*, 3625–3637.
Murphy, A. M., Wong, A. L., & Bezuhly, M. (2015). *Fibrogenesis & Tissue Repair, 8*, 7.
Musaro, A., McCullagh, K., Paul, A., Houghton, L., Dobrowolny, G., Molinaro, M., et al. (2001). *Nature Genetics, 27*, 195–200.
Myllyla, R., Myllyla, V. V., Tolonen, U., & Kivirikko, K. I. (1982). *Archives of Neurology, 39*, 752–755.
Nakae, Y., Dorchies, O. M., Stoward, P. J., Zimmermann, B. F., Ritter, C., & Ruegg, U. T. (2012). *Histochemistry and Cell Biology, 137*, 811–827.
Nakagawa, T., Takeuchi, A., Kakiuchi, R., Lee, T., Yagi, M., Awano, H., et al. (2013). *Clinica Chimica Acta, 423*, 10–14.
Nakamura, H., Fujii, Y., Ohuchi, E., Yamamoto, E., & Okada, Y. (1998). *European Journal of Biochemistry, 253*, 67–75.
Nakamura-Ishizu, A., Okuno, Y., Omatsu, Y., Okabe, K., Morimoto, J., Uede, T., et al. (2012). *Blood, 119*, 5429–5437.
Nguyen, M. H., Cheng, M., & Koh, T. J. (2011). *ScientificWorldJournal, 11*, 1525–1535.
Nicholson, R., Murphy, G., & Breathnach, R. (1989). *Biochemistry, 28*, 5195–5203.
Niethammer, P., Grabher, C., Look, A. T., & Mitchison, T. J. (2009). *Nature, 459*, 996–999.
Niks, E. H., & Aartsma-Rus, A. (2017). *Expert Opinion on Biological Therapy, 17*, 225–236.
Nunan, R., Harding, K. G., & Martin, P. (2014). *Disease Models & Mechanisms, 7*, 1205–1213.
Nunes, A. M., Wuebbles, R. D., Sarathy, A., Fontelonga, T. M., Deries, M., Burkin, D. J., et al. (2017). *Human Molecular Genetics, 26*, 2018–2033.
Nyberg, M., Blackwell, J. R., Damsgaard, R., Jones, A. M., Hellsten, Y., & Mortensen, S. P. (2012). *The Journal of Physiology, 590*, 5361–5370.
Okinaga, T., Mohri, I., Fujimura, H., Imai, K., Ono, J., Urade, Y., et al. (2002). *Acta Neuropathologica, 104*, 377–384.
Painemal, P., Acuna, M. J., Riquelme, C., Brandan, E., & Cabello-Verrugio, C. (2013). *BioFactors, 39*, 467–475.

Palacios, D., Mozzetta, C., Consalvi, S., Caretti, G., Saccone, V., Proserpio, V., et al. (2010). *Cell Stem Cell*, *7*, 455–469.

Paliwal, P., Pishesha, N., Wijaya, D., & Conboy, I. M. (2012). *Aging (Albany NY)*, *4*, 553–566.

Pallafacchina, G., Francois, S., Regnault, B., Czarny, B., Dive, V., Cumano, A., et al. (2010). *Stem Cell Research*, *4*, 77–91.

Pannerec, A., Formicola, L., Besson, V., Marazzi, G., & Sassoon, D. A. (2013). *Development*, *140*, 2879–2891.

Pannerec, A., Marazzi, G., & Sassoon, D. (2012). *Trends in Molecular Medicine*, *18*, 599–606.

Passerieux, E., Hayot, M., Jaussent, A., Carnac, G., Gouzi, F., Pillard, F., et al. (2015). *Free Radical Biology & Medicine*, *81*, 158–169.

Pawlikowski, B., Pulliam, C., Betta, N. D., Kardon, G., & Olwin, B. B. (2015). *Skeletal Muscle*, *5*, 42.

Peltonen, L., Myllyla, R., Tolonen, U., & Myllyla, V. V. (1982). *Archives of Neurology*, *39*, 756–759.

Perdiguero, E., Sousa-Victor, P., Ruiz-Bonilla, V., Jardi, M., Caelles, C., Serrano, A. L., et al. (2011). *The Journal of Cell Biology*, *195*, 307–322.

Peterson, J. M., Bryner, R. W., & Alway, S. E. (2008). *American Journal of Physiology. Cell Physiology*, *295*, C521–8.

Petis, S., Howard, J. L., Lanting, B. L., & Vasarhelyi, E. M. (2015). *Canadian Journal of Surgery*, *58*, 128–139.

Pisconti, A., Banks, G. B., Babaeijandaghi, F., Betta, N. D., Rossi, F. M., Chamberlain, J. S., et al. (2016). *Skeletal Muscle*, *6*, 34.

Pisconti, A., Cornelison, D. D., Olguin, H. C., Antwine, T. L., & Olwin, B. B. (2010). *The Journal of Cell Biology*, *190*, 427–441.

Ponsot, E., Lexell, J., & Kadi, F. (2008). *Muscle & Nerve*, *37*, 467–472.

Price, F. D., von Maltzahn, J., Bentzinger, C. F., Dumont, N. A., Yin, H., Chang, N. C., et al. (2014). *Nature Medicine*, *20*, 1174–1181.

Quarta, M., Brett, J. O., DiMarco, R., De Morree, A., Boutet, S. C., Chacon, R., et al. (2016). *Nature Biotechnology*, *34*, 752–759.

Radley, H. G., Davies, M. J., & Grounds, M. D. (2008). *Neuromuscular Disorders*, *18*, 227–238.

Rafael-Fortney, J. A., Chimanji, N. S., Schill, K. E., Martin, C. D., Murray, J. D., Ganguly, R., et al. (2011). *Circulation*, *124*, 582–588.

Rando, T. A. (2006). *Nature*, *441*, 1080–1086.

Rao, N., Evans, S., Stewart, D., Spencer, K. H., Sheikh, F., Hui, E. E., et al. (2013). *Biomedical Microdevices*, *15*, 161–169.

Rapraeger, A. C., Krufka, A., & Olwin, B. B. (1991). *Science*, *252*, 1705–1708.

Redfern, P. A. (1970). *The Journal of Physiology*, *209*, 701–709.

Renault, V., Thornell, L. E., Eriksson, P. O., Butler-Browne, G., & Mouly, V. (2002). *Aging Cell*, *1*, 132–139.

Reuben, D. B., Judd-Hamilton, L., Harris, T. B., Seeman, T. E., & MacArthur Studies of Successful Aging (2003). *Journal of the American Geriatrics Society*, *51*, 1125–1130.

Rhoads, R. P., Johnson, R. M., Rathbone, C. R., Liu, X., Temm-Grove, C., Sheehan, S. M., et al. (2009). *American Journal of Physiology. Cell Physiology*, *296*, C1321–8.

Richardson, D. K., Kashyap, S., Bajaj, M., Cusi, K., Mandarino, S. J., Finlayson, J., et al. (2005). *The Journal of Biological Chemistry*, *280*, 10290–10297.

Riquelme, C., Larrain, J., Schonherr, E., Henriquez, J. P., Kresse, H., & Brandan, E. (2001). *The Journal of Biological Chemistry*, *276*, 3589–3596.

Rochlin, K., Yu, S., Roy, S., & Baylies, M. K. (2010). *Developmental Biology*, *341*, 66–83.

Rodgers, J. T., King, K. Y., Brett, J. O., Cromie, M. J., Charville, G. W., Maguire, K. K., et al. (2014). *Nature*, *510*, 393–396.

Rodgers, J. T., Schroeder, M. D., Ma, C., & Rando, T. A. (2017). *Cell Reports, 19*, 479–486.
Rooney, J. E., Welser, J. V., Dechert, M. A., Flintoff-Dye, N. L., Kaufman, S. J., & Burkin, D. J. (2006). *Journal of Cell Science, 119*, 2185–2195.
Rozo, M., Li, L., & Fan, C. M. (2016). *Nature Medicine, 22*, 889–896.
Rullman, E., Rundqvist, H., Wagsater, D., Fischer, H., Eriksson, P., Sundberg, C. J., et al. (2007). *Journal of Applied Physiology, 102*, 2346–2351.
Ryan, N. A., Zwetsloot, K. A., Westerkamp, L. M., Hickner, R. C., Pofahl, W. E., & Gavin, T. P. (2006). *Journal of Applied Physiology, 100*, 178–185.
Saclier, M., Yacoub-Youssef, H., Mackey, A. L., Arnold, L., Ardjoune, H., Magnan, M., et al. (2013). *Stem Cells, 31*, 384–396.
Sahenk, Z., & Mendell, J. R. (2011). *Current Rheumatology Reports, 13*, 199–207.
Sajko, S., Kubinova, L., Cvetko, E., Kreft, M., Wernig, A., & Erzen, I. (2004). *The Journal of Histochemistry and Cytochemistry, 52*, 179–185.
Sanders, P. M., Russell, S. T., & Tisdale, M. J. (2005). *British Journal of Cancer, 93*, 425–434.
Sanes, J. R., Marshall, L. M., & McMahan, U. J. (1978). *The Journal of Cell Biology, 78*, 176–198.
Sarelius, I. H., Damon, D. N., & Duling, B. R. (1981). *The American Journal of Physiology, 241*, H317–24.
Scadden, D. T. (2006). *Nature, 441*, 1075–1079.
Schakman, O., Kalista, S., Barbe, C., Loumaye, A., & Thissen, J. P. (2013). *The International Journal of Biochemistry & Cell Biology, 45*, 2163–2172.
Scharner, J., & Zammit, P. S. (2011). *Skeletal Muscle, 1*, 28.
Schmalbruch, H. (1976). *Tissue & Cell, 8*, 673–692.
Schnyder, S., & Handschin, C. (2015). *Bone, 80*, 115–125.
Schofield, R. (1978). *Blood Cells, 4*, 7–25.
Scholz, D., Thomas, S., Sass, S., & Podzuweit, T. (2003). *Molecular and Cellular Biochemistry, 246*, 57–67.
Schultz, E. (1974). *The Anatomical Record, 180*, 589–595.
Schuster-Gossler, K., Cordes, R., & Gossler, A. (2007). *Proceedings of the National Academy of Sciences of the United States of America, 104*, 537–542.
Serrano, A. L., & Munoz-Canoves, P. (2017). *Seminars in Cell & Developmental Biology, 64*, 181–190.
Sessa, W. C. (2009). *Journal of Thrombosis and Haemostasis*, 7(Suppl. 1), 35–37.
Shea, K. L., Xiang, W., LaPorta, V. S., Licht, J. D., Keller, C., Basson, M. A., et al. (2010). *Cell Stem Cell, 6*, 117–129.
Shefer, G., Rauner, G., Yablonka-Reuveni, Z., & Benayahu, D. (2010). *PLoS One, 5*, e13307.
Shefer, G., Van de Mark, D. P., Richardson, J. B., & Yablonka-Reuveni, Z. (2006). *Developmental Biology, 294*, 50–66.
Siegel, A. L., Atchison, K., Fisher, K. E., Davis, G. E., & Cornelison, D. D. (2009). *Stem Cells, 27*, 2527–2538.
Silva, A. M., Shen, W., Heo, M., Gallagher, D., Wang, Z., Sardinha, L. B., et al. (2010). *American Journal of Human Biology, 22*, 76–82.
Singh, P., Carraher, C., & Schwarzbauer, J. E. (2010). *Annual Review of Cell and Developmental Biology, 26*, 397–419.
Sousa-Victor, P., & Munoz-Canoves, P. (2016). *Molecular Aspects of Medicine, 50*, 109–117.
Spurney, C. F., Sali, A., Guerron, A. D., Iantorno, M., Yu, Q., Gordish-Dressman, H., et al. (2011). *Journal of Cardiovascular Pharmacology and Therapeutics, 16*, 87–95.
Stark, D. A., Karvas, R. M., Siegel, A. L., & Cornelison, D. D. (2011). *Development, 138*, 5279–5289.
Stearns-Reider, K. M., D'Amore, A., Beezhold, K., Rothrauff, B., Cavalli, L., Wagner, W. R., et al. (2017). *Aging Cell, 16*, 518–528.

Stewart, C. E., James, P. L., Fant, M. E., & Rotwein, P. (1996). *Journal of Cellular Physiology, 169*, 23–32.
Stoick-Cooper, C. L., Moon, R. T., & Weidinger, G. (2007). *Genes & Development, 21*, 1292–1315.
Suelves, M., Lopez-Alemany, R., Lluis, F., Aniorte, G., Serrano, E., Parra, M., et al. (2002). *Blood, 99*, 2835–2844.
Suelves, M., Vidal, B., Serrano, A. L., Tjwa, M., Roma, J., Lopez-Alemany, R., et al. (2007). *The Journal of Cell Biology, 178*, 1039–1051.
Sun, D., Martinez, C. O., Ochoa, O., Ruiz-Willhite, L., Bonilla, J. R., Centonze, V. E., et al. (2009). *The FASEB Journal, 23*, 382–395.
Sun, G. L., Zhao, S., Li, P., & Jiang, H. K. (2006). *Neuroscience Bulletin, 22*, 85–90.
Tajbakhsh, S. (2009). *Journal of Internal Medicine, 266*, 372–389.
Tatsumi, R., Anderson, J. E., Nevoret, C. J., Halevy, O., & Allen, R. E. (1998). *Developmental Biology, 194*, 114–128.
Thakur, R., & Mishra, D. P. (2016). *Pharmacology & Therapeutics, 168*, 61–74.
Tidball, J. G. (2005). *American Journal of Physiology. Regulatory, Integrative and Comparative Physiology, 288*, R345–53.
Tidball, J. G. (2017). *Nature Reviews. Immunology, 17*, 165–178.
Tidball, J. G., Dorshkind, K., & Wehling-Henricks, M. (2014). *Development, 141*, 1184–1196.
Tidball, J. G., & Villalta, S. A. (2010). *American Journal of Physiology. Regulatory, Integrative and Comparative Physiology, 298*, R1173–87.
Tidball, J. G., & Wehling-Henricks, M. (2005). *Current Opinion in Rheumatology, 17*, 707–713.
Tierney, M. T., Aydogdu, T., Sala, D., Malecova, B., Gatto, S., Puri, P. L., et al. (2014). *Nature Medicine, 20*, 1182–1186.
Tierney, M. T., Gromova, A., Sesillo, F. B., Sala, D., Spenle, C., Orend, G., et al. (2016). *Cell Reports, 14*, 1940–1952.
Tonkin, J., Temmerman, L., Sampson, R. D., Gallego-Colon, E., Barberi, L., Bilbao, D., et al. (2015). *Molecular Therapy, 23*, 1189–1200.
Tonon, E., Ferretti, R., Shiratori, J. H., Santo Neto, H., Marques, M. J., & Minatel, E. (2012). *Nutrition, 28*, 686–690.
Turnbull, J., Powell, A., & Guimond, S. (2001). *Trends in Cell Biology, 11*, 75–82.
Uezumi, A., Fukada, S., Yamamoto, N., Takeda, S., & Tsuchida, K. (2010). *Nature Cell Biology, 12*, 143–152.
Urciuolo, A., Quarta, M., Morbidoni, V., Gattazzo, F., Molon, S., Grumati, P., et al. (2013). *Nature Communications, 4*, 1964.
Vachon, P. H., Xu, H., Liu, L., Loechel, F., Hayashi, Y., Arahata, K., et al. (1997). *The Journal of Clinical Investigation, 100*, 1870–1881.
Varga, T., Mounier, R., Horvath, A., Cuvellier, S., Dumont, F., Poliska, S., et al. (2016). *Journal of Immunology, 196*, 4771–4782.
Varga, T., Mounier, R., Patsalos, A., Gogolak, P., Peloquin, M., Horvath, A., et al. (2016). *Immunity, 45*, 1038–1051.
Vasyutina, E., Lenhard, D. C., Wende, H., Erdmann, B., Epstein, J. A., & Birchmeier, C. (2007). *Proceedings of the National Academy of Sciences of the United States of America, 104*, 4443–4448.
Verdijk, L. B., Dirks, M. L., Snijders, T., Prompers, J. J., Beelen, M., Jonkers, R. A., et al. (2012). *Medicine and Science in Sports and Exercise, 44*, 2322–2330.
Verdijk, L. B., Koopman, R., Schaart, G., Meijer, K., Savelberg, H. H., & van Loon, L. J. (2007). *American Journal of Physiology. Endocrinology and Metabolism, 292*, E151–7.
Vial, C., Gutierrez, J., Santander, C., Cabrera, D., & Brandan, E. (2011). *The Journal of Biological Chemistry, 286*, 24242–24252.

Vial, C., Zuniga, L. M., Cabello-Verrugio, C., Canon, P., Fadic, R., & Brandan, E. (2008). *Journal of Cellular Physiology*, *215*, 410–421.
Vidal, B., Ardite, E., Suelves, M., Ruiz-Bonilla, V., Janue, A., Flick, M. J., et al. (2012). *Human Molecular Genetics*, *21*, 1989–2004.
Villalta, S. A., Deng, B., Rinaldi, C., Wehling-Henricks, M., & Tidball, J. G. (2011). *Journal of Immunology*, *187*, 5419–5428.
Villalta, S. A., Nguyen, H. X., Deng, B., Gotoh, T., & Tidball, J. G. (2009). *Human Molecular Genetics*, *18*, 482–496.
Villalta, S. A., Rinaldi, C., Deng, B., Liu, G., Fedor, B., & Tidball, J. G. (2011). *Human Molecular Genetics*, *20*, 790–805.
Villalta, S. A., Rosenberg, A. S., & Bluestone, J. A. (2015). *Rare Diseases*, *3*, e1010966.
Viloria, K., & Hill, N. J. (2016). *Biomolecular Concepts*, *7*, 117–132.
von Moers, A., Zwirner, A., Reinhold, A., Bruckmann, O., van Landeghem, F., Stoltenburg-Didinger, G., et al. (2005). *Acta Neuropathologica*, *109*, 285–293.
Vracko, R., & Benditt, E. P. (1972). *The Journal of Cell Biology*, *55*, 406–419.
Wagatsuma, A. (2007). *Molecular and Cellular Biochemistry*, *298*, 151–159.
Wang, Y. X., & Rudnicki, M. A. (2011). *Nature Reviews. Molecular Cell Biology*, *13*, 127–133.
Waterlow, J. C. (1984). *Human Nutrition. Clinical Nutrition*, *38*, 151–154.
Watts, R., McAinch, A. J., Dixon, J. B., O'Brien, P. E., & Cameron-Smith, D. (2013). *Obesity (Silver Spring)*, *21*, 525–528.
Webster, M. T., Manor, U., Lippincott-Schwartz, J., & Fan, C. M. (2016). *Cell Stem Cell*, *18*, 243–252.
Wehling, M., Spencer, M. J., & Tidball, J. G. (2001). *The Journal of Cell Biology*, *155*, 123–131.
Wehling-Henricks, M., Lee, J. J., & Tidball, J. G. (2004). *Neuromuscular Disorders*, *14*, 483–490.
Wehling-Henricks, M., Sokolow, S., Lee, J. J., Myung, K. H., Villalta, S. A., & Tidball, J. G. (2008). *Human Molecular Genetics*, *17*, 2280–2292.
Wen, Y., Bi, P., Liu, W., Asakura, A., Keller, C., & Kuang, S. (2012). *Molecular and Cellular Biology*, *32*, 2300–2311.
White, R. B., Bierinx, A. S., Gnocchi, V. F., & Zammit, P. S. (2010). *BMC Developmental Biology*, *10*, 21.
Xian, X., Gopal, S., & Couchman, J. R. (2010). *Cell and Tissue Research*, *339*, 31–46.
Yamada, M., Sankoda, Y., Tatsumi, R., Mizunoya, W., Ikeuchi, Y., Sunagawa, K., et al. (2008). *The International Journal of Biochemistry & Cell Biology*, *40*, 2183–2191.
Yamaguchi, M., Watanabe, Y., Ohtani, T., Uezumi, A., Mikami, N., Nakamura, M., et al. (2015). *Cell Reports*, *13*, 302–314.
Yennek, S., Burute, M., Thery, M., & Tajbakhsh, S. (2014). *Cell Reports*, *7*, 961–970.
Yin, H., Price, F., & Rudnicki, M. A. (2013). *Physiological Reviews*, *93*, 23–67.
Yoshida, T., Galvez, S., Tiwari, S., Rezk, B. M., Semprun-Prieto, L., Higashi, Y., et al. (2013). *The Journal of Biological Chemistry*, *288*, 23823–23832.
Young, H. E., Carrino, D. A., & Caplan, A. I. (1990). *Mechanisms of Ageing and Development*, *53*, 179–193.
Zhang, J., Xiao, Z., Qu, C., Cui, W., Wang, X., & Du, J. (2014). *Journal of Immunology*, *193*, 5149–5160.
Zhou, X., Wang, J. L., Lu, J., Song, Y., Kwak, K. S., Jiao, Q., et al. (2010). *Cell*, *142*, 531–543.
Zou, M. H., Shi, C., & Cohen, R. A. (2002). *The Journal of Clinical Investigation*, *109*, 817–826.
Zuo, L., & Pannell, B. K. (2015). *Frontiers in Physiology*, *6*, 338.

CHAPTER THREE

Translational Control of the Myogenic Program in Developing, Regenerating, and Diseased Skeletal Muscle

Ryo Fujita, Colin Crist[1]

Lady Davis Institute for Medical Research, Jewish General Hospital, McGill University, Montreal, QC, Canada

[1]Corresponding author: e-mail address: colin.crist@mcgill.ca

Contents

Abstract

Translational control of genes that code for protein allows a cell to rapidly respond to changes in its environment, in part because translational control of gene expression does not depend on upstream events required to produce an mRNA molecule. The

Current Topics in Developmental Biology, Volume 126
ISSN 0070-2153
https://doi.org/10.1016/bs.ctdb.2017.08.004

© 2018 Elsevier Inc.
All rights reserved.

importance of translational control has been highlighted by studies concerning muscle development, regeneration, and disease. Translational control of specific mRNAs is achieved by microRNAs and RNA-binding proteins, which are particularly relevant to developmental myogenesis, where they ensure the stepwise differentiation of multipotent progenitors to committed myogenic progenitors that ultimately fuse into slow- or fast-type myofibers that make up skeletal muscle. The importance of translational control is also illustrated in muscle disease, where deregulated microRNA expression accelerates or delays progression of disease. Skeletal muscle is also unique for its remarkable capacity to regenerate after injury, which requires the activity of quiescent muscle stem cells, named satellite cells for their position underneath the basal lamina of the myofiber. Mitotically quiescent satellite cells are primed to activate the cell cycle and myogenic program, a unique feature that requires specific regulation of mRNA translation converging with pathways that regulate global protein synthesis. Emerging concepts in translational control of gene expression have shed light on multiple layers of control over the myogenic program. In parallel, the development and regeneration of skeletal muscle represents a unique, relevant, and highly defined context within which new concepts in translational control of gene expression should emerge.

1. INTRODUCTION

1.1 Regulation of Translation

The translation of mRNAs is necessary for protein encoding gene expression and represents a key regulatory step in the control of gene expression. Mechanisms regulating protein synthesis play important roles in most cellular processes, are critical for maintaining homeostasis of cells, and development of tissues in an organism. As cells progress through a stepwise developmental program, the synthesis rate of proteins that define each sequential gene expression program will depend on not only the cellular concentration of mRNA but also the translational efficiency of mRNA.

The translation of eukaryotic mRNAs begins with the assembly of multiprotein complexes at the 7-methyl-GTP-cap structure at the 5′ end of all eukaryotic mRNAs, in a process collectively termed cap-dependent translation. The eukaryotic initiation factor eIF4G acts as a scaffold to assemble eIF4E and eIF4A to form the eIF4F complex. Meanwhile, the 40S ribosomal subunit is loaded with eIF3, eIF5, and the eIF2-GTP-met-tRNA to form a 43S preinitiation complex. The eIF4F complex positions the 43S preinitiation complex at the 5′ cap, which then initiates scanning along the 5′UTR to recognize an AUG start codon. Multiple cellular mechanisms converge on this multistep process of translation initiation to exert translational control. Global regulation by translational control often affects the assembly of the preinitiation complex, whereas specific regulation of mRNA translation is brought about by *cis*-elements in the mRNA that are bound by microRNA- or RNA-binding proteins (RBPs) to inhibit translation initiation (Hershey, Sonenberg, & Mathews, 2012).

The defined, stepwise differentiation of multipotent progenitors to skeletal muscle fibers makes the myogenic program an excellent paradigm to study translational control of gene expression. In the following sections, we need to first introduce the biological context of the myogenic program. Then, we will introduce mechanisms that regulate specific mRNAs as cells transit through the multistep myogenic program and discuss the implications of deregulation translational control of gene expression in muscle disease. We will also review progress toward understanding translational control of gene expression to reduce global rates of protein synthesis to ensure muscle stem cells remain quiescent, but poised to activate the myogenic program in response injury. We try to highlight conceptual advances that have been made during the study of translational control that have become particularly relevant to the myogenic program.

1.2 Myogenesis

1.2.1 Cell Fate Choices by Multipotent Progenitors

Limb and trunk muscle in vertebrates are derived from myogenic progenitors that are first established in the somites, which are paired blocks of paraxial mesoderm that form along an anterior–posterior axis to establish the metameric patterning of the body axis. Initially cells within newly formed somites express the paired homeodomain transcription factor *Pax3* (Schubert et al., 2001). As somites mature, the dorsal domain of somites maintains *Pax3* expression and becomes the dermomyotome, while the ventral domain downregulates *Pax3* and upregulates *Pax1* and *Pax9* to form the sclerotome. Cells within the sclerotome are progenitors for bone of the vertebral column, ribs, cartilage, and tendons. The dermomyotome is an epithelial structure initially named for being the source of bipotent progenitors that give rise to dermis and skeletal muscle. However, there is growing evidence indicating that cells in the dermomyotome are multipotent progenitors that give rise to vascular smooth muscle, endothelial cells (Ben-Yair & Kalcheim, 2008; Lagha et al., 2009; Mayeuf-Louchart et al., 2014, 2016), and brown fat (Atit et al., 2006).

1.2.2 Multiple Sequential Steps Define the Myogenic Program

At the onset of myogenesis, myogenic progenitors delaminate from the dermomyotome and migrate from the somite to sites of myogenesis such as in the limb, where they also upregulate *Pax7* expression. Meanwhile, *Pax3*/

Pax7-expressing progenitors present within the central dermomyotome drop into the myotome, which is the site of the first developing skeletal muscle in the embryo. At this point, progenitors determined to enter the myogenic lineage begin expressing bHLH transcription factors known as myogenic regulatory factors *Myf5*, *MyoD*, and *Mrf4* (Kassar-Duchossoy et al., 2004; Rudnicki, Schnegelsberg, Stead, & Braun, 1993), while *Pax3* and *Pax7* expression are silenced. The expression of both *Pax3* and *Pax7* must be downregulated for myogenic differentiation to occur (Boutet, Disatnik, Chan, Iori, & Rando, 2007; Olguin & Olwin, 2004; Relaix, Rocancourt, Mansouri, & Buckingham, 2005). The continued expression of *MyoD* and *Mrf4* and the subsequent upregulation of *Myogenin* are later required to maintain the myogenic differentiation program (Hasty et al., 1993; Nabeshima et al., 1993). During the myogenic program, *Myf5*, *MyoD*, and *Myog* are all transiently expressed myogenic regulatory transcription factors, whereas *Mrf4* expression increases sharply during late fetal development and is maintained in myonuclei of adult muscle (Hinterberger, Sassoon, Rhodes, & Konieczny, 1991). The final steps in the myogenic program are concluded when myoblasts stop proliferating and fuse to form myotubes. At this step, differentiation of slow-twitch or fast-twitch myofibers occurs. This step is regulated in part by the expression of a member of the Sox family of transcription factors, *Sox6*, which is expressed in nuclei of fast-type muscle fibers and functions to repress the expression of slow-type fiber-specific genes (Hagiwara, Yeh, & Liu, 2007).

1.2.3 Establishment of the Muscle Stem Cell Pool

The establishment of a muscle stem cell pool, a reservoir of undifferentiated skeletal muscle progenitors required for adult skeletal muscle regeneration, occurs in concert with the development of skeletal muscle. Using electron microscopy to study skeletal muscle, Alexander Mauro was the first to characterize the muscle stem cell, which he has named the "satellite cell" for its position between the basal lamina and the plasma membrane of the muscle fiber. Speculating on the origin and role of the satellite cell, Alexandre Mauro hypothesized:

> *The satellite cells are remnants from the embryonic development of the multinucleate muscle cell which results from the process of fusion of individual myoblasts. Thus the satellite cells are merely dormant myoblasts that failed to fuse with other myoblasts and are ready to recapitulate the embryonic development of skeletal muscle fiber when the main multinucleate cell is damaged.*
>
> **Mauro (1961)**

After five decades of further study, it is understood that satellite cells are the principal cells required for skeletal muscle regeneration (Lepper, Partridge, & Fan, 2011; McCarthy et al., 2011; Murphy, Lawson, Mathew, Hutcheson, & Kardon, 2011; Sambasivan et al., 2011). Satellite cells are also derived from somites, express *Pax7,* and *Pax3* in a subset of trunk muscle, and are first found in skeletal muscle underneath a basal lamina between E16.5 and E18.5 (Relaix et al., 2005). Lineage analyses taking advantage of Cre/lox genetic tools show that the majority of satellite cells have activated the expression of myogenic regulatory factors *Myf5, MyoD, and Mrf4* in their developmental history (Biressi et al., 2013; Kanisicak, Mendez, Yamamoto, Yamamoto, & Goldhamer, 2009; Sambasivan et al., 2013), yet these regulatory factors are not abundant, at least at the protein level, in adult satellite cells. Therefore, satellite cells likely reverse the differentiation program to establish the muscle stem cell pool, one that is poised to activate the myogenic program, while remaining quiescent.

2. MicroRNA REGULATION OF THE MYOGENIC PROGRAM

One of the most exciting aspects that has emerged out of the study of translational control during the last decade is the discovery of microRNAs (Lee, Feinbaum, & Ambros, 1993; Wightman, Ha, & Ruvkun, 1993). MicroRNAs are found throughout the genome and are present in intergenic regions as well as intragenic regions. In canonical microRNA biogenesis, primary microRNAs are transcribed by RNA polymerase II or III, and are cleaved into pre-microRNAs by the DGC8 and Drosha microprocessor complex. Pre-microRNAs are exported to the cytoplasm, where they are further processed by Dicer and TRBP to a functional microRNA that is loaded with Argonaute (Ago2) into the RNA-induced silencing complex (RISC) (Winter, Jung, Keller, Gregory, & Diederichs, 2009). Specificity is achieved by virtue of the microRNA "seed sequence," the first seven to eight nucleotides at the 5′end of the microRNA that bind to sites of complementary sequence in 3′UTRs of target mRNAs. Accumulating evidence indicates that the earliest event in microRNA activity is the inhibition of cap-dependent translation initiation, which is achieved by the RISC targeting the activity of the cap-binding complex eIF4F to prevent translation initiation (Bazzini, Lee, & Giraldez, 2012; Djuranovic, Nahvi, & Green, 2012; Mathonnet, Fabian, Svitkin, & Parsyan, 2007). Thereafter, microRNA-mediated repression of target mRNA translation is consolidated by deadenylation and subsequent degradation of target mRNAs. This step is mediated by the GW182 component of the RISC, which recruits the CCR4-NOT deadenylase machinery (Fabian et al., 2011, 2009).

2.1 MicroRNA Regulation of Embryonic Myogenesis

2.1.1 Dicer Mutants

The importance of microRNAs in developing muscle is highlighted by conditional inactivation of the microRNA-processing enzyme Dicer in developing somites. This is achieved by crossing *Dicer*$^{fl/fl}$ mice with a *Tcre* transgenic strain where Cre recombinase expression is placed under the control of the *T* (*Brachyury*) promoter (Zhang et al., 2011). By embryonic day 9.0 (E9.0), *Dicer* mRNA is no longer detected in somites, which are smaller with large populations of cells undergoing apoptosis within the dermomyotome and sclerotome compartments of the somite. In addition, there are fewer *Pax7*-expressing cells in *Dicer* mutants, which are normally present in the central compartment of the dermomyotome, as they are also apoptotic (Zhang et al., 2011). These results suggest that *Dicer* is required for the survival of *Pax3/Pax7*-expressing multipotent progenitors within the dermomyotome of the developing somite. The identification of microRNAs within the dermomyotome that are specifically required for the survival of *Pax3/Pax7*-expressing multipotent progenitors awaits further investigation. Mutant *Tcre; Dicer*$^{fl/fl}$ embryos begin to die at E12.5, which prevents further investigation of downstream events in the myogenic pathway.

Instead, the essential role of Dicer in developing muscle is demonstrated by crossing *Dicer*$^{fl/fl}$ mice with *MyoD*$^{Cre/+}$ mice (O'Rourke et al., 2007), which would begin to inactivate *Dicer* expression within the myotome of the developing somites between E10.5 and E11.5 (Hinterberger et al., 1991). By E14.5, muscle-specific expression of microRNAs is reduced by 90% (O'Rourke et al., 2007). By E18.5, the amount of muscle in mutant mice is drastically reduced and all skeletal muscle *Dicer* mutants die within minutes following birth (O'Rourke et al., 2007). Interestingly, *Dicer*-defective myoblasts increase their expression of myogenic regulatory factors *Mef2*, *MyoD*, and *Myogenin*, but also undergo apoptosis and contribute poorly to differentiation, ultimately leading to abnormal myofiber morphologies (O'Rourke et al., 2007).

2.1.2 MicroRNA Regulation of Myogenic Determination

The specific roles of myomiRs and their targets during the myogenic program are beginning to be elucidated. We and others have proposed that

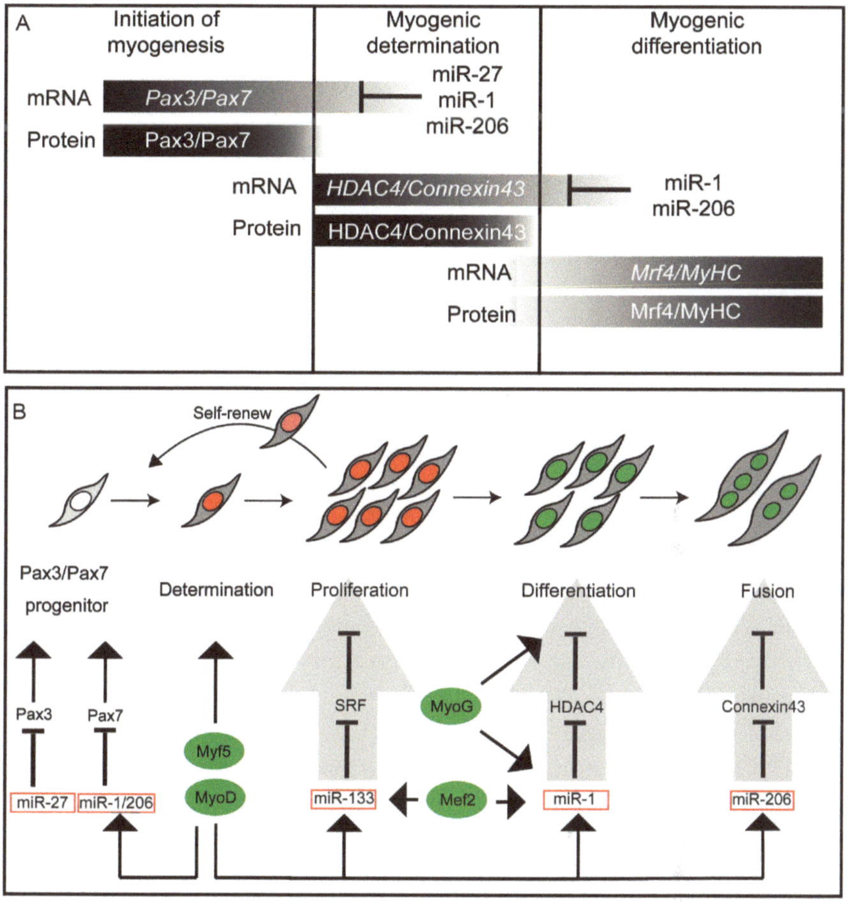

Fig. 1 Integrating microRNA activity within the myogenic program. (A) Selected examples highlighting how microRNAs confer robustness to the myogenic program. Pax3 and Pax7 transcriptionally activate the myogenic program (myogenic determination), and thereafter must be tightly downregulated. This is achieved in part by the activity of miR-27, miR-1, and miR-206 to prevent the translation of *Pax3/Pax7* mRNA that perdures. Myogenic differentiation requires that proliferating myoblasts reduce expression of HDAC4 and Connexin43 prior to their fusion, which is regulated in part by miR-1 and miR-206. (B) The expression of microRNAs (*red boxes*) are regulated by myogenic regulatory transcription factors (*green*) to ensure the stepwise progression of the myogenic program, as well as the balance between proliferating myoblasts (*red*) and their differentiation and fusion into polynucleated myotubes (*green*).

microRNAs ensure robust transitions through the myogenic program by silencing gene expression programs as new myogenic gene expression programs are upregulated (Fig. 1). One of the first events within the myogenic program that illustrates this concept occurs when *Pax3/Pax7*-expressing myogenic progenitors upregulate the expression of *Myf5, Mrf4*, and *MyoD*. This step coincides by the subsequent downregulation of *Pax3/Pax7*, which requires in part the activity of miR-27 to downregulate the translation of *Pax3* transcripts through a single target site within the *Pax3* 3′UTR. Ectopic expression of miR-27 in *Pax3*-expressing myogenic progenitor cells results in downregulation of Pax3 and premature differentiation in vivo (Crist et al., 2009) (Fig. 1).

2.1.3 MyomiRs

When the myogenic program is activated, the balance between committed myogenic progenitors and their differentiated progeny that exit the cell cycle and initiate fusion must also be precisely regulated. Precocious differentiation at the expense of expansion of the committed myogenic progenitor cell pool ultimately results in severe muscle hypotrophy (Schuster-Gossler, Cordes, & Gossler, 2007). The balance between proliferating myoblasts and their subsequent differentiation is regulated by a group of "myomiRs" (miR-1, miR-206, miR-133). While miR-1 and miR-133 are present in skeletal and cardiac muscle, miR-206 is specific to skeletal muscle alone. The expression of myomiRs is directly regulated by the myogenic regulatory factors Mef2, MyoD, and Myogenin (Liu et al., 2007; Rao, Kumar, Farkhondeh, Baskerville, & Lodish, 2006; Sweetman et al., 2008) (Fig. 1B). Even though miR-1, miR-206, and miR-133 originate from bicistronic transcripts on three different chromosomes, they have opposing effects on progression of the myogenic program. Studies in differentiating C2C12 myoblasts show that miR-1 promotes differentiation by targeting *Hdac4*, while miR-133 prevents differentiation and promotes proliferation by targeting *Srf* (Chen et al., 2006). MiR-1 and miR-206 promote myogenic differentiation by targeting *Pax7* to ensure its robust downregulation (Chen et al., 2010), and also target *Connexin43* which is expressed in proliferating myogenic progenitors and requires downregulation to initiate myoblast fusion (Anderson, Catoe, & Werner, 2006). MiR-206 has also been shown to ensure robust downregulation of *Pax3 expression* (Boutet et al., 2012), while also targeting transcripts for follistatin (*Fstl1)* and utrophin (*Utrn*) (Rosenberg, Georges, Asawachaicharn, Analau, & Tapscott, 2006) to promote myogenic differentiation (Fig. 1).

2.2 MicroRNA Regulation of Quiescent Satellite Cells

In adult muscle, the conditional inactivation of Dicer in satellite cells was achieved by crossing *Dicer*$^{fl/fl}$ mice with an inducible *Pax7-CreER* line (Cheung et al., 2012). Six days after the induction of Cre expression in

Pax7-expressing satellite cells by tamoxifen administration, levels of Dicer protein and individual microRNAs were downregulated. In the absence of Dicer, satellite cells activate the cell cycle, but are subsequently prone to apoptosis (Cheung et al., 2012), recapitulating the importance of *Dicer* for cell survival of myogenic progenitors during development (O'Rourke et al., 2007; Zhang et al., 2011).

Normally mitotically quiescent satellite cells activate the cell cycle and the myogenic program in response to injury. Upon activation, satellite cells rapidly activate the myogenic program and enter the cell cycle to give rise to a pool of myogenic progenitors that differentiate to repair muscle, or self-renew to restore the satellite cell pool. The rapid activation of the myogenic program is achieved, in part, because *Myf5* is transcribed to prime the quiescent satellite cell. However, while the cellular concentration of *Myf5* transcripts is high in quiescent satellite cells, the translational efficiency of these mRNAs is low (Crist, Montarras, & Buckingham, 2012; Zismanov et al., 2016). Quiescent satellite cells accumulate miR-31, which has a single target site on the *Myf5* 3′UTR. Overexpression of miR-31 in activated satellite cells, which normally downregulate miR-31, delays activation of the myogenic program and subsequent differentiation by inhibiting the accumulation of Myf5 protein (Crist et al., 2012). The normally quiescent state of satellite cells is also regulated by microRNA silencing of gene expression programs that would otherwise promote the cell cycle. Cell cycle genes *Dlk1*, *Ccnd*, and *Cdc25* are transcriptionally activated in quiescent satellite cells, but miR-489 (Cheung et al., 2012), miR-195, and miR-497 (Sato, Yamamoto, & Sehara-Fujisawa, 2014) prevent their translation.

The activity of microRNA can be tested in vivo by intravenous administration of antagomiRs, which are chemically modified, cholesterol-conjugated antisense oligonucleotides that provide efficient and long-lasting silencing of microRNAs (Krützfeldt et al., 2005). Intravenous administration of antagomiRs against miR-31 and miR-489 to adult mice activates the myogenic program and cell cycle in satellite cells, respectively (Cheung et al., 2012; Crist et al., 2012).

2.3 Intronic MicroRNAs Enforce the Activity of Host Genes Throughout Myogenesis

One concept that has emerged with respect to microRNA regulation of gene expression is the presence of microRNAs in the genome within introns of host genes. In human tissues, intronic microRNAs are often expressed with their host gene mRNA, suggesting that they may ultimately be derived from a common transcript (Baskerville, 2005). Furthermore, emerging evidence indicates that intronic microRNAs support the activity of their host genes by preventing the expression of genes that antagonize the host gene or a biological effect mediated by the host gene. We highlight the concept that intronic microRNAs support host gene activity by a number of examples within the myogenic program below.

The last step of the myogenic program is broadly characterized by differentiation into two distinct myofibers called type I slow twitch and type II fast twitch muscle. These fiber types have different contraction strength, metabolic strategies, and endurance. Type I slow twitch myofibers express predominantly β-MHC (*Myh7*) and *Myh7b*. Within the introns of these genes reside two highly related microRNAs, miR-208b and miR-499. MiR-499 functions to repress the translation of accumulating *Sox6* mRNAs in slow skeletal muscle. The negative correlation between the expression of *Myh7b*, *Myh7* (slow fiber type), and *Sox6* (fast fiber type) is significant because Sox6 is a transcription factor that represses *Myh7* expression (Fig. 2A). Skeletal muscle of $miR\text{-}499^{-/-}$; $miR\text{-}208b^{-/-}$ double knockout mice shows increased *Sox6*

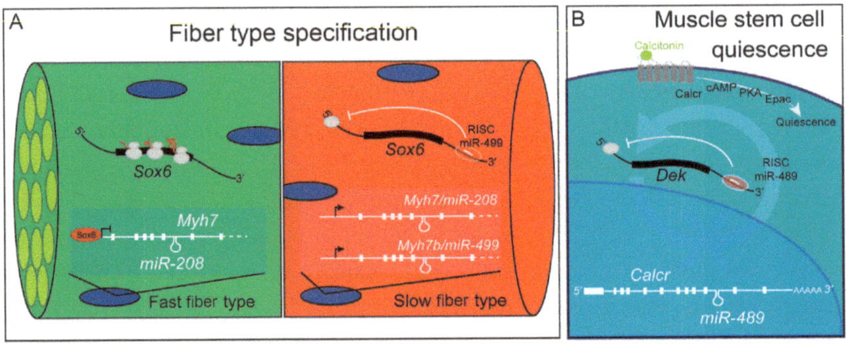

Fig. 2 MicroRNAs reinforce the activity of their host genes throughout myogenesis. (A) MicroRNAs reinforce skeletal muscle fiber type specification. Fast skeletal muscle fibers (*green*) are specified by the activity of *Sox6*, a transcriptional repressor of slow myosin heavy chain *Myh7*. In slow skeletal muscle fibers (*red*), *Myh7* and *Myh7b* are expressed with microRNAs miR-208 and miR-499, respectively. MiR-499 represses the translation of inappropriately expressed *Sox6* transcripts to reinforce the specification of slow myofiber types. (B) MicroRNAs reinforce the quiescent state of the satellite cell. Satellite cells maintain quiescence by the expression of the GPCR calcitonin receptor, which responds to calcitonin to maintain the quiescent state by activation of cAMP/PKA/Epac signaling pathways. miR-489 resides within an intron of *Calcr*, and represses the translation of the cell cycle promoter *Dek* to reinforce satellite cell quiescence.

expression, reduced slow β-MHC (*Myh7*) expression, and a significant loss of type I slow twitch myofibers. Conversely, transgenic mice overexpressing miR-499 in skeletal muscle convert 100% of myofibers to type I slow twitch fiber types (van Rooij et al., 2009). The suppression of Sox6 expression in slow twitch muscle by miR-499 is also confirmed in zebrafish (Wang et al., 2011), providing strong evidence that microRNA regulatory networks are evolutionarily conserved among vertebrates.

A second example concerns the regulation of satellite cell quiescence by miR-489 (described earlier), which is a microRNA found within an intron of the host gene for the G protein-coupled receptor (GPCR), or calcitonin receptor, *Calcr*. Both the mRNA and protein of *Calcr* are highly expressed in quiescent satellite cells and downregulated in activated satellite cells, mirroring the expression of miR-489 (Cheung et al., 2012; Fukada et al., 2007). Calcitonin stimulation of satellite cells delays their activation of both the cell cycle and the myogenic program (Fukada et al., 2007). The stimulation of Calcr appears to maintain satellite cell quiescence by activating both classical cAMP-dependent protein kinase A (PKA) and the exchange protein directly activated by cAMP (Epac) signaling pathways (Yamaguchi et al., 2015). Since miR-489 inhibits the translation of mRNAs for *Dek*, it supports the activity of Calcr/cAMP/PKA/Epac by preventing the translation of mRNAs that would promote the cell cycle (Fig. 2B).

MiR-378 is found within the first intron of the *Ppargc1b* (Pgc1b) gene, which along with Pgc1a, is a member of the PPARγ coactivator-1 (Pgc-1) family of transcriptional coactivators. MyoD binds directly to regulatory sequences upstream of *Pgc1b/miR-378* to regulate the expression of miR-378. MiR-378, in turn, represses the translation of mRNA for *MyoR*, a known repressor of MyoD expression. Therefore, a feedforward loop is established by which MyoD promotes the expression of miR-378 to indirectly downregulate the expression of its repressor *MyoR* (Gagan, Dey, Layer, Yan, & Dutta, 2011). Pgc1b and especially Pgc1a play a role regulating cellular energy metabolic pathways in skeletal muscle, liver, and other organ systems (Finck & Kelly, 2006). In the liver, where miR-378 and Pgc1b are coregulated and coexpressed, miR-378 fine-tunes the activity of Pgc1b (Carrer et al., 2012). In skeletal muscle, Pgc1a coactivates the activity of myogenic regulatory factor Mef2 to drive its own expression in an autoregulatory loop (Handschin, Rhee, Lin, Tarr, & Spiegelman, 2003). While Pgc1b is also expressed in skeletal muscle, defining its role within the myogenic program requires further investigation.

2.4 MicroRNA Dysregulation in Muscle Disease

Following the discovery of microRNAs, one concept that immediately emerged is that microRNAs function to restrain the translation of inappropriately expressed mRNAs. In other words, highly expressed microRNAs specific to a given cell type prevent the expression of genes that specify other cell types. This is also an important concept when differentiated cell types inappropriately acquire microRNA activity to prevent the translation of mRNAs otherwise required to maintain the differentiated cell type. The effect of microRNA gain-of-function mutations is strikingly illustrated by Texel sheep, which exhibit skeletal muscle hypertrophy and are bred for their meat. Quantitative trait locus analysis followed by fine mapping revealed a G-to-A transition in the 3′UTR of *Gdf8* (myostatin), a negative regulator of muscle cell growth and differentiation. The G-to-A transition in the 3′UTR of *Gdf8* in Texel sheep creates a microRNA target site for muscle-specific myomiRs -1 and -206. The resultant microRNA-mediated silencing of *Gdf8* translation contributes to the muscle hypertrophy of Texel sheep (Clop et al., 2006).

The importance of microRNAs to the homeostasis of muscle is also highlighted by their deregulation in diseased muscle (Eisenberg et al., 2007; Greco et al., 2009). MicroRNA profiling of adult healthy and dystrophic muscle from patients or mouse models of Duchenne muscular dystrophy reveals a large numbers of deregulated microRNAs. Deregulated microRNAs may be part of a normal regenerative response, due to chronic muscle degeneration, to the infiltration of skeletal muscle with immune cells and fibroblasts (Cacchiarelli et al., 2010; Greco et al., 2009), or a direct consequence of perturbation of regulatory pathways by the defect in the gene causing the myopathy (Cacchiarelli et al., 2010; Fiorillo et al., 2015).

Duchenne muscular dystrophy is an X-linked recessive disease caused by mutations in the *Dmd* gene, which encodes the dystrophin protein. Dystrophin functions within the dystrophin-associated protein complex (DAPC) to link the cytoskeleton to the membrane of muscle fibers. In the absence of Dystrophin protein, muscle fibers become more prone to damage. The DAPC also plays an important role in the intracellular nitric oxide (NO) pathway to directly regulate histone deacetylases that mediate epigenetic changes in gene expression. Nitric oxide-mediated S-nitrosylation of HDAC2 influences gene expression in skeletal muscle by altering the acetylation status of histones at specific loci, including at regulatory sequences for microRNA. In dystrophic muscle, the absence of S-nitrosylation of HDAC2 results in their continued association with and repression of a subset of microRNAs. These include miR-1 and miR-29, with consequent oxidative stress and fibrosis caused by increased translation of normally targeted *G6PD* and *Col1A1/Eln*, respectively (Cacchiarelli et al., 2010).

Other microRNAs exhibit increased expression in the absence of dystrophin without being regulated by the DAPC. MiR-31 normally targets accumulating *Myf5* transcripts in quiescent satellite cells (Crist et al., 2012), but is also inappropriately overexpressed in dystrophic muscle. Satellite cells normally downregulate miR-31 upon their activation, but miR-31 levels increase in activated satellite cells isolated from muscle of *mdx* mice and prevent their differentiation when cultured ex vivo (Cacchiarelli et al., 2011; Crist et al., 2012). MiR-31 is also more abundant in human DMD biopsies than in healthy skeletal muscle and in skeletal muscle of Becker Muscular Dystrophy patients, which maintain partial dystrophin function. In addition to miR-31 regulation of *Myf5*, the 3′UTR of *Dmd* mRNA also has a functional, validated miR-31 target site. Given recent evidence that dystrophin protein is upregulated in activated satellite cells to regulate polarity and asymmetric cell divisions (Dumont et al., 2015), future studies might ask whether dystrophin protein and miR-31 negatively regulate each other such that (a) the activation of dystrophin expression in activated satellite cells is required to downregulate miR-31 expression and (b) whether miR-31 normally represses the translation of accumulating *Dmd* transcripts in satellite cells.

Deregulated microRNAs in dystrophic muscle may also exert deleterious effects on therapeutic strategies to restore dystrophin expression and muscle function. The majority of mutations in the *Dmd* gene are deletions, duplications, or nonsense mutations that disrupt the open reading frame and cause premature termination of translation. A promising therapeutic strategy for DMD is by exon skipping which aim to restore translation of *Dmd* mRNA by removing the exon harboring the mutation in question without shifting the open reading frame. In 2016, the US Food and Drug Administration granted accelerated approval status to Exondys51, an exon skipping-based drug that aims to restore the translatability of *Dmd* mRNA by skipping exon 51 in defective gene variants (Lim, Maruyama, & Yokota, 2017). Traditional approval of the drug will require that issues related to Exondys51 efficacy be addressed to provide clinical benefit.

Exon skipping strategies do not normally alter the 3′UTR of *Dmd* mRNA and all associated microRNA-binding sites. Strategies to inhibit microRNAs that block the translation of *Dmd* mRNA could work in synergy with exon skipping to restore more Dystrophin protein (Fig. 3). To

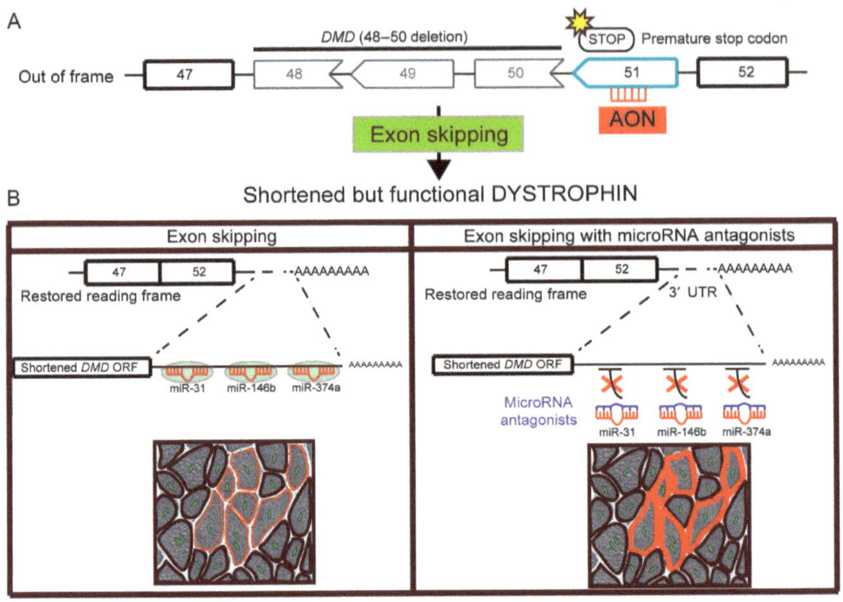

Fig. 3 Increasing the translatability of DMD mRNA by exon skipping therapeutics. (A) Some DMD patients have a deletion in the *DMD* gene spanning exons 48–50 (*gray boxes*, labeled 48, 49, 50), which creates a frameshift and introduces a stop codon in exon 51 (*blue box*, STOP), resulting in no dystrophin expression. Exon skipping strategies aim to influence the splicing machinery to skip exon 51 with antisense oligonucleotides (AON, *red*). The reading frame is restored in the resulting *DMD* mRNA, which is translated into a shorter DYSTROPHIN protein that retains partial function. (B) A scheme illustrating how the low efficacy reported for exon skipping to restore translation of *DMD* mRNA (depicted by low DYSTROPHIN production; *thin red* myofibers, *left panel*) could be improved with strategies to block microRNAs that inhibit *DMD* translation. Multiple microRNAs are overexpressed in dystrophic muscle, of which miR-31, miR-146a, and miR-374b have been validated to target the *DMD* 3′UTR (*green*). Targeting these microRNAs with antisense microRNA inhibitors (*blue*) should further increase the translatability of *DMD* mRNA (*thick red* myofibers, *right panel*).

demonstrate this synergistic strategy, DMD myoblasts expressing high levels of miR-31 were treated with modified antisense oligonucleotides targeting exon 51 of *Dmd* mRNA and a miR-31 sponge. A threefold increase in dystrophin protein was observed when the miR-31 sponge was included, compared to the exon skipping strategy alone (Cacchiarelli et al., 2011). In addition, inflammatory cytokines, in particular TNFα, that are prevalent in dystrophic muscle induce the expression of miR-146b and miR-374a. These inflammatory microRNAs also target the 3′UTR of *Dmd* mRNA and inhibit dystrophin expression ex vivo and in vivo (Fiorillo et al., 2015) (Fig. 3).

While the deregulation of some microRNAs may enhance the progression of disease, the activated expression of other microRNAs that normally occur during a regenerative response may function to delay progression of disease. Amyotrophic lateral sclerosis is characterized by degeneration of motor neurons that leads to muscle atrophy. Comparing microRNA expression in skeletal muscle of wild-type mice and a mouse model of ALS (*G93A-SOD1* transgenic mice), the most dramatically upregulated microRNA is the myomiR miR-206. The upregulation of miR-206 facilitates reinnervation of muscle by preventing the translation of mRNAs for *HDAC4*, a negative repressor of Fgf signaling pathway that normally promotes muscle reinnervation. Mice that are deficient for miR-206 develop neuromuscular synapses normally, but do not regenerate neuromuscular synapses after acute nerve injury. Furthermore, the absence of miR-206 accelerates disease progression in *G93A-SOD1* mice (Williams et al., 2009).

2.5 Regulation of MicroRNA Activity With Long Noncoding RNAs

The inhibition of microRNAs by antagomiRs (Krützfeldt et al., 2005) and other antisense oligonucleotide strategies not only constitutes a new class of drug for muscle pathologies (Brown & Naldini, 2009; Cacchiarelli et al., 2011; Fiorillo et al., 2015; van Rooij & Olson, 2007), but has also been an important experimental tool used to examine the role of individual microRNAs. In addition to inhibition of microRNAs by antisense oligonucleotides, microRNA activity can be inhibited by overexpressing transgenic reporters that contain microRNA-binding sites (Brown et al., 2007), referred to as microRNA "sponges." These strategies provided biochemical evidence that microRNA activity can be sequestered away from their natural targets by antisense oligonucleotides.

Accumulating evidence argues that natural microRNA "sponges" exist in the form of competing endogenous RNAs (ceRNAs), which act as negative regulators of microRNA activity by limiting their activity on target mRNAs (Salmena, Poliseno, Tay, Kats, & Pandolfi, 2011). Experimental evidence supporting the ceRNA hypothesis indicate that many lncRNAs function as ceRNAs, in particular circRNAs (generated by noncanonical splicing to end-join exons into a loop) (Hansen et al., 2013), pseudogene-derived lncRNAs (duplicated genes with acquired mutations) (Wang et al., 2013), and other lncRNAs (Franco-Zorrilla et al., 2007).

One argument against the ceRNA hypothesis concerns the abundance of mRNA targets. Since an individual microRNA is predicted to directly inhibit the expression of hundreds of genes (Selbach et al., 2008), it remains unknown how the expression of a single ceRNA can make a dent in a microRNA regulatory network. For example, miR-122, one of the most abundant microRNAs in mammals, is present at 120,000 molecules per hepatocyte. Taking advantage of adenoviral vectors expressing miR-122 target mRNA *AldoA*, it is estimated that up to 150,000 additional miR-122 target sites are required to increase the expression of the endogenous mRNAs normally repressed by miR-122. The expression of mRNAs and ceRNAs does not reach this level in vivo (Broderick & Zamore, 2014; Denzler, Agarwal, Stefano, Bartel, & Stoffel, 2014). It may be more likely that ceRNAs function to mop up endogenous microRNAs as their expression is being downregulated as a cell transits through a developmental pathway. Furthermore, it cannot be excluded that multiple ceRNAs work in sync to collectively dampen the activity of a microRNA.

Evidence for the ability of 3′UTRs to impact the myogenic program was first described by Blau and colleagues, almost 20 years before the ceRNA hypothesis. In an experiment designed to identify molecular regulators of the myogenic program by genetic complementation of a differentiation-defective myoblast mutant, cDNAs for structural genes *troponin I* and *tropomyosin* were identified. Interestingly, the activity of these cDNAs to activate the expression of *Myogenin* was mapped to their 3′UTRs. Of several mechanisms by which the 3′UTRs could promote differentiation, the group hypothesized that 3′UTRs sequester *trans*-acting factors to alter the stability of other mRNAs (Rastinejad & Blau, 1993). The identification of these factors, which might be microRNAs or other RBPs, awaits further experimentation.

More recently, additional ceRNAs affecting myogenesis have been identified. MyoD directly regulates the expression of *linc-MD1* as mouse and human myoblasts differentiate ex vivo. *Linc-MD1* is also expressed in regenerating dystrophic muscle in vivo (Cesana et al., 2011). *Linc-MD1* contains target sites for the myomiR miR-133 and miR-135, which regulate the translation of mRNAs for myogenic regulatory factor *Mef2c* and its cofactor *Maml1*. Overexpression of *linc-MD1* promotes muscle differentiation, while siRNAs directed against *lnc-MD1* delay differentiation (Cesana et al., 2011). Another ceRNA that functions to regulate the myogenic program is *lnc-mg*, which was determined to inhibit the activity of miR-125b to alleviate the translational silencing of insulin growth factor 2 (*Igf2*) mRNAs to promote myogenesis (Zhu et al., 2017).

When considering the function of lncRNAs, including ceRNAs, we also highlight recent studies demonstrating that annotated mouse and human lncRNAs harbor ORFs that are translated into functional peptides regulating muscle fusion and performance (Anderson et al., 2015; Bi et al., 2017). Further experimentation might lead to our improved understanding of whether lncRNAs that harbor ORFs have dual RNA/peptide functionality and whether previously reported ceRNAs harbor ORFs with additional function as micropeptides.

2.6 Future Directions

Over the last 10 years, significant progress has been made to place microRNAs within the regulatory networks that oversee myogenesis. Key roles for microRNA have been described to maintain primed satellite cells quiescent and to ensure the stepwise progression of the myogenic program (Fig. 1).

Nevertheless, little is known about microRNA specification of the myogenic program, the process in the developing embryo by which multipotent progenitors marked by Pax3-expression first acquire myogenic identity. This developmental context should be a rich paradigm to explore the concept that microRNAs specific to a given cell type function to prevent the expression of genes that specify other cell types. The regulation of cell fate specification by myomiRs has been more clearly demonstrated in adult muscle and brown fat, where the expression of *Prdm16* drives myogenic and white fat precursors to adopt brown adipocyte fate (Seale et al., 2008). Muscle-enriched miR-133 prevents the translation of *Prdm16*. However after cold exposure, miR-133 is downregulated to permit the differentiation of brown adipocytes (Trajkovski, Ahmed, Esau, & Stoffel, 2012; Yin et al., 2013).

It is expected that an individual microRNA targets hundreds of mRNAs and each mRNA is potentially regulated by several microRNAs (Bartel, 2009; Selbach et al., 2008). The impact of microRNAs on the proteome can be addressed by taking advantage of quantitative proteomics. Stable isotope labeling of amino acids in culture (SILAC) is a straightforward approach to quantitatively determine changes in protein abundance and is well suited to strategies to inhibit or promote microRNA activity (Selbach et al., 2008). When combined with analysis of changes in mRNA and target prediction algorithms that search for microRNA seed sequences in 3′UTRs, putative targets can be rapidly identified for validation. The utility of SILAC to identify miR-1 targets has been performed in HeLa cells (Selbach et al., 2008), a cervical cancer cell line that does not provide further insight into miR-1 regulation of the myogenic program. Alternative SILAC analysis of microRNA regulation of the myogenic program might rather take advantage of recent advances in cell culture conditions that promote the ex vivo expansion of satellite cells (Gilbert et al., 2010; Quarta et al., 2016; Zismanov et al., 2016), or the derivation of myogenic progenitors from pluripotent stem cells (Chal et al., 2015; Darabi et al., 2008; Shelton et al., 2014).

3. REGULATION OF THE MYOGENIC PROGRAM BY RNA-BINDING PROTEINS (RBPs)

While microRNAs are the best-studied example of specific regulation of mRNA translation, their role to regulate the translation of specific mRNAs may also be achieved by RBPs. This can also be true mechanistically, since some RBPs that regulate the translation of specific mRNAs do so by sharing components of the RISC that function to prevent translation initiation and degrade target mRNA (Fabian et al., 2013).

RBPs that exhibit sequence specificity to target specific mRNAs include a family of adenylate uridylate-rich element (ARE)-binding proteins, of which HuR, Aufl, and TTP are prominent players. AREs commonly exist in the 3′UTRs of mammalian mRNAs, and exist in up to 8% of human transcripts. These mRNAs regulate proliferation, RNA metabolism, transcriptional regulation, signaling, response to stress, and developmental processes.

During myogenic differentiation HuR shuttles from the nucleus to the cytoplasm to simultaneously stabilize transcripts required for differentiation, while promoting the decay of transcripts that are associated with proliferation of myogenic progenitors. Cytoplasmic HuR binds to AREs in 3′UTRs of mRNAs for *MyoD*, *Myogenin*, and *p21* to ensure their stability and translation. Knockdown of *HuR* reduces *MyoD* and *Myogenin* expression to delay myogenic differentiation. Overexpression of HuR causes precocious differentiation and early myotube fusion (Figueroa et al., 2003; van der Giessen, Di Marco, Clair, & Gallouzi, 2003). HuR also binds to U-rich elements in the 3′UTR of the mRNA for *NPM* (nucleophosmin), leading to mRNA degradation. The degradation of *NPM* depends on HuR recruitment of an additional RBP, KSRP, which recruits RNA degradation machinery to degrade *NPM* transcripts (Cammas et al., 2014). The activity of HuR to simultaneously stabilize some transcripts, while degrading others, illustrates that posttranscriptional silencing complexes can be heterogeneous with respect to composition and function.

Like *Myf5*, the transcription of *MyoD* is also activated in quiescent satellite cells, thereby priming them to activate the myogenic program. MyoD protein, on the other hand, is not present. Instead, MyoD accumulates within hours after satellite cell activation. The translational efficiency of accumulating *MyoD* transcripts is regulated in part by phosphorylation status of the ARE-binding protein tristetraprolin (TTP) (Hausburg et al., 2015). In the unphosphorylated, active state, TTP binds to AREs in the 3′UTR of target transcripts to mediate their degradation. This is achieved in part by recruitment of the Ccr4-Not deadenylase complex (Lykke-Andersen & Wagner, 2005; Sandler, Kreth, Timmers, & Stoecklin, 2011). Mouse TTP is phosphorylated, and inactivated by Mapk-activated protein kinase 2, which is also known to be required for the asymmetric cell divisions that ensure both proliferative expansion and self-renewal of satellite cells (Troy et al., 2012). Phosphorylated TTP is unable to bind Ccr4-Not and deadenylate mRNAs (Marchese et al., 2010), which allows the translation and the rapid accumulation of MyoD protein. The importance of TTP function in the regulation of satellite cells is illustrated by conditional knockout of *Ttp* in satellite cells, which results in their activation of the myogenic program and fusion to myofibers (Hausburg et al., 2015).

Satellite cell activity to regenerate muscle is also regulated by the ARE-binding protein Auf1. Specifically, satellite cells have reduced capacity to self-renew in *Auf1*$^{-/-}$ mice, due to the increased translation of Auf1 target transcripts like matrix metalloprotease *Mmp9*. Mmp9 degrades extracellular matrix proteins, including laminin, that are otherwise components of the satellite cell niche. Deregulation of Mmp9 activity in the absence of Auf1 leads to depletion of ECM components that are required for satellite cells to remodel their niche and return to quiescence after a course of regeneration. Aging *Auf1*$^{-/-}$ mice have accelerated loss of satellite cell capacity to regenerate muscle and also exhibit muscle atrophy. These findings are also relevant to muscle pathology, since the loss or mutation of *AUF1* is related to the development of limb girdle muscular dystrophy (LGMD). Multiple families with LGMD have a mutation at the 4q21 locus, a chromosomal segment that contains the *AUF1* gene, and one family has a mutation in a poorly described *AUF1* homolog *HNRNPDL* (Chenette et al., 2016).

3.1 Cooperation Between MicroRNAs and RBP Regulation of Gene Expression

Future studies should ask whether the canonical mechanism of microRNA silencing (i.e., inhibition of translation initiation consolidated by mRNA decay) remains true in different cell contexts. Emerging evidence from *Drosophila melanogaster* and *Caenorhabditis elegans* indicate that microRNA silencing also occurs in the absence of GW182 (Fukaya & Tomari, 2012; Jannot et al., 2016; Wu, Isaji, & Carthew, 2013), which is a component of the RISC that is otherwise required for recruitment of CCR4-NOT deadenylase complex (Fabian et al., 2009).

The heretofore analysis and validation of singularly acting microRNAs or RBPs within the myogenic program does not accurately provide a picture by which a given 3′UTR is likely bound by multiple microRNA/RISC complexes and multiple RBPs. The interaction of these factors should positively or negatively impact rates of polyadenlyation/deadenylation, stability, translation, transport, and localization (Mazumder, Seshadri, & Fox, 2003). The interaction of RBPs with microRNA/RISC has been demonstrated in the context of cell recovery from stress, in which HuR relieves repression of *CAT-1* translation by miR-122, showing that microRNA silencing is reversible (Bhattacharyya, Habermacher, Martine, Closs, & Filipowicz, 2006). In neurons, fragile X mental retardation protein (FMRP) interacts with the RISC to deliver microRNA-targeted mRNAs required for local protein synthesis at the synapse (Muddashetty et al., 2011) and for axon elongation (Wang et al., 2015). Quiescent satellite cells require FMRP for reversible miR-31 silencing of *Myf5* (Crist et al., 2012). Mechanistic insight into reversible microRNA silencing within these different cellular contexts requires further experimentation. Reversible microRNA silencing implies that the target mRNA pool is not significantly degraded, yet the underlying mechanisms that protect the mRNA from microRNA-mediated decay have not been made clear.

4. SATELLITE CELLS ARE REGULATED BY TRANSLATIONAL CONTROL MECHANISMS IMPACTING GLOBAL PROTEIN SYNTHESIS

4.1 Regulation of Global Protein Synthesis

Protein synthesis requires substantial amounts of cellular material and energy. Cell growth and proliferation requires that cells have adequate resources available to drive protein synthesis. Therefore, regulating global protein synthesis is particularly relevant to quiescent satellite cells, which have few mitochondria and limited energy production. Quiescent satellite cells must use their limited energy resources to survive such that they remain present to ensure homeostasis of the tissue.

4.2 The Phosphorylation of eIF2α Is a Translational Control Mechanism Responding to Various Cellular Stresses

A main pathway regulating global protein synthesis limits the efficiency by which the methionine-loaded tRNA (Met-tRNA) is recruited to the start codon to initiate translation. This step is regulated by the phosphorylation of eIF2α, a component of the eIF2 complex that recruits met-tRNA to the start codon in a GTP-dependent manner. When eIF2α is phosphorylated, it no longer prevents the catalysis of GDP to GTP to facilitate recycling of eIF2-GTP-Met-tRNA ternary complexes that are needed to facilitate translation reinitiation (Koromilas, 2015). The phosphorylation of eIF2α normally occurs in response to various cellular stresses, and is mediated by stress-associated kinases PKR, PERK, GCN2, and HRI. The phosphorylation of eIF2α serves as a "checkpoint" mechanism in cells, allowing cells to reduce protein synthesis and arrest the cell cycle in order to recuperate from the stress or be eliminated if the stress is persistent and damage is beyond repair (Koromilas, 2015).

Satellite cells have coopted PERK phosphorylation of eIF2α to maintain low levels of protein synthesis and the quiescent state (Zismanov et al., 2016). Genetic manipulations that inactivate eIF2α phosphorylation specifically in satellite cells result in their increased protein synthesis, activation of the myogenic program, entry into the cell cycle, and fusion with myofibers. Polysome analysis of mRNAs for *Myf5*, *MyoD*, and *Dek* shows their shift toward efficient translation when eIF2α phosphorylation is inactivated. While phosphorylation of eIF2α is not required for differentiation of satellite cells, it is required for their self-renewal. Furthermore, culture of satellite cells in the presence of sal003, a derivative of salubrinal that blocks eIF2α dephosphorlation, promotes their expansion in a naïve state such that they retain their stem cell capacity to regenerate muscle and contribute to the satellite cell pool after engraftment into a preclinical mouse model of Duchenne muscular dystrophy (Zismanov et al., 2016).

Along with the emergence of eIF2α phosphorylation as a prominent regulator of global protein synthesis, the role played by the 5′ untranslated region (5′UTR) of mRNAs to regulate the efficiency of their translation is of equal importance (Hinnebusch, Ivanov, & Sonenberg, 2016). When eIF2α is phosphorylated, some mRNAs are selectively translated by their upstream open reading frames (uORFs). These uORFs are sites of translation initiation and termination that prevent the translation of downstream ORFs. Ribosome readthrough of these uORFs that occurs when eIF2α is phosphorylated permits translation initiation at the main ORF. High-resolution ribosome footprinting indicates that uORFs are prevalent in vertebrate transcriptomes, are commonly associated with ribosomes, and potently regulate the translation of mRNAs with a similar magnitude to microRNAs (Johnstone, Bazzini, & Giraldez, 2016). Nevertheless, direct evidence that uORFs inhibit the translation of downstream ORFs exists for only a small number of genes, and furthermore, many mRNAs that are selectively translated when eIF2α is phosphorylated do not contain uORFs, suggesting that other 5′UTR features may exert control (Baird et al., 2014). For a review of 5′UTR elements regulating translation efficiency, see Hinnebusch et al. (2016).

The regulatory role of 5′UTRs should also play a substantial role within the context of a quiescent satellite cell. We expect accumulating mRNAs that are sensitive to eIF2α phosphorylation prime the satellite cell for activation (Zismanov et al., 2016). Meanwhile, a subset of mRNAs that are selectively translated will likely be important for satellite cell quiescence and self-renewal (Fig. 4). Future experiments should identify these mRNAs and identify mechanisms of their selective translation.

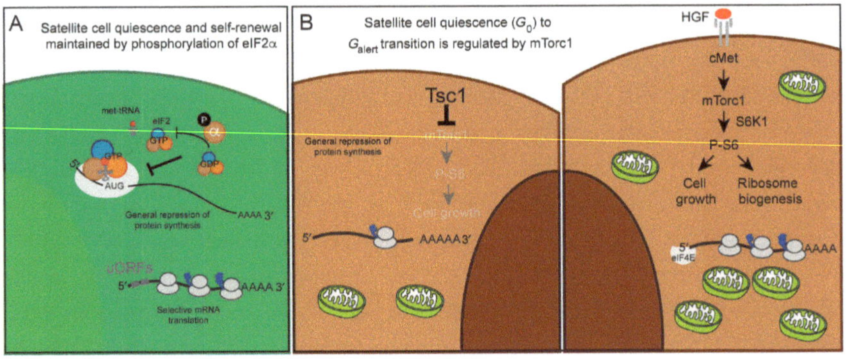

Fig. 4 Pathways regulating global protein synthesis regulate satellite cell activity. (A) A simplified scheme of the multistep process of translation initiation to highlight phosphorylation of eIF2α regulation of muscle stem cell quiescence and self-renewal. The phosphorylation (*black* P) of eIF2α (*brown*) prevents recycling of eIF2-GTP-met-tRNA ternary complexes required to reinitiate translation, resulting in a general repression of protein synthesis. Elements within 5′UTRs of certain mRNAs, such as uORFs (*dark gray*), prevent the translation of the main ORF when eIF2-GTP-met-tRNA is abundant. However, when eIF2α is phosphorylated, limiting availability of eIF2-GTP-met-tRNA causes ribosome readthrough of uORFs to permit translation initiation at the main ORF. (B) Satellite cell transition to a G_{alert} state after an acute injury at a distant site is mediated by the mTorc1 pathway. Genetic strategies to eliminate an inhibitor of mTorc1 signaling (Tsc1) illustrate the importance of mTorc1 inhibition to prevent cell growth and maintain muscle stem cell homeostasis (*left*). After acute injury (*right*) circulating HGF (*red*) activates the c-Met tyrosine kinase (*gray*), which activates mTorc1 to phosphorylate 40S ribosomal subunit protein S6 (ribosome biogenesis, cell growth). G_{alert} satellite cells exhibit increased cell growth (*right*), increased ATP, increased mitochondrial content (*green* mitochondria), and exhibit enhanced regenerative capacity.

4.3 Regulation of Satellite Cells by mTorc1

A second major pathway regulating protein synthesis occurs through mTor, which is a central nutrient sensor complex that signals the cell to grow or proliferate. mTor interacts with different proteins to form functionally distinct mTorc1 and mTorc2 complexes. mTorc1 regulates translation by phosphorylating S6K1 (ribosome biogenesis, cell growth) and 4E-BP (translation initiation, cell proliferation). Evidence that satellite cells are sensitive to mTor signaling arose from studying their activity in skeletal muscle at sites contralateral to an acute injury. A distant acute injury results in the activation of circulating inflammatory signals, leading to activation of hepatocyte growth factor (HGF). HGF activation of the c-Met tyrosine kinase leads to activation of mTorc1, phosphorylation of S6, increased levels of ATP and mitochondria, and increased cell size. These characteristics are recapitulated by genetic strategies to activate mTorc1 by inactivation of the mTorc1 inhibitor *Tsc1* (Rodgers et al., 2015; Rodgers, Schroeder, Ma, & Rando, 2017). Interestingly, most satellite cells (>90%) at sites distant to an acute muscle injury remain quiescent, while a small fraction initiate proliferation. Phosphorylation of 4E-BP by mTorc1 mediates cell proliferation (Dowling et al., 2010), but it remains unclear the extent to which 4E-BP plays a role in maintaining satellite cell quiescence.

5. CONCLUDING REMARKS

In this chapter, we have highlighted how conceptual advances made through the study of translational control have shed light on an additional layer controlling gene expression within the myogenic program. The first major advance in this regard took place after the discovery of microRNAs, which function to prevent the translation of specific mRNAs through complementary sequences in their 3′UTRs. The importance of the microRNA pathway to myogenesis is illustrated by conditional knockout of the *Dicer* component of the microRNA biogenesis machinery, which causes defective myogenesis as muscle progenitor or satellite cells undergo apoptosis. MicroRNAs regulate each aspect of the myogenic program, initially from the cell fate decision to enter the myogenic program, to myogenic determination, differentiation, and fiber type specification. The expression of a group of muscle-specific "myomiRs" is directly activated in skeletal muscle by the myogenic regulatory transcription factors to regulate the balance between proliferating myogenic progenitors and differentiating myoblasts.

Despite significant progress, there remains gaps in our knowledge of translational control of the myogenic program. In particular, new studies should ask how microRNAs regulate cell fate decisions required for muscle development. This is particularly relevant to how *Pax3*-expressing multipotent progenitors give rise to myogenic progenitors. Knowledge of microRNA participation in this process, and subsequent validation of their targets, will reinforce the concept that microRNAs specific to a cell type function to repress the translation of mRNAs that specify other cell types. Knowledge of microRNAs that participate in this process, and their manipulation, might lead to the derivation of myogenic progenitors from pluripotent stem cells at greater efficiencies.

The concept that microRNAs function to repress the translation of inappropriately expressed mRNAs is also highlighted when this activity is deregulated in muscle disease. Defining microRNAs that are deregulated in muscular dystrophy and amyotrophic lateral sclerosis has led to new avenues for therapeutic opportunities. Within this theme, we have particularly highlighted inappropriately expressed microRNAs in dystrophic muscle, which may not only contribute to the progression of disease (for example, proinflammatory or profibrotic microRNAs) but also impede the efficacy of treatment strategies. Blocking inappropriately expressed microRNAs that have target sites on *DMD* mRNA should increase the efficacy of exon skipping strategies that aim to restore the translatability of DMD mRNAs.

The study of satellite cells also helped to reinforce emerging concepts in microRNA regulation. Reversible microRNA silencing, first characterized in cells recovering from stress, is also conceptually important to the quiescent satellite cell, which have primed the activation of genes that are required to rapidly activate the myogenic program and the cell cycle. Transcripts for *Myf5* and *Dek* are present in quiescent satellite cells, but silenced by the activity of microRNAs. Reversible silencing of accumulating *MyoD* transcripts also occurs, but mediated instead by the phosphorylation of TTP.

We also understand that satellite cells not only regulate the translation of specific transcripts in a microRNA- and RBP-dependent manner but also employ pathways that regulate global protein synthesis to maintain quiescence, self-renew, and become alert to sites of distant injury. Within this unique cellular context, how some transcripts are repressed while others are selectively translated will undoubtedly become an emerging theme. We also predict that "priming" satellite cells for activation requires a convergence of mechanisms that regulate specific transcripts with mechanisms that regulate global protein synthesis. For example, whether mechanisms regulating miR-dependent regulation of translation initiation, deadenylation, and mRNA degradation are influenced when eIF2α is phosphorylated remains unclear.

ACKNOWLEDGMENTS

C.C. thanks the Canadian Institutes for Health Research (CIHR 286519), the Muscular Dystrophy Association (MDA 351259), and the Stem Cell Network for their support.

REFERENCES

Anderson, D. M., Anderson, K. M., Chang, C.-L., Makarewich, C. A., Nelson, B. R., McAnally, J. R., et al. (2015). A micropeptide encoded by a putative long noncoding RNA regulates muscle performance. *Cell, 160*, 595–606. https://doi.org/10.1016/j.cell.2015.01.009.

Anderson, C., Catoe, H., & Werner, R. (2006). MIR-206 regulates connexin 43 expression during skeletal muscle development. *Nucleic Acids Research, 34*, 5863–5871. https://doi.org/10.1093/nar/gkl743.

Atit, R., Sgaier, S. K., Mohamed, O. A., Taketo, M. M., Dufort, D., Joyner, A. L., et al. (2006). Beta-catenin activation is necessary and sufficient to specify the dorsal dermal fate in the mouse. *Developmental Biology, 296*, 164–176. https://doi.org/10.1016/j.ydbio.2006.04.449.

Baird, T. D., Palam, L. R., Fusakio, M. E., Willy, J. A., Davis, C. M., McClintick, J. N., et al. (2014). Selective mRNA translation during eIF2 phosphorylation induces expression of IBTK. *Molecular Biology of the Cell, 25*, 1686–1697. https://doi.org/10.1091/mbc.E14-02-0704.

Bartel, D. P. (2009). Micro RNAs: Target recognition and regulatory functions. *Cell, 136*, 215–233. https://doi.org/10.1016/j.cell.2009.01.002.

Baskerville, S. (2005). Microarray profiling of micro RNAs reveals frequent coexpression with neighboring mi RNAs and host genes. *RNA, 11*, 241–247. https://doi.org/10.1261/rna.7240905.

Bazzini, A. A., Lee, M. T., & Giraldez, A. J. (2012). Ribosome profiling shows that mi R-430 reduces translation before causing mRNA decay in zebrafish. *Science, 336*, 233–237. https://doi.org/10.1126/science.1215704.

Ben-Yair, R., & Kalcheim, C. (2008). Notch and bone morphogenetic protein differentially act on dermomyotome cells to generate endothelium, smooth, and striated muscle. *The Journal of Cell Biology, 180*, 607–618. https://doi.org/10.1083/jcb.200707206.

Bhattacharyya, S. N., Habermacher, R., Martine, U., Closs, E. I., & Filipowicz, W. (2006). Relief of micro RNA-mediated translational repression in human cells subjected to stress. *Cell, 125*, 1111–1124. https://doi.org/10.1016/j.cell.2006.04.031.

Bi, P., Ramirez-Martinez, A., Li, H., Cannavino, J., McAnally, J. R., Shelton, J. M., et al. (2017). Control of muscle formation by the fusogenic micropeptide myomixer. *Science, 356*, 323–327. https://doi.org/10.1126/science.aam9361.

Biressi, S., Bjornson, C. R. R., Carlig, P. M. M., Nishijo, K., Keller, C., & Rando, T. A. (2013). Myf5 expression during fetal myogenesis defines the developmental progenitors of adult satellite cells. *Developmental Biology, 379*, 195–207. https://doi.org/10.1016/j.ydbio.2013.04.021.

Boutet, S. C., Cheung, T. H., Quach, N. L., Liu, L., Prescott, S. L., Edalati, A., et al. (2012). Alternative polyadenylation mediates micro RNA regulation of muscle stem cell function. *Cell Stem Cell, 10*, 327–336. https://doi.org/10.1016/j.stem.2012.01.017.

Boutet, S. C., Disatnik, M.-H., Chan, L. S., Iori, K., & Rando, T. A. (2007). Regulation of Pax 3 by proteasomal degradation of monoubiquitinated protein in skeletal muscle progenitors. *Cell, 130*, 349–362. https://doi.org/10.1016/j.cell.2007.05.044.

Broderick, J. A., & Zamore, P. D. (2014). Competitive endogenous RNAs cannot alter micro RNA function in vivo. *Molecular Cell, 54*, 711–713. https://doi.org/10.1016/j.molcel.2014.05.023.

Brown, B. D., Gentner, B., Cantore, A., Colleoni, S., Amendola, M., Zingale, A., et al. (2007). Endogenous micro RNA can be broadly exploited to regulate transgene expression according to tissue, lineage and differentiation state. *Nature Biotechnology, 25*, 1457–1467. https://doi.org/10.1038/nbt1372.

Brown, B. D., & Naldini, L. (2009). Exploiting and antagonizing micro RNA regulation for therapeutic and experimental applications. *Nature Reviews. Genetics, 10*, 578–585. https://doi.org/10.1038/nrg2628.

Cacchiarelli, D., Incitti, T., Martone, J., Cesana, M., Cazzella, V., Santini, T., et al. (2011). mi R-31 modulates dystrophin expression: New implications for Duchenne muscular dystrophy therapy. *EMBO Reports, 12*, 136–141. https://doi.org/10.1038/embor.2010.208.

Cacchiarelli, D., Martone, J., Girardi, E., Cesana, M., Incitti, T., Morlando, M., et al. (2010). Micro RNAs involved in molecular circuitries relevant for the Duchenne muscular dystrophy pathogenesis are controlled by the dystrophin/nNOS pathway. *Cell Metabolism, 12*, 341–351. https://doi.org/10.1016/j.cmet.2010.07.008.

Cammas, A., Sanchez, B. J., Lian, X. J., Dormoy-Raclet, V., van der Giessen, K., López de Silanes, I., et al. (2014). Destabilization of nucleophosmin mRNA by the HuR/KSRP complex is required for muscle fibre formation. *Nature Communications, 5*, 4190. https://doi.org/10.1038/ncomms5190.

Carrer, M., Liu, N., Grueter, C. E., Williams, A. H., Frisard, M. I., Hulver, M. W., et al. (2012). Control of mitochondrial metabolism and systemic energy homeostasis by micro RNAs 378 and 378. *Proceedings of the National Academy of Sciences of the United States of America, 109*, 15330–15335. https://doi.org/10.1073/pnas.1207605109.

Cesana, M., Cacchiarelli, D., Legnini, I., Santini, T., Sthandier, O., Chinappi, M., et al. (2011). A long noncoding RNA controls muscle differentiation by functioning as a competing endogenous RNA. *Cell, 147*, 358–369. https://doi.org/10.1016/j.cell.2011.09.028.

Chal, J., Oginuma, M., Al Tanoury, Z., Gobert, B., Sumara, O., Hick, A., et al. (2015). Differentiation of pluripotent stem cells to muscle fiber to model Duchenne muscular dystrophy. *Nature Biotechnology, 33*, 962–969. https://doi.org/10.1038/nbt.3297.

Chen, J.-F., Mandel, E. M., Thomson, J. M., Wu, Q., Callis, T. E., Hammond, S. M., et al. (2006). The role of micro RNA-1 and micro RNA-133 in skeletal muscle proliferation and differentiation. *Nature Genetics, 38*, 228–233. https://doi.org/10.1038/ng1725.

Chen, J.-F., Tao, Y., Li, J., Deng, Z., Yan, Z., Xiao, X., et al. (2010). Micro RNA-1 and micro RNA-206 regulate skeletal muscle satellite cell proliferation and differentiation by repressing Pax 7. *The Journal of Cell Biology, 190*, 867–879. https://doi.org/10.1083/jcb.200911036.

Chenette, D. M., Cadwallader, A. B., Antwine, T. L., Larkin, L. C., Wang, J., Olwin, B. B., et al. (2016). Targeted mRNA decay by RNA binding protein AUF1 regulates adult muscle stem cell fate, promoting skeletal muscle integrity. *Cell Reports, 16*, 1379–1390. https://doi.org/10.1016/j.celrep.2016.06.095.

Cheung, T. H., Quach, N. L., Charville, G. W., Liu, L., Park, L., Edalati, A., et al. (2012). Maintenance of muscle stem-cell quiescence by micro RNA-489. *Nature, 482*, 524–528. https://doi.org/10.1038/nature10834.

Clop, A., Marcq, F., Takeda, H., Pirottin, D., Tordoir, X., Bibé, B., et al. (2006). A mutation creating a potential illegitimate micro RNA target site in the myostatin gene affects muscularity in sheep. *Nature Genetics, 38*, 813–818. https://doi.org/10.1038/ng1810.

Crist, C. G., Montarras, D., & Buckingham, M. (2012). Muscle satellite cells are primed for myogenesis but maintain quiescence with sequestration of Myf 5 mRNA targeted by micro RNA-31 in mRNP granules. *Cell Stem Cell, 11*, 118–126. https://doi.org/10.1016/j.stem.2012.03.011.

Crist, C. G., Montarras, D., Pallafacchina, G., Rocancourt, D., Cumano, A., Conway, S. J., et al. (2009). Muscle stem cell behavior is modified by micro RNA-27 regulation of Pax 3 expression. *Proceedings of the National Academy of Sciences of the United States of America, 106*, 13383–13387. https://doi.org/10.1073/pnas.0900210106.

Darabi, R., Gehlbach, K., Bachoo, R. M., Kamath, S., Osawa, M., Kamm, K. E., et al. (2008). Functional skeletal muscle regeneration from differentiating embryonic stem cells. *Nature Medicine, 14*, 134–143. https://doi.org/10.1038/nm1705.

Denzler, R., Agarwal, V., Stefano, J., Bartel, D. P., & Stoffel, M. (2014). Assessing the ceRNA hypothesis with quantitative measurements of mi RNA and target abundance. *Molecular Cell, 54*, 766–776. https://doi.org/10.1016/j.molcel.2014.03.045.

Djuranovic, S., Nahvi, A., & Green, R. (2012). miRNA-mediated gene silencing by translational repression followed by mRNA deadenylation and decay. *Science, 336*, 237–240. https://doi.org/10.1126/science.1215691.

Dowling, R. J. O., Topisirovic, I., Alain, T., Bidinosti, M., Fonseca, B. D., Petroulakis, E., et al. (2010). mTORC1-mediated cell proliferation, but not cell growth, controlled by the 4E-BPs. *Science, 328*, 1172–1176. https://doi.org/10.1126/science.1187532.

Dumont, N. A., Wang, Y. X., von Maltzahn, J., Pasut, A., Bentzinger, C. F., Brun, C. E., et al. (2015). Dystrophin expression in muscle stem cells regulates their polarity and asymmetric division. *Nature Medicine*, 1–11. https://doi.org/10.1038/nm.3990.

Eisenberg, I., Eran, A., Nishino, I., Moggio, M., Lamperti, C., Amato, A. A., et al. (2007). Distinctive patterns of micro RNA expression in primary muscular disorders. *Proceedings of the National Academy of Sciences of the United States of America, 104*, 17016–17021. https://doi.org/10.1073/pnas.0708115104.

Fabian, M. R., Cieplak, M. K., Frank, F., Morita, M., Green, J., Srikumar, T., et al. (2011). miRNA-mediated deadenylation is orchestrated by GW182 through two conserved motifs that interact with CCR4-NOT. *Nature Structural & Molecular Biology, 18*, 1211–1217. https://doi.org/10.1038/nsmb.2149.

Fabian, M. R., Frank, F., Rouya, C., Siddiqui, N., Lai, W. S., Karetnikov, A., et al. (2013). Structural basis for the recruitment of the human CCR4-NOT deadenylase complex by tristetraprolin. *Nature Structural & Molecular Biology, 20*, 735–739. https://doi.org/10.1038/nsmb.2572.

Fabian, M. R., Mathonnet, G., Sundermeier, T., Mathys, H., Zipprich, J. T., Svitkin, Y. V., et al. (2009). Mammalian mi RNA RISC recruits CAF1 and PABP to affect PABP-dependent deadenylation. *Molecular Cell, 35*, 868–880. https://doi.org/10.1016/j.molcel.2009.08.004.

Figueroa, A., Cuadrado, A., Fan, J., Atasoy, U., Muscat, G. E., Muñoz-Cánoves, P., et al. (2003). Role of HuR in skeletal myogenesis through coordinate regulation of muscle differentiation genes. *Molecular and Cellular Biology, 23*, 4991–5004. https://doi.org/10.1128/MCB.23.14.4991-5004.2003.

Finck, B. N., & Kelly, D. P. (2006). PGC-1 coactivators: Inducible regulators of energy metabolism in health and disease. *The Journal of Clinical Investigation, 116*, 615–622. https://doi.org/10.1172/JCI27794.

Fiorillo, A. A., Heier, C. R., Novak, J. S., Tully, C. B., Brown, K. J., Uaesoontrachoon, K., et al. (2015). TNF-α-induced micro RNAs control dystrophin expression in Becker muscular dystrophy. *Cell Reports, 12*, 1678–1690. https://doi.org/10.1016/j.celrep.2015.07.066.

Franco-Zorrilla, J. M., Valli, A., Todesco, M., Mateos, I., Puga, M. I., Rubio-Somoza, I., et al. (2007). Target mimicry provides a new mechanism for regulation of micro RNA activity. *Nature Genetics, 39*, 1033–1037. https://doi.org/10.1038/ng2079.

Fukada, S.-I., Uezumi, A., Ikemoto, M., Masuda, S., Segawa, M., Tanimura, N., et al. (2007). Molecular signature of quiescent satellite cells in adult skeletal muscle. *Stem Cells, 25*, 2448–2459. https://doi.org/10.1634/stemcells.2007-0019.

Fukaya, T., & Tomari, Y. (2012). Micro RNAs mediate gene silencing via multiple different pathways in drosophila. *Molecular Cell, 48*, 825–836. https://doi.org/10.1016/j.molcel.2012.09.024.

Gagan, J., Dey, B. K., Layer, R., Yan, Z., & Dutta, A. (2011). Micro RNA-378 targets the myogenic repressor Myo R during myoblast differentiation. *The Journal of Biological Chemistry, 286*, 19431–19438. https://doi.org/10.1074/jbc.M111.219006.

Gilbert, P. M., Havenstrite, K. L., Magnusson, K. E. G., Sacco, A., Leonardi, N. A., Kraft, P., et al. (2010). Substrate elasticity regulates skeletal muscle stem cell self-renewal in culture. *Science, 329*, 1078–1081. https://doi.org/10.1126/science.1191035.

Greco, S., De Simone, M., Colussi, C., Zaccagnini, G., Fasanaro, P., Pescatori, M., et al. (2009). Common micro-RNA signature in skeletal muscle damage and regeneration induced by Duchenne muscular dystrophy and acute ischemia. *The FASEB Journal, 23*, 3335–3346. https://doi.org/10.1096/fj.08-128579.

Hagiwara, N., Yeh, M., & Liu, A. (2007). Sox 6 is required for normal fiber type differentiation of fetal skeletal muscle in mice. *Developmental Dynamics, 236*, 2062–2076. https://doi.org/10.1002/dvdy.21223.

Handschin, C., Rhee, J., Lin, J., Tarr, P. T., & Spiegelman, B. M. (2003). An autoregulatory loop controls peroxisome proliferator-activated receptor gamma coactivator 1alpha expression in muscle. *Proceedings of the National Academy of Sciences of the United States of America, 100*, 7111–7116. https://doi.org/10.1073/pnas.1232352100.

Hansen, T. B., Jensen, T. I., Clausen, B. H., Bramsen, J. B., Finsen, B., Damgaard, C. K., et al. (2013). Natural RNA circles function as efficient micro RNA sponges. *Nature, 495*, 384–388. https://doi.org/10.1038/nature11993.

Hasty, P., Bradley, A., Morris, J. H., Edmondson, D. G., Venuti, J. M., Olson, E. N., et al. (1993). Muscle deficiency and neonatal death in mice with a targeted mutation in the myogenin gene. *Nature, 364*, 501–506. https://doi.org/10.1038/364501a0.

Hausburg, M. A., Doles, J. D., Clement, S. L., Cadwallader, A. B., Hall, M. N., Blackshear, P. J., et al. (2015). Post-transcriptional regulation of satellite cell quiescence by TTP-mediated mRNA decay. *eLife, 4*, e03390. https://doi.org/10.7554/eLife.03390.

Hershey, J. W. B., Sonenberg, N., & Mathews, M. B. (2012). Principles of translational control: An overview. *Cold Spring Harbor Perspectives in Biology, 4*, 1–10. https://doi.org/10.1101/cshperspect.a011528.

Hinnebusch, A. G., Ivanov, I. P., & Sonenberg, N. (2016). Translational control by 5′-untranslated regions of eukaryotic mRNAs. *Science, 352*, 1413–1416. https://doi.org/10.1126/science.aad9868.

Hinterberger, T. J., Sassoon, D. A., Rhodes, S. J., & Konieczny, S. F. (1991). Expression of the muscle regulatory factor MRF4 during somite and skeletal myofiber development. *Developmental Biology, 147*, 144–156.

Jannot, G., Michaud, P., Quévillon Huberdeau, M., Morel-Berryman, L., Brackbill, J. A., Piquet, S., et al. (2016). GW182-free micro RNA silencing complex controls post-transcriptional gene expression during Caenorhabditis elegans embryogenesis. *PLoS Genetics, 12*, e1006484. https://doi.org/10.1371/journal.pgen.1006484.

Johnstone, T. G., Bazzini, A. A., & Giraldez, A. J. (2016). Upstream ORFs are prevalent translational repressors in vertebrates. *The EMBO Journal*. https://doi.org/10.15252/embj.201592759.

Kanisicak, O., Mendez, J. J., Yamamoto, S., Yamamoto, M., & Goldhamer, D. J. (2009). Progenitors of skeletal muscle satellite cells express the muscle determination gene, Myo D. *Developmental Biology, 332*, 131–141. https://doi.org/10.1016/j.ydbio.2009.05.554.

Kassar-Duchossoy, L., Gayraud-Morel, B., Gomès, D., Rocancourt, D., Buckingham, M., Shinin, V., et al. (2004). Mrf 4 determines skeletal muscle identity in Myf 5:Myod double-mutant mice. *Nature, 431*, 466–471. https://doi.org/10.1038/nature02876.

Koromilas, A. E. (2015). Roles of the translation initiation factor eIF2α serine 51 phosphorylation in cancer formation and treatment. *Biochimica et Biophysica Acta, 1849*, 871–880. https://doi.org/10.1016/j.bbagrm.2014.12.007.

Krützfeldt, J., Rajewsky, N., Braich, R., Rajeev, K. G., Tuschl, T., Manoharan, M., et al. (2005). Silencing of micro RNAs in vivo with "antagomirs". *Nature, 438*, 685–689. https://doi.org/10.1038/nature04303.

Lagha, M., Brunelli, S., Messina, G., Cumano, A., Kume, T., Relaix, F., et al. (2009). Pax 3: Foxc 2 reciprocal repression in the somite modulates muscular versus vascular cell fate choice in multipotent progenitors. *Developmental Cell, 17*, 892–899. https://doi.org/10.1016/j.devcel.2009.10.021.

Lee, R. C., Feinbaum, R. L., & Ambros, V. (1993). The C. elegans heterochronic gene lin-4 encodes small RNAs with antisense complementarity to lin-14. *Cell, 75*, 843–854.

Lepper, C., Partridge, T. A., & Fan, C.-M. (2011). An absolute requirement for Pax 7-positive satellite cells in acute injury-induced skeletal muscle regeneration. *Development, 138*, 3639–3646. https://doi.org/10.1242/dev.067595.

Lim, K. R. Q., Maruyama, R., & Yokota, T. (2017). Eteplirsen in the treatment of Duchenne muscular dystrophy. *Drug Design, Development and Therapy, 11*, 533–545. https://doi.org/10.2147/DDDT.S97635.

Liu, N., Williams, A. H., Kim, Y., McAnally, J., Bezprozvannaya, S., Sutherland, L. B., et al. (2007). An intragenic MEF2-dependent enhancer directs muscle-specific expression of micro RNAs 1 and 133. *Proceedings of the National Academy of Sciences of the United States of America, 104*, 20844–20849. https://doi.org/10.1073/pnas.0710558105.

Lykke-Andersen, J., & Wagner, E. (2005). Recruitment and activation of mRNA decay enzymes by two ARE-mediated decay activation domains in the proteins TTP and BRF-1. *Genes & Development, 19*, 351–361. https://doi.org/10.1101/gad.1282305.

Marchese, F. P., Aubareda, A., Tudor, C., Saklatvala, J., Clark, A. R., & Dean, J. L. E. (2010). MAPKAP kinase 2 blocks tristetraprolin-directed mRNA decay by inhibiting CAF1 deadenylase recruitment. *Journal of Biological Chemistry, 285*, 27590–27600. https://doi.org/10.1074/jbc.M110.136473.

Mathonnet, G., Fabian, M. R., Svitkin, Y. V., & Parsyan, A. (2007). MicroRNA inhibition of translation initiation in vitro by targeting the cap-binding complex eIF4F. *Science, 317*, 1760–1764. https://doi.org/10.1126/science.1147939.

Mauro, A. (1961). Satellite cell of skeletal muscle fibers. *The Journal of Biophysical and Biochemical Cytology, 9*, 493–495.

Mayeuf-Louchart, A., Lagha, M., Danckaert, A., Rocancourt, D., Relaix, F., Vincent, S. D., et al. (2014). Notch regulation of myogenic versus endothelial fates of cells that migrate from the somite to the limb. *Proceedings of the National Academy of Sciences of the United States of America, 111*, 8844–8849. https://doi.org/10.1073/pnas.1407606111.

Mayeuf-Louchart, A., Montarras, D., Bodin, C., Kume, T., Vincent, S. D., & Buckingham, M. (2016). Endothelial cell specification in the somite is compromised in Pax 3-positive progenitors of Foxc 1/2 conditional mutants, with loss of forelimb myogenesis. *Development, 143*, 872–879. https://doi.org/10.1242/dev.128017.

Mazumder, B., Seshadri, V., & Fox, P. L. (2003). Translational control by the 3′-UTR: The ends specify the means. *Trends in Biochemical Sciences, 28*, 91–98. https://doi.org/10.1016/S0968-0004(03)00002-1.

McCarthy, J. J., Mula, J., Miyazaki, M., Erfani, R., Garrison, K., Farooqui, A. B., et al. (2011). Effective fiber hypertrophy in satellite cell-depleted skeletal muscle. *Development, 138*, 3657–3666. https://doi.org/10.1242/dev.068858.

Muddashetty, R. S., Nalavadi, V. C., Gross, C., Yao, X., Xing, L., Laur, O., et al. (2011). Reversible inhibition of PSD-95 mRNA translation by miR-125a, FMRP phosphorylation, and mGluR signaling. *Molecular Cell*, *42*, 673–688. https://doi.org/10.1016/j.molcel.2011.05.006.

Murphy, M. M., Lawson, J. A., Mathew, S. J., Hutcheson, D. A., & Kardon, G. (2011). Satellite cells, connective tissue fibroblasts and their interactions are crucial for muscle regeneration. *Development*, *138*, 3625–3637. https://doi.org/10.1242/dev.064162.

Nabeshima, Y., Hanaoka, K., Hayasaka, M., Esumi, E., Li, S., & Nonaka, I. (1993). Myogenin gene disruption results in perinatal lethality because of severe muscle defect. *Nature*, *364*, 532–535. https://doi.org/10.1038/364532a0.

Olguin, H. C., & Olwin, B. B. (2004). Pax-7 up-regulation inhibits myogenesis and cell cycle progression in satellite cells: A potential mechanism for self-renewal. *Developmental Biology*, *275*, 375–388. https://doi.org/10.1016/j.ydbio.2004.08.015.

O'Rourke, J. R., Georges, S. A., Seay, H. R., Tapscott, S. J., McManus, M. T., Goldhamer, D. J., et al. (2007). Essential role for dicer during skeletal muscle development. *Developmental Biology*, *311*, 359–368. https://doi.org/10.1016/j.ydbio.2007.08.032.

Quarta, M., Brett, J. O., DiMarco, R., De Morree, A., Boutet, S. C., Chacon, R., et al. (2016). An artificial niche preserves the quiescence of muscle stem cells and enhances their therapeutic efficacy. *Nature Biotechnology*, *34*, 752–759. https://doi.org/10.1038/nbt.3576.

Rao, P. K., Kumar, R. M., Farkhondeh, M., Baskerville, S., & Lodish, H. F. (2006). Myogenic factors that regulate expression of muscle-specific micro RNAs. *Proceedings of the National Academy of Sciences of the United States of America*, *103*, 8721–8726. https://doi.org/10.1073/pnas.0602831103.

Rastinejad, F., & Blau, H. M. (1993). Genetic complementation reveals a novel regulatory role for 3′ untranslated regions in growth and differentiation. *Cell*, *72*, 903–917.

Relaix, F., Rocancourt, D., Mansouri, A., & Buckingham, M. (2005). A Pax 3/Pax 7-dependent population of skeletal muscle progenitor cells. *Nature*, *435*, 948–953. https://doi.org/10.1038/nature03594.

Rodgers, J. T., King, K. Y., Brett, J. O., Cromie, M. J., Charville, G. W., Maguire, K. K., et al. (2015). mTORC1 controls the adaptive transition of quiescent stem cells from G0 to G(Alert). *Nature*, *510*, 393–396. https://doi.org/10.1038/nature13255.

Rodgers, J. T., Schroeder, M. D., Ma, C., & Rando, T. A. (2017). HGFA is an injury-regulated systemic factor that induces the transition of stem cells into GAlert. *Cell Reports*, *19*, 479–486. https://doi.org/10.1016/j.celrep.2017.03.066.

Rosenberg, M. I., Georges, S. A., Asawachaicharn, A., Analau, E., & Tapscott, S. J. (2006). Myo D inhibits Fstl 1 and Utrn expression by inducing transcription of mi R-206. *The Journal of Cell Biology*, *175*, 77–85. https://doi.org/10.1083/jcb.200603039.

Rudnicki, M. A., Schnegelsberg, P., Stead, R. H., & Braun, T. (1993). Myo D or Myf-5 is required for the formation of skeletal muscle. *Cell*, *75*, 1351–1359. https://doi.org/10.1016/0092-8674(93)90621-V.

Salmena, L., Poliseno, L., Tay, Y., Kats, L., & Pandolfi, P. P. (2011). A ceRNA hypothesis: The Rosetta stone of a hidden RNA language? *Cell*, *146*, 353–358. https://doi.org/10.1016/j.cell.2011.07.014.

Sambasivan, R., Comai, G., Le Roux, I., Gomès, D., Konge, J., Dumas, G., et al. (2013). Embryonic founders of adult muscle stem cells are primed by the determination gene Mrf 4. *Developmental Biology*, *381*, 241–255. https://doi.org/10.1016/j.ydbio.2013.04.018.

Sambasivan, R., Yao, R., Kissenpfennig, A., Van Wittenberghe, L., Paldi, A., Gayraud-Morel, B., et al. (2011). Pax 7-expressing satellite cells are indispensable for adult skeletal muscle regeneration. *Development*, *138*, 4333. https://doi.org/10.1242/dev.073601.

Sandler, H., Kreth, J., Timmers, H. T. M., & Stoecklin, G. (2011). Not 1 mediates recruitment of the deadenylase Caf 1 to mRNAs targeted for degradation by tristetraprolin. *Nucleic Acids Research*, *39*, 4373–4386. https://doi.org/10.1093/nar/gkr011.

Sato, T., Yamamoto, T., & Sehara-Fujisawa, A. (2014). miR-195/497 induce postnatal quiescence of skeletal muscle stem cells. *Nature Communications*, *5*, 1–11. https://doi.org/10.1038/ncomms5597.

Schubert, F. R., Tremblay, P., Mansouri, A., Faisst, A. M., Kammandel, B., Lumsden, A., et al. (2001). Early mesodermal phenotypes in splotch suggest a role for Pax 3 in the formation of epithelial somites. *Developmental Dynamics*, *222*, 506–521. https://doi.org/10.1002/dvdy.1211.

Schuster-Gossler, K., Cordes, R., & Gossler, A. (2007). Premature myogenic differentiation and depletion of progenitor cells cause severe muscle hypotrophy in Delta 1 mutants. *Proceedings of the National Academy of Sciences of the United States of America*, *104*, 537–542. https://doi.org/10.1073/pnas.0608281104.

Seale, P., Bjork, B., Yang, W., Kajimura, S., Chin, S., Kuang, S., et al. (2008). PRDM16 controls a brown fat/skeletal muscle switch. *Nature*, *454*, 961–967. https://doi.org/10.1038/nature07182.

Selbach, M., Schwanhäusser, B., Thierfelder, N., Fang, Z., Khanin, R., & Rajewsky, N. (2008). Widespread changes in protein synthesis induced by micro RNAs. *Nature*, *455*, 58–63. https://doi.org/10.1038/nature07228.

Shelton, M., Metz, J., Liu, J., Carpenedo, R. L., Demers, S.-P., Stanford, W. L., et al. (2014). Derivation and expansion of PAX7-positive muscle progenitors from human and mouse embryonic stem cells. *Stem Cell Reports*, *3*, 516–529. https://doi.org/10.1016/j.stemcr.2014.07.001.

Sweetman, D., Goljanek, K., Rathjen, T., Oustanina, S., Braun, T., Dalmay, T., et al. (2008). Specific requirements of MRFs for the expression of muscle specific micro RNAs, mi R-1, mi R-206 and mi R-133. *Developmental Biology*, *321*, 491–499. https://doi.org/10.1016/j.ydbio.2008.06.019.

Trajkovski, M., Ahmed, K., Esau, C. C., & Stoffel, M. (2012). Myomi R-133 regulates brown fat differentiation through Prdm 16. *Nature Cell Biology*, *14*, 1330–1335. https://doi.org/10.1038/ncb2612.

Troy, A., Cadwallader, A. B., Fedorov, Y., Tyner, K., Tanaka, K. K., & Olwin, B. B. (2012). Coordination of satellite cell activation and self-renewal by Par-complex-dependent asymmetric activation of p38α/β MAPK. *Cell Stem Cell*, *11*, 541–553. https://doi.org/10.1016/j.stem.2012.05.025.

van der Giessen, K., Di Marco, S., Clair, E., & Gallouzi, I.-E. (2003). RNAi-mediated HuR depletion leads to the inhibition of muscle cell differentiation. *The Journal of Biological Chemistry*, *278*, 47119–47128. https://doi.org/10.1074/jbc.M308889200.

van Rooij, E., & Olson, E. N. (2007). MicroRNAs: Powerful new regulators of heart disease and provocative therapeutic targets. *The Journal of Clinical Investigation*, *117*, 2369–2376. https://doi.org/10.1172/JCI33099.

van Rooij, E., Quiat, D., Johnson, B. A., Sutherland, L. B., Qi, X., Richardson, J. A., et al. (2009). A family of microRNAs encoded by myosin genes governs myosin expression and muscle performance. *Developmental Cell*, *17*, 662–673. https://doi.org/10.1016/j.devcel.2009.10.013.

Wang, L., Guo, Z.-Y., Zhang, R., Xin, B., Chen, R., Zhao, J., et al. (2013). Pseudogene OCT4-pg4 functions as a natural micro RNA sponge to regulate OCT4 expression by competing for miR-145 in hepatocellular carcinoma. *Carcinogenesis*, *34*, 1773–1781. https://doi.org/10.1093/carcin/bgt139.

Wang, X., Ono, Y., Tan, S. C., Chai, R. J., Parkin, C., & Ingham, P. W. (2011). Prdm1a and miR-499 act sequentially to restrict Sox6 activity to the fast-twitch muscle lineage in the zebrafish embryo. *Development, 138*, 4399–4404. https://doi.org/10.1242/dev.070516.

Wang, B., Pan, L., Wei, M., Wang, Q., Liu, W.-W., Wang, N., et al. (2015). FMRP-mediated axonal delivery of miR-181d regulates axon elongation by locally targeting Map1b and Calm1. *Cell Reports, 13*, 2794–2807. https://doi.org/10.1016/j.celrep.2015.11.057.

Wightman, B., Ha, I., & Ruvkun, G. (1993). Posttranscriptional regulation of the heterochronic gene lin-14 by lin-4 mediates temporal pattern formation in C. elegans. *Cell, 75*, 855–862.

Williams, A. H., Valdez, G., Moresi, V., Qi, X., McAnally, J., Elliott, J. L., et al. (2009). MicroRNA-206 delays ALS progression and promotes regeneration of neuromuscular synapses in mice. *Science, 326*, 1549–1554. https://doi.org/10.1126/science.1181046.

Winter, J., Jung, S., Keller, S., Gregory, R. I., & Diederichs, S. (2009). Many roads to maturity: MicroRNA biogenesis pathways and their regulation. *Nature Cell Biology, 11*, 228–234. https://doi.org/10.1038/ncb0309-228.

Wu, P.-H., Isaji, M., & Carthew, R. W. (2013). Functionally diverse microRNA effector complexes are regulated by extracellular signaling. *Molecular Cell, 52*, 113–123. https://doi.org/10.1016/j.molcel.2013.08.023.

Yamaguchi, M., Watanabe, Y., Ohtani, T., Uezumi, A., Mikami, N., Nakamura, M., et al. (2015). Calcitonin receptor signaling inhibits muscle stem cells from escaping the quiescent state and the niche. *Cell Reports, 13*, 302–314. https://doi.org/10.1016/j.celrep.2015.08.083.

Yin, H., Pasut, A., Soleimani, V. D., Bentzinger, C. F., Antoun, G., Thorn, S., et al. (2013). MicroRNA-133 controls brown adipose determination in skeletal muscle satellite cells by targeting Prdm16. *Cell Metabolism, 17*, 210–224. https://doi.org/10.1016/j.cmet.2013.01.004.

Zhang, Z., O'Rourke, J. R., McManus, M. T., Lewandoski, M., Harfe, B. D., & Sun, X. (2011). The microRNA-processing enzyme dicer is dispensable for somite segmentation but essential for limb bud positioning. *Developmental Biology, 351*, 254–265. https://doi.org/10.1016/j.ydbio.2011.01.005.

Zhu, M., Liu, J., Xiao, J., Yang, L., Cai, M., Shen, H., et al. (2017). Lnc-mg is a long noncoding RNA that promotes myogenesis. *Nature Communications, 8*. 14718. https://doi.org/10.1038/ncomms14718.

Zismanov, V., Chichkov, V., Colangelo, V., Jamet, S., Wang, S., Syme, A., et al. (2016). Phosphorylation of eIF2α is a translational control mechanism regulating muscle stem cell quiescence and self-renewal. *Cell Stem Cell, 18*, 79–90. https://doi.org/10.1016/j.stem.2015.09.020.

CHAPTER FOUR

The Composition, Development, and Regeneration of Neuromuscular Junctions

Wenxuan Liu[*,2], **Joe V. Chakkalakal**[*,†,‡,1]
[*]Center for Musculoskeletal Research, University of Rochester Medical Center, Rochester, NY, United States
[†]Stem Cell and Regenerative Medicine Institute, University of Rochester Medical Center, Rochester, NY, United States
[‡]The Rochester Aging Research Center, University of Rochester Medical Center, Rochester, NY, United States
[1]Corresponding author: e-mail address: joe_chakkalakal@urmc.rochester.edu

Contents

Abstract

The neuromuscular junction (NMJ) is the specialized site that connects the terminal of a motor neuron axon to skeletal muscle. As a synapse NMJ integrity is essential for transducing motor neuron signals that initiate skeletal muscle contraction. Many diseases and skeletal muscle aging are linked to impaired NMJ function and the associated muscle wasting. In this chapter we review the components of an NMJ and, the processes of NMJ development, maturation, and regeneration. Also, we briefly discuss the cellular and molecular mechanisms of NMJ decline in the context of disease and aging.

1. NEUROMUSCULAR JUNCTION COMPOSITION AND FUNCTION

First described by the German physiologist Wilhelm Friedrich Kühne (1837–1900) at the light microscopy level through examination of many species, the neuromuscular junction (NMJ) is the communication site

[2] Current address: Genea Biocells US Inc., San Diego, California, USA.

Current Topics in Developmental Biology, Volume 126
ISSN 0070-2153
https://doi.org/10.1016/bs.ctdb.2017.08.005
© 2018 Elsevier Inc.
All rights reserved.

between α-motor neurons and multinucleated contractile muscle cells, myofibers (Fig. 1; Sanes & Lichtman, 1999; Wu, Xiong, & Mei, 2010). He not only coined the term "neuromuscular junction" but also recognized its functional significance as the site of signal transmission from nerve to skeletal muscle (Hong & Etherington, 2011; Kühne, 1862). The NMJ is composed of different specialized cell types: motor neurons, skeletal myofibers, terminal Schwann cells, and possibly a fourth cell type, the newly discovered kranocytes (Fig. 2; Court et al., 2008; Griffin & Thompson, 2008; Hong & Etherington, 2011). In adults, each contractile myofiber is innervated by a single axon from a motor neuron at the NMJ. Axons from a pool of motor neurons branch within a given skeletal muscle to innervate multiple myofibers. A motor unit is a motor neuron and all the myofibers it innervates (Larsson & Ansved, 1995; Sanes & Lichtman, 1999). Myofibers within the same motor unit are typically distributed over a relatively wide area within the muscle, presumably to ensure even distribution of the contractile force. Although motor neurons and myofibers have principle roles in the initiation and transduction of signals for generating contractile force, terminal Schwann cells and perhaps kranocytes are also critical supporting components of a NMJ. Terminal Schwann cells are found in close proximity to motor nerve terminals. They express trophic, adhesive, and growth-associated factors and provide the nerve terminals with trophic sustenance and growth stimulus during NMJ development and regeneration. In addition, they are involved

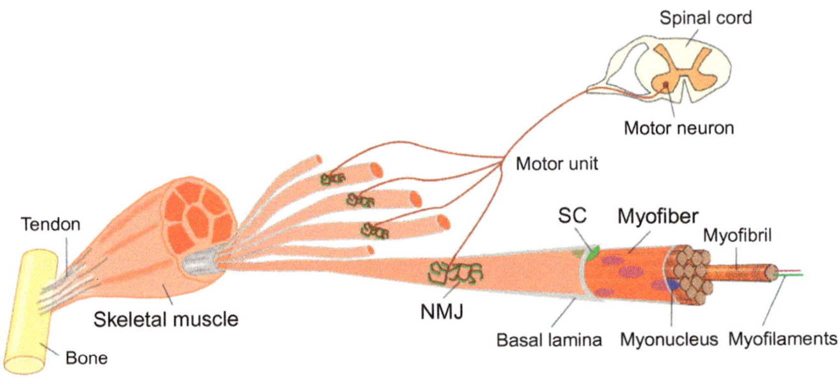

Fig. 1 The components of a motor unit. Skeletal muscles are composed predominantly of multinucleated myofibers. Axons from motor neurons innervate myofibers through the NMJs. A motor neuron and all the myofibers it innervates is called a motor unit. Satellite cells (SCs) are located between the basal lamina and the plasma membrane of myofibers. Each myofiber consists of hundreds of myofibrils, which contain thick and thin myofilaments.

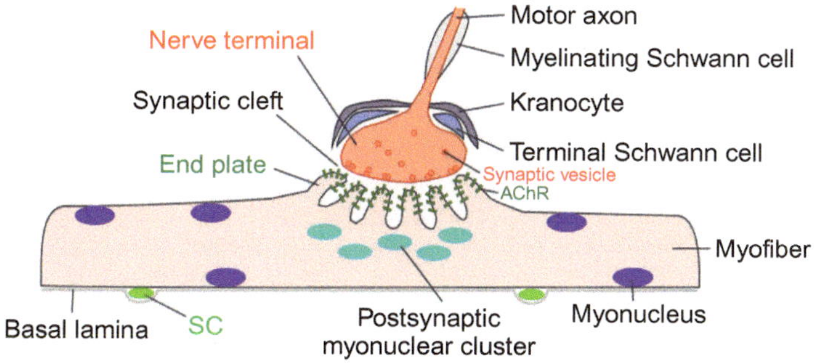

Fig. 2 Cellular constituents of a NMJ. The NMJ is composed of three elements: presynaptic motor nerve terminal, synaptic cleft, and postsynaptic membrane (end plate). Motor neurons and myofibers have principle roles in the initiation and transduction of signals for force generation, whereas terminal Schwann cells and kranocytes play critical supporting roles. The neurotransmitter acetylcholine (ACh) is packaged in synaptic vesicles. When the action potential reaches the motor nerve terminal, ACh is released into the synaptic cleft, binds to AChRs, and triggers membrane potential propagation to initiate muscle contraction. The cluster of postsynaptic myonuclei is essential for NMJ function.

in modulating neuromuscular transmission and long-term maintenance of NMJ structure and function (Barik, Li, Sathyamurthy, Xiong, & Mei, 2016; Griffin & Thompson, 2008; Kang, Tian, Mikesh, Lichtman, & Thompson, 2014; Reddy, Koirala, Sugiura, Herrera, & Ko, 2003). Kranocytes are also reported to respond to neuromuscular insults and may play a permissive role in NMJ regeneration (Court et al., 2008).

The distal end of a motor neuron axon is the presynaptic nerve terminal, which is specialized for neurotransmitter release (Fig. 3). The terminal contains a large number of small (50-nm-diameter) lipid bilayer structures called synaptic vesicles with quanta of neurotransmitters packaged inside (Deutch, 2013). Many readily releasable vesicles tend to anchor at areas on the terminal membrane, active zones, which are marked by high densities of proteins related to transmitter release (Hong & Etherington, 2011; Sanes & Lichtman, 1999). Tightly aligned with the presynaptic terminal is the postsynaptic region of the myofibers. This region is also called the motor end plate because of the invaginations (junctional folds) in the postsynaptic membrane, which raise the postsynaptic apparatus above the rest of the myofiber membrane. Junctional folds are positioned directly opposite the active zones and bear a high concentration of neurotransmitter receptors at the crests, which allow for highly effective synaptic transmission (Sanes & Lichtman, 1999). The space that spans ~50 nm between the presynaptic terminal and the postsynaptic membrane is the synaptic cleft. Filled

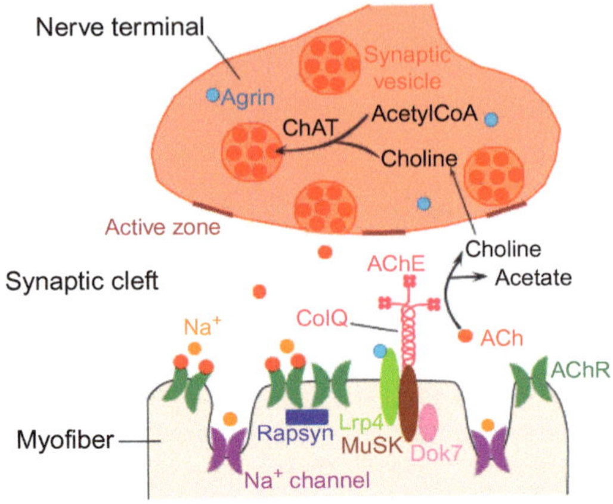

Fig. 3 Critical molecules for NMJ function. Neurotransmitter ACh is synthesized by ChAT from acetyl coenzyme A and choline in the presynaptic nerve terminal. Readily releasable synaptic vesicles with quanta of ACh packaged inside tend to anchor at active zones on the terminal membrane and release ACh into the synaptic cleft through exocytosis. ACh binds to the postsynaptic AChRs and triggers muscle contraction. Neurotransmission is rapidly terminated when ACh is degraded by AChE, which is anchored to the basal lamina in the synaptic cleft by a collagen-like tail ColQ. Agrin/Lrp4/MuSK pathway is critical for NMJ formation and maintenance. During development, nerve-derived Agrin binds to the transmembrane protein Lrp4, which triggers the autophosphorylation and activation of MuSK. MuSK activation leads to AChR clustering through Rapsyn. Genetic mutations in these genes or antibodies against them lead to sever muscle diseases.

with large molecular complexes that assure ultrastructural NMJ arrangement and signal transduction, the synaptic cleft allows for rapid diffusion and degradation of neurotransmitter (Fagerlund & Eriksson, 2009).

The NMJ only occupies approximately 0.1% of the surface area of a myofiber but initiates action potential propagation required for force generation and myofiber maintenance (Bassel-Duby & Olson, 2006; Sanes & Lichtman, 1999; Schiaffino & Reggiani, 2011; Wu et al., 2010). It is a chemical synapse and uses different neurotransmitters in different species: acetylcholine (ACh) in vertebrates, glutamate in *Drosophila*, and both ACh and γ-butyric acid in *Caenorhabditis elegans* (Wu et al., 2010). The chemical transmission of nerve impulses was discovered by Otto Loewi in 1921; this first neurotransmitter ever discovered was later confirmed to be ACh, which was identified by Sir Henry Hallett Dale in 1914 (Dale, 1914; Dale & Dudley, 1929). In vertebrate NMJs, ACh is synthesized from acetyl coenzyme A and choline by the enzyme choline acetyltransferase (ChAT) (Fig. 3). When the action potential from a motor neuron reaches the nerve terminal, voltage-gated calcium channels open, synaptic vesicles fuse with the plasma membrane, and ACh is released from the presynaptic nerve terminal into the synaptic cleft through exocytosis (Hong & Etherington, 2011).

Acetylcholine binds to the acetylcholine receptors (AChRs) on the postsynaptic membrane and triggers skeletal muscle contraction. The postsynaptic AChRs on the myofiber membrane are nicotinic; they bind to the addictive drug nicotine and components of certain snake venoms such as bungarotoxin (BTX). Nicotinic AChR was the first membrane receptor of a neurotransmitter and ion channel to be characterized. The receptor concept was initially proposed by Langley as a "receptive substance" in the muscle which "receives the stimulus from the nerve and transmits it to the effector cell" (Langley, 1905). In 1970, by using both tissue from the electric eel (containing large amounts of soluble cholinergic receptors) and the snake venom toxin BTX, Changeux et al. biochemically characterized AChR for the first time (Changeux, Kasai, & Lee, 1970). The current understanding of AChR structure and function largely comes from Unwin's studies using the AChR-enriched electric ray *Torpedo* and electron crystallography, advanced by recent X-ray crystallographic studies of proteins with similar structure such as nicotinic AChR prokaryotic homologues and ACh-binding proteins (Chen, 2010; Miyazawa, Fujiyoshi, & Unwin, 2003; Unwin, 1995, 2005; Unwin & Fujiyoshi, 2012).

The skeletal muscle AChR is a heteropentamer composed of four types of polypeptide chains: two α subunits and one of each β, γ, and δ subunit (in adults, the embryonic γ subunit is replaced by the homologous ε subunit). ACh binds to the α subunits of the receptor at their interfaces with neighboring γ and δ subunits, triggering the transient opening of the cation channel of the receptor and the influx of small cations (mainly sodium) (Unwin & Fujiyoshi, 2012). The flow of cations into the muscle cell through AChRs triggers a series of reactions in myofibers that leads to muscle contraction (Fig. 4), including: (1) cation influx gives rise to a synaptic potential; (2) this localized change in the membrane potential activates the voltage-gated sodium channels, which elicits an action potential that propagates along the myofiber; (3) when the action potential reaches transverse-tubules (T-tubules, extensions of the myofiber membrane containing large numbers of ion channels that penetrate into the center of myofibers), the voltage-dependent calcium channels open and trigger ryanodine receptors in the sarcoplasmic reticulum (SR) to open and release calcium; (4) calcium then binds to troponin (part of the thin filament of the myofibril together with tropomyosin and actin) which induces a conformational change in the

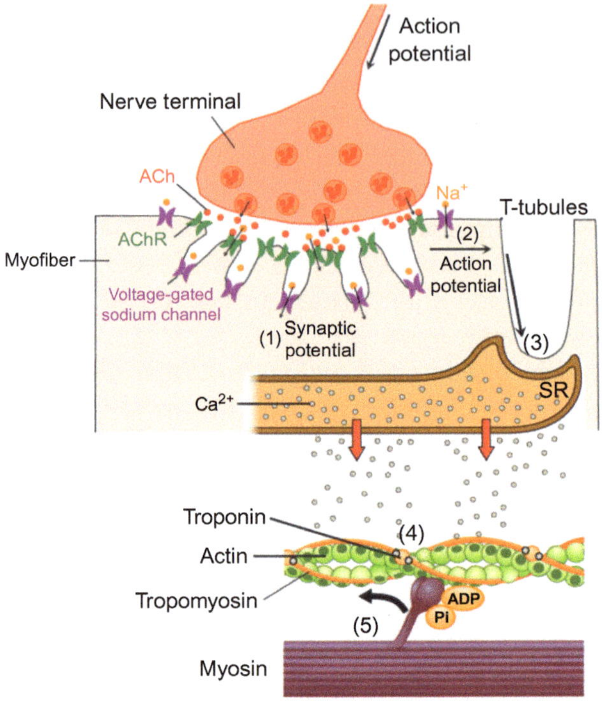

Fig. 4 Summary of excitation–contraction coupling. When the action potential reaches the presynaptic nerve terminal, ACh is released into the synaptic cleft. The binding of ACh to its receptor triggers the transient opening of the cation channel of the receptor and the influx of small cations (mainly Na^+). The flow of cations into the muscle cell through AChRs triggers a series of reactions in myofibers that leads to muscle contraction: (1) cation influx gives rise to a synaptic potential; (2) this localized change in the membrane potential activates the voltage-gated sodium channels, which elicits an action potential that propagates along the myofiber; (3) when the action potential reaches the T-tubules, SR releases calcium; (4) calcium then binds to troponin, exposing binding sites for myosin on the actin filament; (5) myosin bound by ADP and Pi can then form cross-bridges with actin; the release of ADP and Pi produces the power stroke that drives the sliding of the thin filament past the thick filament, thereby eliciting myofiber contraction. *Adapted from BC Open Textbooks https://opentextbc.ca/anatomyandphysiology/chapter/10-3-muscle-fiber-contraction-and-relaxation/.*

troponin complex, exposing binding sites for myosin (the thick filament of the myofibril) on the actin filament; (5) myosin bound by adenosine diphosphate (ADP) and inorganic phosphate (Pi) can then form cross-bridges with actin; the release of ADP and Pi produces the power stroke that drives the sliding of the thin filament past the thick filament, thereby eliciting myofiber contraction (Kuo & Ehrlich, 2015). Neurotransmission is rapidly terminated when ACh is degraded by acetylcholinesterase (AChE), which is anchored to the basal lamina (a layer of extracellular matrix, ECM) in the synaptic cleft by a collagen-like tail called ColQ (Hong & Etherington, 2011). Without the nerve impulse, calcium is transported back into the SR; the low calcium concentration releases myosin from actin and results in myofiber relaxation (Kuo & Ehrlich, 2015). Disruption in these events, including impairments in NMJ integrity, often triggers abnormal intracellular calcium handling, which leads to myofiber organelle dysfunction, protein catabolism, atrophy, sarcolemma disruption, or degeneration (Berchtold, Brinkmeier, & Muntener, 2000; Cisterna, Cardozo, & Saez, 2014).

The NMJ is a critical player in transducing motor neuron signals and triggering skeletal muscle contraction in a rapid, reliable, and precise manner. The essential function of the NMJ is ensured by specializations at this synapse: (1) large numbers of active zones in terminals, leading to excess neurotransmitter release; (2) clustering of AChRs at the crest of postsynaptic junctional folds and their alignment with the active zones, which increases efficiency of AChR activation and synaptic potential generation; (3) the rapid degradation of ACh by AChE after binding to AChR, which allows the temporal fidelity of transmission (Tintignac, Brenner, & Ruegg, 2015).

2. NMJ DEVELOPMENT AND MATURATION

The development and maturation of NMJs require spatially restricted signaling between presynaptic and postsynaptic cells and a series of complex interactions between motor neurons, myofibers, and Schwann cells to form a highly specialized motor nerve terminal and postsynaptic membrane.

During development, the major cell types of the NMJ must migrate over long distances to meet and form a functional synapse. Motor neurons are produced by multipotent progenitors located in the ventricular zone of the neural tube. In the spinal cord motor neurons are arranged into longitudinal columns that innervate the same target area; each motor column is organized as multiple motor pools, which contain cells that innervate the same target muscle (Jessell, 2000). Skeletal myofibers are derived from the mesoderm (the middle layer of an embryo in early development). More specifically, trunk and limb muscles originate from somites (the segmented paraxial mesoderm), whereas head muscles originate from the unsegmented cranial paraxial mesoderm and the prechordal mesoderm (Buckingham & Rigby, 2014). Committed myogenic cells migrate to sites of muscle formation, where they differentiate into myoblasts and further into myocytes before fusing to form immature myofibers, myotubes (Bentzinger, Wang, & Rudnicki, 2012; Buckingham & Rigby, 2014). Following guidance cues from the environment surrounding the axon trajectory and the target muscle, axons of motor neurons extend into the target and innervate myofibers (Bonanomi & Pfaff, 2010).

Prior to the arrival of motor neuron axons, primitive clusters of AChRs form in the center of myotubes in a nerve-independent process known as prepatterning (Fig. 5). This muscle-autonomous process indicates that muscle plays an active role in the initial formation of NMJs (Kummer, Misgeld, & Sanes, 2006). The growth cone of motor neuron axons reaches to some, but not all of the prepatterned AChR clusters. Subsequently the innervated clusters enlarge, whereas the uninnervated prepatterned sites eventually disperse (Lin et al., 2001; Vock, Ponomareva, & Rimer, 2008; Yang et al., 2001). At this stage, AChR clusters are innervated by two or more motor axons. NMJs are morphologically and functionally immature before birth, and NMJ maturation is required to ensure efficient neurotransmission for inducing muscle contractions in a refined and precise manner (Darabid, Perez-Gonzalez, & Robitaille, 2014).

As the NMJ matures postnatally, the postsynaptic membrane invaginates to form junctional folds; AChRs become concentrated at the crests of the folds, which change the shape of the NMJ from a flat plaque shape to a characteristic pretzel-like appearance (Fig. 6; Sanes & Lichtman, 2001). The AChR undergoes a shift in its subunit composition from $\alpha2\beta\delta\gamma$ to $\alpha2\beta\delta\varepsilon$. Also, the density of synaptic AChRs increases ~10-fold from approximately $1000/\mu m^2$ to more than $10,000/\mu m^2$, while the extrasynaptic density falls to $<10/\mu m^2$. Mechanisms that contribute to the high density of

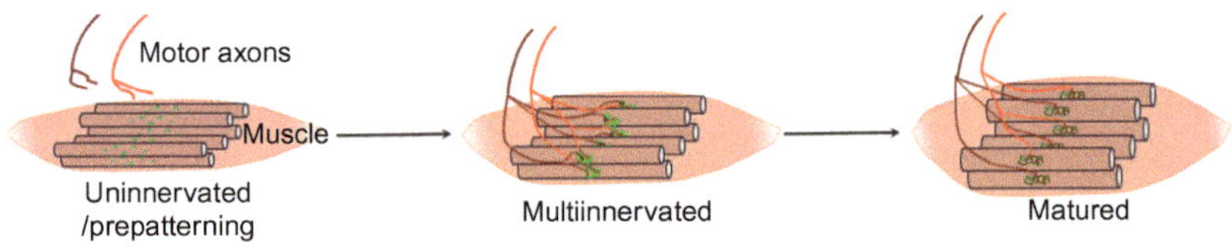

Fig. 5 Development and consolidation of NMJs. Prior to the arrival of motor neuron axons, primitive clusters of AChRs form in the myotubes in a nerve-independent process called prepatterning. Motor neuron axons reach to some AChR clusters, and the uninnervated prepatterned sites eventually disperse. At this stage, AChR clusters are multiinnervated by two or more motor axons. NMJs are morphologically and functionally immature. Extraaxonal inputs are later removed from the NMJ through synapse elimination, which decreases the number of nerve terminals in contact with a myofiber to one.

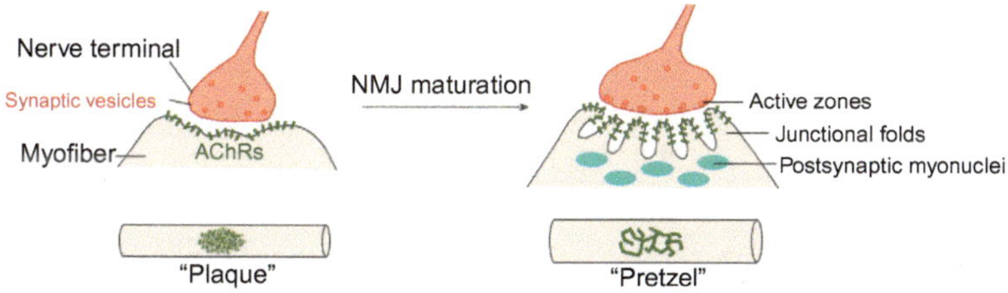

Fig. 6 Postnatal NMJ maturation. As the NMJ matures postnatally, the postsynaptic membrane invaginates to form junctional folds; AChRs become concentrated at the crests of the folds, which changes the shape of the NMJ from a flat plaque shape to a characteristic pretzel-like appearance. The postsynaptic myonuclei are transcriptionally specialized, expressing AChR subunits and other postsynaptic component genes at a higher rate than the extrasynaptic nuclei. Presynaptic nerve terminals also undergo major changes: the number of active zones (aligned with the AChRs on the crest of the junctional folds) are increased; extraaxonal inputs are removed; and uninnervated AChRs are removed thus achieving a perfect alignment between the presynaptic nerve terminal and AChR clusters.

AChR clusters at the postsynaptic membrane include the redistribution of AChRs from the extrasynaptic region due to lateral diffusion and anchorage to the synaptic pool, and the higher stability of postnatal AChRs in comparison to embryonic AChRs (Salpeter & Loring, 1985; Young & Poo, 1983). In addition, myonuclei associated with the postsynaptic membrane, postsynaptic myonuclei, are transcriptionally specialized, expressing AChR subunits and other postsynaptic component genes at a higher rate compared to the extrasynaptic nuclei (Merlie & Sanes, 1985; Schaeffer, de Kerchove d'Exaerde, & Changeux, 2001; Simon, Hoppe, & Burden, 1992). In mice that lack synaptic nuclear envelope protein 1 (Syne1), myonuclei fail to anchor at the postsynaptic membrane, which significantly impairs motor neuron innervation (Grady, Starr, Ackerman, Sanes, & Han, 2005; Zhang et al., 2007).

Postsynaptic rearrangement during NMJ maturation is mediated by key molecular mechanisms, among which the best characterized is the Agrin/lipoprotein receptor-related protein 4 (Lrp4)/muscle-specific tyrosine kinase receptor (MuSK) pathway (Wu et al., 2010). Agrin, a large heparan sulfate proteoglycan, is synthesized mostly by the motor neurons and essential for stabilizing AChR clusters at synaptic sites (Campanelli, Hoch, Rupp, Kreiner, & Scheller, 1991; Gautam et al., 1996). Agrin binds to the transmembrane protein Lrp4, which triggers the autophosphorylation of the tyrosine kinase receptor MuSK (Kim et al., 2008). The activation of MuSK leads to AChR clustering through Rapsyn (a membrane protein that interacts with cytoskeletal proteins and anchors AChR to the membrane) and AChR cluster stabilization by stimulating AChR phosphorylation (Apel, Glass, Moscoso, Yancopoulos, & Sanes, 1997; Gautam et al., 1995; Glass et al., 1996). While the Agrin/Lrp4/MuSK signaling pathway stabilizes AChRs at the synaptic membrane, ACh-induced muscle depolarization suppresses AChR subunit gene transcription and increases AChR degradation in other regions of the myofiber (Lin et al., 2005; Misgeld et al., 2002). The integration of the positive and negative signals results in repression of AChRs in extrasynaptic areas and a high density of AChRs at the postsynaptic membrane (Ferraro, Molinari, & Berghella, 2012).

Presynaptic nerve terminals also undergo major changes during NMJ maturation. Specifically, the number of active zones and the levels of associated proteins are increased; extraaxonal inputs are removed from the NMJ through synapse elimination, which decreases the number of nerve terminals in contact with a single myofiber (Balice-Gordon & Lichtman, 1993; Walsh & Lichtman, 2003). Multiple retrograde signals released by myofibers play significant roles in presynaptic differentiation and maturation: fibroblast growth factors (specifically FGF7/10/22) are involved in vesicle clustering and presynaptic differentiation (Fox et al., 2007); ECM proteins (such as laminins β2, α4 and α5, Collagens IV and XIII) are important for regulating active zone development and the proper alignment between pre- and postsynaptic membranes (Fox et al., 2007; Nishimune, Sanes, & Carlson, 2004; Nishimune et al., 2008; Singhal & Martin, 2011); glial cell-derived neurotrophic factor (GDNF) and brain-derived neurotrophic factor (BDNF) also regulate maturation of the presynaptic terminal and synapse elimination (Baudet et al., 2008; Je et al., 2013; Nguyen, Parsadanian, Snider, & Lichtman, 1998). Moreover, terminal Schwann cells produce and release synaptogenic factors required for NMJ maturation such as transforming growth factor β (TGFβ). They are particularly important for synaptic competition and elimination, during which they clear retracting axon endings and eliminate superfluous nerve terminals (Feng & Ko, 2008; Fuentes-Medel et al., 2012).

The development and maturation of the NMJ rely on the coordination between nerve- and muscle-derived molecules with concomitant contribution from terminal Schwann cells. The end result of neuromuscular synaptogenesis is the formation of a highly efficient communication site, where a myofiber is innervated by a single motor axon and postsynaptic receptors are perfectly aligned with presynaptic active zones.

3. NMJ REGENERATION

The regeneration of NMJs in response to peripheral nerve lesions can occur; however, the duration and quality of recovery depends on the severity of injury (Buti et al., 1996; Gutmann, Guttmann, Medawar, & Young, 1942). Shortly after nerve injury, the distal end of the axon undergoes Wallerian degeneration. The axolemma (cell membrane of the axon) rapidly degrades, followed by granular disintegration of the axonal cytoskeleton and organelles (Gillingwater & Ribchester, 2001). The myelin sheath then separates from the axon. Activated Schwann cells proliferate, phagocytose the detached axon/myelin debris, and produce cytokines/chemokines to recruit macrophages. The resident and recruited macrophages further clear the axon/myelin debris and produce factors that facilitate Schwann cell migration and axon regeneration (Chen, Piao, & Bonaldo, 2015). At the NMJ, terminal Schwann cells extend processes to actively invade into the synaptic cleft, accompanied by the fragmentation and phagocytosis of the presynaptic nerve terminal (Miledi & Slater, 1970). While the distal portion of the axon degenerates, the growth and regeneration occur in the proximal portion of the axon. Activating Schwann cells form longitudinal cell strands termed Bands of Büngner to provide structural guidance for axon regeneration and also release trophic factors to promote axon growth (Büngner, 1890; Heumann, Korsching, Bandtlow, & Thoenen, 1987). During the period of denervation, terminal Schwann cells extend elaborate processes, which eventually serve as both a guide and a growth promoter for regenerating axons to reinnervate myofibers (Kang et al., 2014; Son & Thompson, 1995). Regenerating axons follow their previously occupied endoneurial tubes to reinnervate the original synaptic site when the injury is minimal; innervation of a new ectopic site can occur if the nerve trauma is more severe (Nguyen, Sanes, & Lichtman, 2002). In addition, axon sprouts may arise from nerve terminals of neighboring NMJs and extend toward the denervated myofiber. After axons arrive at the synaptic site, there is commonly a remodeling of these contacts to eliminate multiple innervations and reorganize the postsynaptic AChRs (Kang et al., 2014; Rich & Lichtman, 1989).

Skeletal muscles also play a part in NMJ regeneration. In the absence of myofibers the initial reinnervation of synaptic basal lamina by motor axons can proceed; however, the prolonged maintenance of reinnervated NMJs on basal lamina ghosts eventually declines (Sanes & Lichtman, 1999;

Sanes, Marshall, & McMahan, 1978). These observations indicate the requirement of muscle-derived contribution for the continued differentiation and maintenance of regenerating NMJs. Other studies also suggested that proper signaling in muscle is essential for NMJ regeneration (Fig. 7). For instance, muscle-derived microRNA (miR)-206 was shown to be required for efficient regeneration of NMJs after nerve injury. In the absence of miR-206, reinnervation of denervated muscles was delayed after nerve crush or transection surgeries (Williams et al., 2009). Muscle-derived secretory factors such as GDNF were suggested to promote retrograde signaling that facilitates NMJ regeneration (Baudet et al., 2008). A recent study showed that the absence of Hippo signaling factor Yap in muscles prevented reinnervation of denervated myofibers after nerve transplantation or nerve crush injury (Zhao et al., 2017).

Adult skeletal muscle stem cells, satellite cells (SCs), were also revealed to be an essential muscle component for NMJ regeneration. Numerous studies have provided evidence that SCs, derivatives of embryonic and postnatal myogenesis, are the primary source for the nuclei of myofibers (myonuclei) and indispensable for skeletal muscle regeneration (Brack & Rando, 2012; Lepper, Partridge, & Fan, 2011; McCarthy et al., 2011; Murphy, Lawson, Mathew, Hutcheson, & Kardon, 2011; Sambasivan et al., 2011). Previously SCs were considered to have limited, if any, roles during muscle atrophy upon nerve insults, given the little myonuclear turnover and myogenic progenitor fusion observed in models of chronic denervation (where reinnervation is prevented) (Bruusgaard & Gundersen, 2008). Recently the fates and roles of SCs were reexamined in a model of peripheral nerve injury that enables NMJ regeneration. Increased SC activity and fusion of indelibly

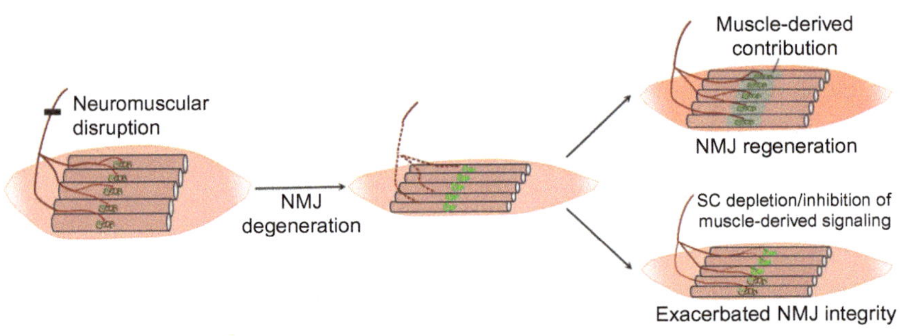

Fig. 7 Muscle-derived contribution to NMJ regeneration. Proper signaling in muscle is essential for NMJ regeneration. Inhibition of muscle-derived miR-206, GDNF–Ret signaling, and Yap signaling pathways have been shown to be detrimental to NMJ regeneration upon neuromuscular disruption. SCs also contribute to the regeneration of NMJs, depletion of which leads to impaired reinnervation and postsynaptic integrity.

labeled SC-derived progenitors to myofibers were demonstrated in the vicinity of regenerating NMJs. Depletion of SCs led to deficits in NMJ reinnervation, reductions in postsynaptic morphology, and loss of postsynaptic myonuclei, demonstrating a role for SCs in the regeneration of NMJs. Moreover, SC depletion was also associated with reduced myofiber size and further declines in muscle force-generation capacity, indicating the necessity for SC-mediated NMJ regeneration in the regulation of skeletal muscle integrity upon neuromuscular disruption (Liu, Wei-LaPierre, Klose, Dirksen, & Chakkalakal, 2015).

Consistent with a vital role for NMJ integrity in skeletal muscle maintenance, failed NMJ regeneration after injury leads to skeletal muscle loss and functional decline. Two major protein degradation pathways, the ubiquitin–proteasome and the autophagy–lysosome systems, are activated during skeletal muscle denervation atrophy (Bonaldo & Sandri, 2013). In response to denervation, Myogenin and FoxO transcription factors are upregulated to trigger the expression of E3 ubiquitin ligases such as MuRF1, Atrogin1, and Fbxo30, promoting muscle proteolysis and atrophy (Bodine et al., 2001; Moresi et al., 2010; Sartori et al., 2013; Winbanks et al., 2013). In addition, upon denervation interaction of TNF-like weak inducer of apoptosis (TWEAK) with its receptor Fn14 induces the activation of NF-κB, which increases the expression of MuRF1 and leads to muscle wasting and weakness (Mittal et al., 2010; Paul et al., 2010). Autophagy, which plays a crucial role in the turnover of cell components, is also activated following denervation and contributes to the loss of muscle mass (O'Leary, Vainshtein, Carter, Zhang, & Hood, 2012; Zhao et al., 2007). Skeletal muscle function decline occurs concurrently with muscle atrophy. Fibrillation activity where myofiber contraction is triggered by a spontaneous action potential without nerve input can be observed in denervated muscles, possibly due to the expanded expression of AChRs along the myofiber and disruption of the SR. Also, upon stimulation force-generation capacity is dramatically reduced in the denervated muscles, accompanied by modified contraction kinetics (Dulhunty, 1985; Lewis, 1972; Liu et al., 2015). Following the dramatic decline of muscle mass and loss of voluntary function, longer-term denervation can further lead to degeneration of myofibers and infiltration of fibrosis or fat tissue (Carlson, 2014).

4. NMJs AND DISEASES/AGING

Many diseases have been linked to impaired NMJ function and associated muscle wasting. Congenital myasthenic syndromes (CMS) are a group of genetic diseases characterized by compromised neurotransmission and skeletal muscle weakness/fatigability (Engel, 2012; Lorenzoni, Scola, Kay, & Werneck, 2012). Based on the subtypes the symptoms of CMS are manifested in different muscle groups involving ocular, bulbar, and limb muscles, which leads to ptosis (droopy eyelids), slurred speech, or difficulties with movements such as chewing, swallowing, and walking. The severity and symptomatic onset of diseases vary greatly. Some patients present muscle weakness in infant stage or childhood, whereas others are not symptomatic until adolescence or adulthood. Many mutations identified to be associated with CMS are NMJ development or function-related genes (Hantai, Nicole, & Eymard, 2013). Based on the location of primary defect within the NMJ, CMS can be classified as presynaptic, synaptic, or postsynaptic. Mutations in gene *CHAT*, which affects the synthesis of ACh in the presynaptic nerve terminal, are associated with the presynaptic subtype of CMS. The most common form of synaptic CMS is attributed to mutations in *COLQ*, which results in failed binding of AChE to the basal lamina thus an end plate AChE deficiency and prolonged exposure of AChRs to ACh stimulation. Rare cases with mutations in *LAMB2*, which encodes laminins β2, were also reported to cause synaptic CMS. Diverse mutations have been identified to be associated with postsynaptic CMS such as mutations in genes encoding AChR subunits, Agrin, MuSK, Rapsyn, downstream of tyrosine kinase 7 (Dok7), Lrp4, and $Na_v1.4$. These mutations lead to AChR deficiency, kinetic abnormalities of the AChR, or excitability of the voltage-gated sodium channel, thus defective neurotransmission. Current therapies for CMS include AChE inhibitors such as pyridostigmine, AChR open-channel blockers such as fluoxetine and quinidine sulfate, and adrenergic agonists such as salbutamol and ephedrine. The genetic cause of the disease is still unknown in many CMS patients. Further research in the future may identify additional mutations that have causal links to CMS, which may help identify genes involved in NMJ function or maintenance (Tintignac et al., 2015).

The autoimmune disease myasthenia gravis is another neuromuscular disorder characterized by defective neurotransmission. This defect is typically caused by antibodies against AChRs, or less frequently Agrin, MuSK, and Lrp4 (Berrih-Aknin, Frenkian-Cuvelier, & Eymard, 2014). Similar to CMS, the severity of varies markedly among patients, ranging from localized eye muscle weakness to more severe/generalized forms in which multiple voluntary muscles, even diaphragm, are affected. The treatment of myasthenia gravis usually involves AChE inhibitors to reduce ACh breakdown, immunosuppressive drugs, or thymectomy (Gilhus & Verschuuren, 2015). Lambert–Eaton myasthenic syndrome (LEMS) is another autoimmune disorder and mostly results from antibodies against voltage-gated calcium channels at the presynaptic nerve terminal (Hulsbrink & Hashemolhosseini, 2014). Muscle weakness in this disorder mostly starts from lower limb muscles. LEMS often occurs as a paraneoplastic syndrome with small cell lung carcinoma, though it can also be idiopathic. An effective symptomatic treatment for LEMS is 3,4-diaminopyridine (3,4-DAP), which inhibits presynaptic voltage-gated potassium channels and counteracts the diminished quantal release of acetylcholine at the NMJ.

Studies have also suggested that NMJs are early and important pathological targets for neuromuscular diseases such as amyotrophic lateral sclerosis (ALS) and spinal muscular atrophy (SMA) (Boyer, Ferrier, & Kothary, 2013; Murray, Talbot, & Gillingwater, 2010). ALS is an adult-onset and age-related lethal neuromuscular disease, characterized by progressive motor neuron death and atrophy/weakening of denervated muscles (Logroscino et al., 2015). The affected individuals generally die from respiratory failure within 3–5 years of the onset of the disease. Most ALS cases are sporadic, and 10% or less of them are inherited. About 20% of the familial ALS cases are linked to mutations in the Cu/Zn superoxide dismutase (SOD1) gene; an expansion of the intronic hexanucleotide repeat sequence in C9ORF72 has been reported to account for about 46% of the familial ALS cases (Renton et al., 2011). Riluzole remains the only Food and Drug Administration (FDA)-approved disease-modifying drug, which prolongs the survival of the patients by a few months. Studies have suggested that the changes at the NMJ occur prior to the loss of motor neurons and symptom onset (Dupuis & Loeffler, 2009; Fischer et al., 2004; Kennel, Finiels, Revah, & Mallet, 1996). Specific manipulation of gene expression in muscle has been shown to affect the progression of the disease, indicating the therapeutic potential of targeting the NMJ and skeletal muscle (Dobrowolny et al., 2005; Dupuis et al., 2009; Wong & Martin, 2010). SMA is a predominantly childhood onset neuromuscular disease caused by mutations in the survival motor neuron gene (*SMN*). Spinraza, a *SMN2*-derected antisense oligonucleotide drug, was approved by FDA recently for SMA therapy. Similar to ALS, motor neuron death and muscle denervation/atrophy lead to paralysis of patients. Retrograde signals from skeletal muscles and NMJs have also been demonstrated to contribute to the pathogenesis of SMA (Boyer et al., 2013). NMJ alterations including defective/delayed maturation, denervation, and neurofilament accumulation occur at early stages of SMA (Boido & Vercelli, 2016). The study of NMJ and muscle deficits is critical for better understanding of these neuromuscular diseases and could provide more therapeutic options.

In addition to the aforementioned pathological conditions, sarcopenia, the progressive decline of skeletal muscle with age has also been linked to defects of the NMJ. Previous studies have shown that human and rodent NMJs undergo marked morphological remodeling during aging, including axonal swellings, sprouting, partial or complete denervation, and fragmentation of the postsynaptic apparatus (Chai, Vukovic, Dunlop, Grounds, & Shavlakadze, 2011; Oda, 1984; Valdez et al., 2010; Wokke et al., 1990). Functional changes in aged skeletal muscle connected to NMJ degeneration include a marked drop (~45%) of peak tetanic force generated by the extensor digitorum longus muscle of aged rats compared with young adult controls (Ibebunjo et al., 2013). Global gene expression profiles of aged mouse or rat skeletal muscles showed perturbed genes/pathways associated with NMJ patency, providing molecular evidence of sarcopenia-related NMJ remodeling and dysfunction (Barns et al., 2014; Ibebunjo et al., 2013; Liu et al., 2017).

The principal cause of age-related NMJ degeneration is still unknown. A commonly accepted theory is that aging is associated with motor neuron loss. Upon the death of a motor neuron, myofibers previously innervated by the neuron become denervated; axons from residual neurons then sprout to reinnervate these myofibers. With aging, it is thought that the adaptive axonal sprouting capacity declines and leads to eventual myofiber denervation and atrophy (Gonzalez-Freire, de Cabo, Studenski, & Ferrucci, 2014). Some studies showed modest declines in the number of motor neurons or axons in spinal ventral roots in aged human and rodents (Aagaard, Suetta, Caserotti, Magnusson, & Kjaer, 2010; Rowan et al., 2012; Tomlinson & Irving, 1977; Valdez et al., 2010). Nevertheless, causal links between motor neuron death, failure of denervation–reinnervation, and age-related NMJ degeneration have mostly been inferred rather than examined. Although geriatric mice have a high proportion of partially innervated NMJs where some postsynaptic folds are devoid of nerve terminals, no evidence of functional denervation was found by physiological criteria (Banker, Kelly, & Robbins, 1983; Willadt, Nash, & Slater, 2016). Moreover, recent mouse studies showed that although NMJs and their interactions with myofibers are greatly altered during aging, little if any loss of motor neurons occurs based on serial section analysis of lumbar spinal cord regions (Chai et al., 2011).

Fig. 8 Deterioration of the NMJ with aging. NMJs deteriorate with age, demonstrated by reduced nerve terminal area and loss of postsynaptic complexity. Aging is associated with SC loss, and SC depletion accelerates age-related NMJ deterioration.

Considering that NMJs respond to impacts from both the motor neuron and the myofiber, age-related NMJ degeneration may be rooted from changes in the aged skeletal muscle. Autophagy declines in aged skeletal muscles; inhibition or rescue of autophagy flux was shown to exacerbate or improve aged NMJ integrity, respectively (Carnio et al., 2014). Although vital imaging of the sternomastoid muscle in aging mice found rare local degenerative–regenerative events associated with persistent NMJ changes, indelible labeling experiments have demonstrated SC-derived contributions to sedentary adult myofibers with negligible degeneration (Keefe et al., 2015; Li, Lee, & Thompson, 2011). Nonetheless, such observations suggest an interrelationship between age-related declines in SC number and function and NMJ integrity (Chakkalakal, Jones, Basson, & Brack, 2012; Relaix & Zammit, 2012; Sousa-Victor, Garcia-Prat, Serrano, Perdiguero, & Munoz-Canoves, 2015). Indeed, fate tracking of indelibly labeled SCs demonstrates notable contribution to adult NMJs that is lost after middle age (Liu et al., 2017; Fig. 8). Furthermore, loss of SCs is sufficient to accelerate age-related NMJ degeneration, whereas gain of SCs is associated with attenuation of NMJ deterioration (Liu et al., 2017).

5. CONCLUSION AND FUTURE DIRECTIONS

The NMJ is a critical connection point between the nervous system and the skeletal muscles. Proper development, maturation, and maintenance of NMJs are crucial for the functionality of the neuromuscular system. Assaults from both the neuron and muscle sides, such as nerve injury or skeletal muscle damages, can affect NMJ function and trigger NMJ deterioration. The reinnervation and reorganization of NMJs after nerve injury are critical for denervated skeletal muscle recovery. Genetic mutations or autoimmune targeting in NMJ components lead to severe muscle weakness/atrophy in CMS and myasthenia gravis. Studies on neuromuscular diseases such as ALS or SMA have provided evidence that NMJ vulnerability and intrinsic muscle defects lead to motor neuron degeneration in a "dying-back" manner (Boyer et al., 2013; Moloney, de Winter, & Verhaagen, 2014; Murray et al., 2010). During aging, NMJ deterioration occurs concomitantly with progressive decline in skeletal muscle regenerative potential.

Given that SCs from animal models or human biopsies for both ALS and SMA showed defective myogenic potential, it is plausible that SC dysfunction may contribute to the progression of these diseases (Iascone, Henderson, & Lee, 2015; Loeffler, Picchiarelli, Dupuis, & Gonzalez De Aguilar, 2016). Further research will be needed to determine whether SCs play a pathogenic role in these diseases or act as a mediator between motor neuron degeneration and skeletal muscle atrophy. Recently some exciting results were published where genome editing by a clustered regularly interspaced short palindromic repeats (CRISPR)/CRISPR-associated nuclease (Cas) 9 system was utilized to correct the gene mutation in the Duchenne muscular dystrophy (DMD) mouse models (Long et al., 2016; Nelson et al., 2016; Tabebordbar et al., 2016). The corrected gene expression was shown to increase over time, and the degree of muscle phenotypic rescue even exceeded the efficiency of gene correction, indicating the significance of achieving editing efficiency in SCs or muscle progenitors (Long et al., 2016, 2014). Similar to DMD, NMJ mutation-associated diseases such as CMS are also monogenic disorders (Engel, Ohno, & Sine, 2003; Tintignac et al., 2015). Correcting CMS-associated mutations through CRISPR/Cas9-based gene-editing system could provide a potential means for treating these diseases. Considering the cellular turnover around the NMJ and SC-derived contribution to NMJs, gene-editing efficiency in SCs needs to be considered to ensure extended restoration of corrected gene expression.

Previously sarcopenia has been considered as a neurogenic syndrome considering the crucial influence of neuronal activity on skeletal muscle function (Kwan, 2013). However, there is a lack of consensus if prevailing motor neuron death occurs before the development of sarcopenia-related features (Banker et al., 1983; Chai et al., 2011; Tomlinson & Irving, 1977). Moreover, skeletal muscle-resident stem and progenitor cell populations such as SCs are beginning to be appreciated for roles in NMJ regeneration and life-long maintenance. Skeletal muscle is a source of additional resident mesenchymal stem and progenitor populations, such as, PW1+/Pax7− interstitial cells, and myoendothelial cells (Huang, Schulz, Beauvais, Tseng, & Gussoni, 2014; Joe et al., 2010; Malecova & Puri, 2012; Mitchell et al., 2010). Understanding the extent to which these resident skeletal muscle stem and progenitor populations contribute either individually or in concert to the regeneration and maintenance of NMJs should provide important insights into the etiology of NMJ degenerative disorders and sarcopenia.

REFERENCES

Aagaard, P., Suetta, C., Caserotti, P., Magnusson, S. P., & Kjaer, M. (2010). Role of the nervous system in sarcopenia and muscle atrophy with aging: Strength training as a countermeasure. *Scandinavian Journal of Medicine & Science in Sports*, *20*, 49–64.

Apel, E. D., Glass, D. J., Moscoso, L. M., Yancopoulos, G. D., & Sanes, J. R. (1997). Rapsyn is required for MuSK signaling and recruits synaptic components to a MuSK-containing scaffold. *Neuron*, *18*, 623–635.

Balice-Gordon, R. J., & Lichtman, J. W. (1993). In vivo observations of pre- and postsynaptic changes during the transition from multiple to single innervation at developing neuromuscular junctions. *The Journal of Neuroscience: The Official Journal of the Society for Neuroscience*, *13*, 834–855.

Banker, B. Q., Kelly, S. S., & Robbins, N. (1983). Neuromuscular transmission and correlative morphology in young and old mice. *The Journal of Physiology*, *339*, 355–377.

Barik, A., Li, L., Sathyamurthy, A., Xiong, W. C., & Mei, L. (2016). Schwann cells in neuromuscular junction formation and maintenance. *The Journal of Neuroscience: The Official Journal of the Society for Neuroscience*, *36*, 9770–9781.

Barns, M., Gondro, C., Tellam, R. L., Radley-Crabb, H. G., Grounds, M. D., & Shavlakadze, T. (2014). Molecular analyses provide insight into mechanisms underlying sarcopenia and myofibre denervation in old skeletal muscles of mice. *The International Journal of Biochemistry & Cell Biology*, *53*, 174–185.

Bassel-Duby, R., & Olson, E. N. (2006). Signaling pathways in skeletal muscle remodeling. *Annual Review of Biochemistry*, *75*, 19–37.

Baudet, C., Pozas, E., Adameyko, I., Andersson, E., Ericson, J., & Ernfors, P. (2008). Retrograde signaling onto Ret during motor nerve terminal maturation. *The Journal of Neuroscience: The Official Journal of the Society for Neuroscience*, *28*, 963–975.

Bentzinger, C. F., Wang, Y. X., & Rudnicki, M. A. (2012). Building muscle: Molecular regulation of myogenesis. *Cold Spring Harbor Perspectives in Biology*, *4*, a008342.

Berchtold, M. W., Brinkmeier, H., & Muntener, M. (2000). Calcium ion in skeletal muscle: Its crucial role for muscle function, plasticity, and disease. *Physiological Reviews*, *80*, 1215–1265.

Berrih-Aknin, S., Frenkian-Cuvelier, M., & Eymard, B. (2014). Diagnostic and clinical classification of autoimmune myasthenia gravis. *Journal of Autoimmunity*, *48–49*, 143–148.

Bodine, S. C., Latres, E., Baumhueter, S., Lai, V. K., Nunez, L., Clarke, B. A., et al. (2001). Identification of ubiquitin ligases required for skeletal muscle atrophy. *Science*, *294*, 1704–1708.

Boido, M., & Vercelli, A. (2016). Neuromuscular junctions as key contributors and therapeutic targets in spinal muscular atrophy. *Frontiers in Neuroanatomy*, *10*, 6.

Bonaldo, P., & Sandri, M. (2013). Cellular and molecular mechanisms of muscle atrophy. *Disease Models & Mechanisms*, *6*, 25–39.

Bonanomi, D., & Pfaff, S. L. (2010). Motor axon pathfinding. *Cold Spring Harbor Perspectives in Biology*, *2*, a001735.

Boyer, J. G., Ferrier, A., & Kothary, R. (2013). More than a bystander: The contributions of intrinsic skeletal muscle defects in motor neuron diseases. *Frontiers in Physiology*, *4*, 356.
Brack, A. S., & Rando, T. A. (2012). Tissue-specific stem cells: Lessons from the skeletal muscle satellite cell. *Cell Stem Cell*, *10*, 504–514.
Bruusgaard, J. C., & Gundersen, K. (2008). In vivo time-lapse microscopy reveals no loss of murine myonuclei during weeks of muscle atrophy. *The Journal of Clinical Investigation*, *118*, 1450–1457.
Buckingham, M., & Rigby, P. W. (2014). Gene regulatory networks and transcriptional mechanisms that control myogenesis. *Developmental Cell*, *28*, 225–238.
Büngner, O. (1890). *Ueber die degenerations- und Regenerationsvorgänge am Nerven nach Verletzungen*. Marburg, Germany: Universität Marburg.
Buti, M., Verdu, E., Labrador, R. O., Vilches, J. J., Fores, J., & Navarro, X. (1996). Influence of physical parameters of nerve chambers on peripheral nerve regeneration and reinnervation. *Experimental Neurology*, *137*, 26–33.
Campanelli, J. T., Hoch, W., Rupp, F., Kreiner, T., & Scheller, R. H. (1991). Agrin mediates cell contact-induced acetylcholine receptor clustering. *Cell*, *67*, 909–916.
Carlson, B. M. (2014). The biology of long-term denervated skeletal muscle. *European Journal of Translational Myology*, *24*, 3293.
Carnio, S., LoVerso, F., Baraibar, M. A., Longa, E., Khan, M. M., Maffei, M., et al. (2014). Autophagy impairment in muscle induces neuromuscular junction degeneration and precocious aging. *Cell Reports*, *8*, 1509–1521.
Chai, R. J., Vukovic, J., Dunlop, S., Grounds, M. D., & Shavlakadze, T. (2011). Striking denervation of neuromuscular junctions without lumbar motoneuron loss in geriatric mouse muscle. *PloS One*, *6*, e28090.
Chakkalakal, J. V., Jones, K. M., Basson, M. A., & Brack, A. S. (2012). The aged niche disrupts muscle stem cell quiescence. *Nature*, *490*, 355–360.
Changeux, J. P., Kasai, M., & Lee, C. Y. (1970). Use of a snake venom toxin to characterize the cholinergic receptor protein. *Proceedings of the National Academy of Sciences of the United States of America*, *67*, 1241–1247.
Chen, L. (2010). In pursuit of the high-resolution structure of nicotinic acetylcholine receptors. *The Journal of Physiology*, *588*, 557–564.
Chen, P., Piao, X., & Bonaldo, P. (2015). Role of macrophages in Wallerian degeneration and axonal regeneration after peripheral nerve injury. *Acta Neuropathologica*, *130*, 605–618.
Cisterna, B. A., Cardozo, C., & Saez, J. C. (2014). Neuronal involvement in muscular atrophy. *Frontiers in Cellular Neuroscience*, *8*, 405.
Court, F. A., Gillingwater, T. H., Melrose, S., Sherman, D. L., Greenshields, K. N., Morton, A. J., et al. (2008). Identity, developmental restriction and reactivity of extralaminar cells capping mammalian neuromuscular junctions. *Journal of Cell Science*, *121*, 3901–3911.
Dale, H. H. (1914). The action of certain esters and ethers of choline, and their relation to muscarine. *Journal of Pharmacology and Experimental Therapeutics*, *6*, 147–190.
Dale, H. H., & Dudley, H. W. (1929). The presence of histamine and acetylcholine in the spleen of the ox and the horse. *The Journal of Physiology*, *68*, 97–123.
Darabid, H., Perez-Gonzalez, A. P., & Robitaille, R. (2014). Neuromuscular synaptogenesis: Coordinating partners with multiple functions. *Nature Reviews. Neuroscience*, *15*, 703–718.
Deutch, A. Y. (2013). Chapter 6—Neurotransmitters A2. In L. R. Squire, D. Berg, F. E. Bloom, S. D. Lac, A. Ghosh, & N. C. Spitzer (Eds.), *Fundamental neuroscience* (4th ed., pp. 117–138). San Diego: Academic Press.

Dobrowolny, G., Giacinti, C., Pelosi, L., Nicoletti, C., Winn, N., Barberi, L., et al. (2005). Muscle expression of a local Igf-1 isoform protects motor neurons in an ALS mouse model. *The Journal of Cell Biology, 168*, 193–199.

Dulhunty, A. F. (1985). Excitation-contraction coupling and contractile properties in denervated rat EDL and soleus muscles. *Journal of Muscle Research and Cell Motility, 6*, 207–225.

Dupuis, L., Gonzalez de Aguilar, J. L., Echaniz-Laguna, A., Eschbach, J., Rene, F., Oudart, H., et al. (2009). Muscle mitochondrial uncoupling dismantles neuromuscular junction and triggers distal degeneration of motor neurons. *PloS One, 4*, e5390.

Dupuis, L., & Loeffler, J. P. (2009). Neuromuscular junction destruction during amyotrophic lateral sclerosis: Insights from transgenic models. *Current Opinion in Pharmacology, 9*, 341–346.

Engel, A. G. (2012). Current status of the congenital myasthenic syndromes. *Neuromuscular Disorders: NMD, 22*, 99–111.

Engel, A. G., Ohno, K., & Sine, S. M. (2003). Sleuthing molecular targets for neurological diseases at the neuromuscular junction. *Nature Reviews. Neuroscience, 4*, 339–352.

Fagerlund, M. J., & Eriksson, L. I. (2009). Current concepts in neuromuscular transmission. *British Journal of Anaesthesia, 103*, 108–114.

Feng, Z., & Ko, C. P. (2008). Schwann cells promote synaptogenesis at the neuromuscular junction via transforming growth factor-beta1. *The Journal of Neuroscience: The Official Journal of the Society for Neuroscience, 28*, 9599–9609.

Ferraro, E., Molinari, F., & Berghella, L. (2012). Molecular control of neuromuscular junction development. *Journal of Cachexia, Sarcopenia and Muscle, 3*, 13–23.

Fischer, L. R., Culver, D. G., Tennant, P., Davis, A. A., Wang, M., Castellano-Sanchez, A., et al. (2004). Amyotrophic lateral sclerosis is a distal axonopathy: Evidence in mice and man. *Experimental Neurology, 185*, 232–240.

Fox, M. A., Sanes, J. R., Borza, D. B., Eswarakumar, V. P., Fassler, R., Hudson, B. G., et al. (2007). Distinct target-derived signals organize formation, maturation, and maintenance of motor nerve terminals. *Cell, 129*, 179–193.

Fuentes-Medel, Y., Ashley, J., Barria, R., Maloney, R., Freeman, M., & Budnik, V. (2012). Integration of a retrograde signal during synapse formation by glia-secreted TGF-beta ligand. *Current Biology: CB, 22*, 1831–1838.

Gautam, M., Noakes, P. G., Moscoso, L., Rupp, F., Scheller, R. H., Merlie, J. P., et al. (1996). Defective neuromuscular synaptogenesis in agrin-deficient mutant mice. *Cell, 85*, 525–535.

Gautam, M., Noakes, P. G., Mudd, J., Nichol, M., Chu, G. C., Sanes, J. R., et al. (1995). Failure of postsynaptic specialization to develop at neuromuscular junctions of rapsyn-deficient mice. *Nature, 377*, 232–236.

Gilhus, N. E., & Verschuuren, J. J. (2015). Myasthenia gravis: Subgroup classification and therapeutic strategies. *The Lancet Neurology, 14*, 1023–1036.

Gillingwater, T. H., & Ribchester, R. R. (2001). Compartmental neurodegeneration and synaptic plasticity in the Wlds mutant mouse. *The Journal of Physiology, 534*, 627–639.

Glass, D. J., DeChiara, T. M., Stitt, T. N., DiStefano, P. S., Valenzuela, D. M., & Yancopoulos, G. D. (1996). The receptor tyrosine kinase MuSK is required for neuromuscular junction formation and is a functional receptor for agrin. *Cold Spring Harbor Symposia on Quantitative Biology, 61*, 435–444.

Gonzalez-Freire, M., de Cabo, R., Studenski, S. A., & Ferrucci, L. (2014). The neuromuscular junction: Aging at the crossroad between nerves and muscle. *Frontiers in Aging Neuroscience, 6*, 208.

Grady, R. M., Starr, D. A., Ackerman, G. L., Sanes, J. R., & Han, M. (2005). Syne proteins anchor muscle nuclei at the neuromuscular junction. *Proceedings of the National Academy of Sciences of the United States of America, 102*, 4359–4364.

Griffin, J. W., & Thompson, W. J. (2008). Biology and pathology of nonmyelinating Schwann cells. *Glia, 56*, 1518–1531.
Gutmann, E., Guttmann, L., Medawar, P. B., & Young, J. Z. (1942). The rate of regeneration of nerve. *Journal of Experimental Biology, 19*, 14–44.
Hantai, D., Nicole, S., & Eymard, B. (2013). Congenital myasthenic syndromes: An update. *Current Opinion in Neurology, 26*, 561–568.
Heumann, R., Korsching, S., Bandtlow, C., & Thoenen, H. (1987). Changes of nerve growth factor synthesis in nonneuronal cells in response to sciatic nerve transection. *The Journal of Cell Biology, 104*, 1623–1631.
Hong, I. H. K., & Etherington, S. J. (2011). Neuromuscular junction. In: *Encyclopedia of Life Sciences (ELS)*. Chichester: John Wiley & Sons, Ltd.
Huang, P., Schulz, T. J., Beauvais, A., Tseng, Y. H., & Gussoni, E. (2014). Intramuscular adipogenesis is inhibited by myo-endothelial progenitors with functioning Bmpr1a signalling. *Nature Communications, 5*, 4063.
Hulsbrink, R., & Hashemolhosseini, S. (2014). Lambert-Eaton myasthenic syndrome—Diagnosis, pathogenesis and therapy. *Clinical Neurophysiology, 125*, 2328–2336.
Iascone, D. M., Henderson, C. E., & Lee, J. C. (2015). Spinal muscular atrophy: From tissue specificity to therapeutic strategies. *F1000Prime Reports*, 7, 04.
Ibebunjo, C., Chick, J. M., Kendall, T., Eash, J. K., Li, C., Zhang, Y., et al. (2013). Genomic and proteomic profiling reveals reduced mitochondrial function and disruption of the neuromuscular junction driving rat sarcopenia. *Molecular and Cellular Biology, 33*, 194–212.
Je, H. S., Yang, F., Ji, Y., Potluri, S., Fu, X. Q., Luo, Z. G., et al. (2013). ProBDNF and mature BDNF as punishment and reward signals for synapse elimination at mouse neuromuscular junctions. *The Journal of Neuroscience: The Official Journal of the Society for Neuroscience, 33*, 9957–9962.
Jessell, T. M. (2000). Neuronal specification in the spinal cord: Inductive signals and transcriptional codes. *Nature Reviews. Genetics, 1*, 20–29.
Joe, A. W., Yi, L., Natarajan, A., Le Grand, F., So, L., Wang, J., et al. (2010). Muscle injury activates resident fibro/adipogenic progenitors that facilitate myogenesis. *Nature Cell Biology, 12*, 153–163.
Kang, H., Tian, L., Mikesh, M., Lichtman, J. W., & Thompson, W. J. (2014). Terminal Schwann cells participate in neuromuscular synapse remodeling during reinnervation following nerve injury. *The Journal of Neuroscience: The Official Journal of the Society for Neuroscience, 34*, 6323–6333.
Keefe, A. C., Lawson, J. A., Flygare, S. D., Fox, Z. D., Colasanto, M. P., Mathew, S. J., et al. (2015). Muscle stem cells contribute to myofibres in sedentary adult mice. *Nature Communications, 6*, 7087.
Kennel, P. F., Finiels, F., Revah, F., & Mallet, J. (1996). Neuromuscular function impairment is not caused by motor neurone loss in FALS mice: An electromyographic study. *Neuroreport*, 7, 1427–1431.
Kim, N., Stiegler, A. L., Cameron, T. O., Hallock, P. T., Gomez, A. M., Huang, J. H., et al. (2008). Lrp4 is a receptor for Agrin and forms a complex with MuSK. *Cell, 135*, 334–342.
Kühne, W. (1862). *Über die peripherischen Endorgane der motorischen Nerven*. Leipzig, Germany: W. Engelmann.
Kummer, T. T., Misgeld, T., & Sanes, J. R. (2006). Assembly of the postsynaptic membrane at the neuromuscular junction: Paradigm lost. *Current Opinion in Neurobiology, 16*, 74–82.
Kuo, I. Y., & Ehrlich, B. E. (2015). Signaling in muscle contraction. *Cold Spring Harbor Perspectives in Biology*, 7, a006023.
Kwan, P. (2013). Sarcopenia, a neurogenic syndrome? *Journal of Aging Research, 2013*. 791679.

Langley, J. N. (1905). On the reaction of cells and of nerve-endings to certain poisons, chiefly as regards the reaction of striated muscle to nicotine and to curari. *The Journal of Physiology, 33*, 374–413.

Larsson, L., & Ansved, T. (1995). Effects of ageing on the motor unit. *Progress in Neurobiology, 45*, 397–458.

Lepper, C., Partridge, T. A., & Fan, C. M. (2011). An absolute requirement for Pax7-positive satellite cells in acute injury-induced skeletal muscle regeneration. *Development, 138*, 3639–3646.

Lewis, D. M. (1972). The effect of denervation on the mechanical and electrical responses of fast and slow mammalian twitch muscle. *The Journal of Physiology, 222*, 51–75.

Li, Y., Lee, Y., & Thompson, W. J. (2011). Changes in aging mouse neuromuscular junctions are explained by degeneration and regeneration of muscle fiber segments at the synapse. *The Journal of Neuroscience: The Official Journal of the Society for Neuroscience, 31*, 14910–14919.

Lin, W., Burgess, R. W., Dominguez, B., Pfaff, S. L., Sanes, J. R., & Lee, K. F. (2001). Distinct roles of nerve and muscle in postsynaptic differentiation of the neuromuscular synapse. *Nature, 410*, 1057–1064.

Lin, W., Dominguez, B., Yang, J., Aryal, P., Brandon, E. P., Gage, F. H., et al. (2005). Neurotransmitter acetylcholine negatively regulates neuromuscular synapse formation by a Cdk5-dependent mechanism. *Neuron, 46*, 569–579.

Liu, W., Klose, A., Forman, S., Paris, N. D., Wei-LaPierre, L., Cortes-Lopez, M., et al. (2017). Loss of adult skeletal muscle stem cells drives age-related neuromuscular junction degeneration. *eLife, 6*, e26464.

Liu, W., Wei-LaPierre, L., Klose, A., Dirksen, R. T., & Chakkalakal, J. V. (2015). Inducible depletion of adult skeletal muscle stem cells impairs the regeneration of neuromuscular junctions. *eLife, 4*, e09221.

Loeffler, J. P., Picchiarelli, G., Dupuis, L., & Gonzalez De Aguilar, J. L. (2016). The role of skeletal muscle in amyotrophic lateral sclerosis. *Brain Pathology, 26*, 227–236.

Logroscino, G., Tortelli, R., Rizzo, G., Marin, B., Preux, P. M., & Malaspina, A. (2015). Amyotrophic lateral sclerosis: An aging-related disease. *Current Geriatrics Reports, 4*, 142–153.

Long, C., Amoasii, L., Mireault, A. A., McAnally, J. R., Li, H., Sanchez-Ortiz, E., et al. (2016). Postnatal genome editing partially restores dystrophin expression in a mouse model of muscular dystrophy. *Science, 351*, 400–403.

Long, C., McAnally, J. R., Shelton, J. M., Mireault, A. A., Bassel-Duby, R., & Olson, E. N. (2014). Prevention of muscular dystrophy in mice by CRISPR/Cas9-mediated editing of germline DNA. *Science, 345*, 1184–1188.

Lorenzoni, P. J., Scola, R. H., Kay, C. S., & Werneck, L. C. (2012). Congenital myasthenic syndrome: A brief review. *Pediatric Neurology, 46*, 141–148.

Malecova, B., & Puri, P. L. (2012). "Mix of Mics"—Phenotypic and biological heterogeneity of "multipotent" muscle interstitial cells (MICs). *Journal of Stem Cell Research & Therapy. 004,* (Suppl 11).

McCarthy, J. J., Mula, J., Miyazaki, M., Erfani, R., Garrison, K., Farooqui, A. B., et al. (2011). Effective fiber hypertrophy in satellite cell-depleted skeletal muscle. *Development, 138*, 3657–3666.

Merlie, J. P., & Sanes, J. R. (1985). Concentration of acetylcholine receptor mRNA in synaptic regions of adult muscle fibres. *Nature, 317*, 66–68.

Miledi, R., & Slater, C. R. (1970). On the degeneration of rat neuromuscular junctions after nerve section. *The Journal of Physiology, 207*, 507–528.

Misgeld, T., Burgess, R. W., Lewis, R. M., Cunningham, J. M., Lichtman, J. W., & Sanes, J. R. (2002). Roles of neurotransmitter in synapse formation: Development of neuromuscular junctions lacking choline acetyltransferase. *Neuron, 36*, 635–648.

Mitchell, K. J., Pannerec, A., Cadot, B., Parlakian, A., Besson, V., Gomes, E. R., et al. (2010). Identification and characterization of a non-satellite cell muscle resident progenitor during postnatal development. *Nature Cell Biology*, *12*, 257–266.
Mittal, A., Bhatnagar, S., Kumar, A., Lach-Trifilieff, E., Wauters, S., Li, H., et al. (2010). The TWEAK-Fn14 system is a critical regulator of denervation-induced skeletal muscle atrophy in mice. *The Journal of Cell Biology*, *188*, 833–849.
Miyazawa, A., Fujiyoshi, Y., & Unwin, N. (2003). Structure and gating mechanism of the acetylcholine receptor pore. *Nature*, *423*, 949–955.
Moloney, E. B., de Winter, F., & Verhaagen, J. (2014). ALS as a distal axonopathy: Molecular mechanisms affecting neuromuscular junction stability in the presymptomatic stages of the disease. *Frontiers in Neuroscience*, *8*, 252.
Moresi, V., Williams, A. H., Meadows, E., Flynn, J. M., Potthoff, M. J., McAnally, J., et al. (2010). Myogenin and class II HDACs control neurogenic muscle atrophy by inducing E3 ubiquitin ligases. *Cell*, *143*, 35–45.
Murphy, M. M., Lawson, J. A., Mathew, S. J., Hutcheson, D. A., & Kardon, G. (2011). Satellite cells, connective tissue fibroblasts and their interactions are crucial for muscle regeneration. *Development*, *138*, 3625–3637.
Murray, L. M., Talbot, K., & Gillingwater, T. H. (2010). Review: Neuromuscular synaptic vulnerability in motor neurone disease: Amyotrophic lateral sclerosis and spinal muscular atrophy. *Neuropathology and Applied Neurobiology*, *36*, 133–156.
Nelson, C. E., Hakim, C. H., Ousterout, D. G., Thakore, P. I., Moreb, E. A., Castellanos Rivera, R. M., et al. (2016). In vivo genome editing improves muscle function in a mouse model of Duchenne muscular dystrophy. *Science*, *351*, 403–407.
Nguyen, Q. T., Parsadanian, A. S., Snider, W. D., & Lichtman, J. W. (1998). Hyperinnervation of neuromuscular junctions caused by GDNF overexpression in muscle. *Science*, *279*, 1725–1729.
Nguyen, Q. T., Sanes, J. R., & Lichtman, J. W. (2002). Pre-existing pathways promote precise projection patterns. *Nature Neuroscience*, *5*, 861–867.
Nishimune, H., Sanes, J. R., & Carlson, S. S. (2004). A synaptic laminin-calcium channel interaction organizes active zones in motor nerve terminals. *Nature*, *432*, 580–587.
Nishimune, H., Valdez, G., Jarad, G., Moulson, C. L., Muller, U., Miner, J. H., et al. (2008). Laminins promote postsynaptic maturation by an autocrine mechanism at the neuromuscular junction. *The Journal of Cell Biology*, *182*, 1201–1215.
Oda, K. (1984). Age changes of motor innervation and acetylcholine receptor distribution on human skeletal muscle fibres. *Journal of the Neurological Sciences*, *66*, 327–338.
O'Leary, M. F., Vainshtein, A., Carter, H. N., Zhang, Y., & Hood, D. A. (2012). Denervation-induced mitochondrial dysfunction and autophagy in skeletal muscle of apoptosis-deficient animals. *American Journal of Physiology. Cell Physiology*, *303*, C447–454.
Paul, P. K., Gupta, S. K., Bhatnagar, S., Panguluri, S. K., Darnay, B. G., Choi, Y., et al. (2010). Targeted ablation of TRAF6 inhibits skeletal muscle wasting in mice. *The Journal of Cell Biology*, *191*, 1395–1411.
Reddy, L. V., Koirala, S., Sugiura, Y., Herrera, A. A., & Ko, C. P. (2003). Glial cells maintain synaptic structure and function and promote development of the neuromuscular junction in vivo. *Neuron*, *40*, 563–580.
Relaix, F., & Zammit, P. S. (2012). Satellite cells are essential for skeletal muscle regeneration: The cell on the edge returns centre stage. *Development*, *139*, 2845–2856.
Renton, A. E., Majounie, E., Waite, A., Simón-Sánchez, J., Rollinson, S., Gibbs, J. R., et al. (2011). A hexanucleotide repeat expansion in C9ORF72 is the cause of chromosome 9p21-linked ALS-FTD. *Neuron*, *72*, 257–268.
Rich, M. M., & Lichtman, J. W. (1989). In vivo visualization of pre- and postsynaptic changes during synapse elimination in reinnervated mouse muscle. *The Journal of Neuroscience: The Official Journal of the Society for Neuroscience*, *9*, 1781–1805.

Rowan, S. L., Rygiel, K., Purves-Smith, F. M., Solbak, N. M., Turnbull, D. M., & Hepple, R. T. (2012). Denervation causes fiber atrophy and myosin heavy chain co-expression in senescent skeletal muscle. *PloS One*, *7*, e29082.

Salpeter, M. M., & Loring, R. H. (1985). Nicotinic acetylcholine receptors in vertebrate muscle: Properties, distribution and neural control. *Progress in Neurobiology*, *25*, 297–325.

Sambasivan, R., Yao, R., Kissenpfennig, A., Van Wittenberghe, L., Paldi, A., Gayraud-Morel, B., et al. (2011). Pax7-expressing satellite cells are indispensable for adult skeletal muscle regeneration. *Development*, *138*, 3647–3656.

Sanes, J. R., & Lichtman, J. W. (1999). Development of the vertebrate neuromuscular junction. *Annual Review of Neuroscience*, *22*, 389–442.

Sanes, J. R., & Lichtman, J. W. (2001). Induction, assembly, maturation and maintenance of a postsynaptic apparatus. *Nature Reviews. Neuroscience*, *2*, 791–805.

Sanes, J. R., Marshall, L. M., & McMahan, U. J. (1978). Reinnervation of muscle fiber basal lamina after removal of myofibers. Differentiation of regenerating axons at original synaptic sites. *The Journal of Cell Biology*, *78*, 176–198.

Sartori, R., Schirwis, E., Blaauw, B., Bortolanza, S., Zhao, J., Enzo, E., et al. (2013). BMP signaling controls muscle mass. *Nature Genetics*, *45*, 1309–1318.

Schaeffer, L., de Kerchove d'Exaerde, A., & Changeux, J. P. (2001). Targeting transcription to the neuromuscular synapse. *Neuron*, *31*, 15–22.

Schiaffino, S., & Reggiani, C. (2011). Fiber types in mammalian skeletal muscles. *Physiological Reviews*, *91*, 1447–1531.

Simon, A. M., Hoppe, P., & Burden, S. J. (1992). Spatial restriction of AChR gene expression to subsynaptic nuclei. *Development*, *114*, 545–553.

Singhal, N., & Martin, P. T. (2011). Role of extracellular matrix proteins and their receptors in the development of the vertebrate neuromuscular junction. *Developmental Neurobiology*, *71*, 982–1005.

Son, Y. J., & Thompson, W. J. (1995). Schwann cell processes guide regeneration of peripheral axons. *Neuron*, *14*, 125–132.

Sousa-Victor, P., Garcia-Prat, L., Serrano, A. L., Perdiguero, E., & Munoz-Canoves, P. (2015). Muscle stem cell aging: Regulation and rejuvenation. *Trends in Endocrinology and Metabolism: TEM*, *26*, 287–296.

Tabebordbar, M., Zhu, K., Cheng, J. K., Chew, W. L., Widrick, J. J., Yan, W. X., et al. (2016). In vivo gene editing in dystrophic mouse muscle and muscle stem cells. *Science*, *351*, 407–411.

Tintignac, L. A., Brenner, H. R., & Ruegg, M. A. (2015). Mechanisms regulating neuromuscular junction development and function and causes of muscle wasting. *Physiological Reviews*, *95*, 809–852.

Tomlinson, B. E., & Irving, D. (1977). The numbers of limb motor neurons in the human lumbosacral cord throughout life. *Journal of the Neurological Sciences*, *34*, 213–219.

Unwin, N. (1995). Acetylcholine receptor channel imaged in the open state. *Nature*, *373*, 37–43.

Unwin, N. (2005). Refined structure of the nicotinic acetylcholine receptor at 4A resolution. *Journal of Molecular Biology*, *346*, 967–989.

Unwin, N., & Fujiyoshi, Y. (2012). Gating movement of acetylcholine receptor caught by plunge-freezing. *Journal of Molecular Biology*, *422*, 617–634.

Valdez, G., Tapia, J. C., Kang, H., Clemenson, G. D., Jr., Gage, F. H., Lichtman, J. W., et al. (2010). Attenuation of age-related changes in mouse neuromuscular synapses by caloric restriction and exercise. *Proceedings of the National Academy of Sciences of the United States of America*, *107*, 14863–14868.

Vock, V. M., Ponomareva, O. N., & Rimer, M. (2008). Evidence for muscle-dependent neuromuscular synaptic site determination in mammals. *The Journal of Neuroscience: The Official Journal of the Society for Neuroscience*, *28*, 3123–3130.

Walsh, M. K., & Lichtman, J. W. (2003). In vivo time-lapse imaging of synaptic takeover associated with naturally occurring synapse elimination. *Neuron, 37*, 67–73.
Willadt, S., Nash, M., & Slater, C. R. (2016). Age-related fragmentation of the motor endplate is not associated with impaired neuromuscular transmission in the mouse diaphragm. *Scientific Report, 6*, 24849.
Williams, A. H., Valdez, G., Moresi, V., Qi, X., McAnally, J., Elliott, J. L., et al. (2009). MicroRNA-206 delays ALS progression and promotes regeneration of neuromuscular synapses in mice. *Science, 326*, 1549–1554.
Winbanks, C. E., Chen, J. L., Qian, H., Liu, Y., Bernardo, B. C., Beyer, C., et al. (2013). The bone morphogenetic protein axis is a positive regulator of skeletal muscle mass. *The Journal of Cell Biology, 203*, 345–357.
Wokke, J. H., Jennekens, F. G., van den Oord, C. J., Veldman, H., Smit, L. M., & Leppink, G. J. (1990). Morphological changes in the human end plate with age. *Journal of the Neurological Sciences, 95*, 291–310.
Wong, M., & Martin, L. J. (2010). Skeletal muscle-restricted expression of human SOD1 causes motor neuron degeneration in transgenic mice. *Human Molecular Genetics, 19*, 2284–2302.
Wu, H., Xiong, W. C., & Mei, L. (2010). To build a synapse: Signaling pathways in neuromuscular junction assembly. *Development, 137*, 1017–1033.
Yang, X., Arber, S., William, C., Li, L., Tanabe, Y., Jessell, T. M., et al. (2001). Patterning of muscle acetylcholine receptor gene expression in the absence of motor innervation. *Neuron, 30*, 399–410.
Young, S. H., & Poo, M. M. (1983). Rapid lateral diffusion of extrajunctional acetylcholine receptors in the developing muscle membrane of Xenopus tadpole. *The Journal of Neuroscience: The Official Journal of the Society for Neuroscience, 3*, 225–231.
Zhang, X., Xu, R., Zhu, B., Yang, X., Ding, X., Duan, S., et al. (2007). Syne-1 and Syne-2 play crucial roles in myonuclear anchorage and motor neuron innervation. *Development, 134*, 901–908.
Zhao, J., Brault, J. J., Schild, A., Cao, P., Sandri, M., Schiaffino, S., et al. (2007). FoxO3 coordinately activates protein degradation by the autophagic/lysosomal and proteasomal pathways in atrophying muscle cells. *Cell Metabolism, 6*, 472–483.
Zhao, K., Shen, C., Lu, Y., Huang, Z., Li, L., Rand, C. D., et al. (2017). Muscle yap is a regulator of neuromuscular junction formation and regeneration. *The Journal of Neuroscience: The Official Journal of the Society for Neuroscience, 37*, 3465–3477.

CHAPTER FIVE

Cellular Biomechanics in Skeletal Muscle Regeneration

Edward W. Li*, Olivia C. McKee-Muir*, Penney M. Gilbert*,†,‡,1
*Institute of Biomaterials and Biomedical Engineering, University of Toronto, Toronto, ON, Canada
†University of Toronto, Toronto, ON, Canada
‡Terrence Donnelly Centre for Cellular and Biomolecular Research, Toronto, ON, Canada
1Corresponding author: e-mail address: penney.gilbert@utoronto.ca

Contents

Abstract

Satellite cells, adult stem cells in skeletal muscle tissue, reside within a mechanically dynamic three-dimensional microenvironment. With each contraction–relaxation cycle, a satellite cell is expected to experience tensile, shear, and compressive stresses, and through cell–extracellular matrix interactions, also gauge the stiffness of the niche. Via mechanoreceptors, cells can sense these biophysical parameters of the niche, which serve to physically induce conformational changes that impact biomolecule activity, and thereby alter downstream signal transduction pathways and ultimately cell fate.

Current Topics in Developmental Biology, Volume 126
ISSN 0070-2153
https://doi.org/10.1016/bs.ctdb.2017.08.007
© 2018 Elsevier Inc.
All rights reserved.

An emerging body of literature supports the notion that myogenic cells, too, integrate biochemical factors together with biomechanical stresses and that this may serve to provide spatio-temporal control of cell fate in the complicated three-dimensional niche. Further, skeletal muscle regenerative medicine therapies are being improved by applying this fresh insight. In this focused chapter, the progression of skeletal muscle regeneration is dissected into a dynamic conversation between muscle progenitor cells and the mechanical properties of the extracellular matrix. The significance of biophysical regulation to myogenic repair is reinforced by the exaggerative influences of extrinsic mechanical stresses and the pathological implications of ECM dysregulation. Additional fundamental studies that further define the satellite cell biophysical environment in health, regeneration, aging, and disease may serve to close knowledge gaps and bolster skeletal muscle regenerative medicine.

1. INTRODUCTION

Skeletal muscle, given the wide variety of traumatic insults and degenerative injuries it may sustain, possesses a remarkable ability to replace damaged or destroyed muscle fibers (Ceafalan, Popescu, & Hinescu, 2014; Stilhano, Martins, Ingham, Pesquero, & Huard, 2015; Tedesco, Dellavalle, Diaz-Manera, Messina, & Cossu, 2010). The regenerative capacity of skeletal muscle is largely dependent on the contributions of satellite cells. These tissue-resident stem cells, in their quiescent form, are characterized by the expression of the transcription factor paired-box 7 (Pax7) and remain anchored to the plasma membrane, or sarcolemma, of mature myofibers throughout adulthood (Collins et al., 2005; Gillies & Lieber, 2011; Sambasivan et al., 2011; Seale et al., 2000). On the opposite side of and surrounding these satellite cells is the basal lamina, a thin sheet-like layer of proteoglycans, type IV collagen, laminin, and other carefully regulated proteins which give structure to the stem cell "niche." Further outward are sheaths of connective tissue, containing more fibrillar proteins that serve to bundle individual myofibers together. These structural components form part of an expansive extracellular matrix (ECM) which contains several of the biochemical cues necessary for the activation of myogenesis and the ensuing stages of regeneration (Kuang, Gillespie, & Rudnicki, 2008; Morrissey, Cheng, Davoudi, & Gilbert, 2015; Osses & Brandan, 2002; Stern et al., 2009). Less appreciated is that the biomechanical features of the ECM play a supportive, if not occasionally crucial, role in converting naïve satellite cells into committed myoblasts and, eventually, into multinucleated myofibers.

In this chapter, we review the mechanisms by which the structure and composition of the ECM regulate the proliferation, differentiation, migration, and fusion of muscle progenitor cells during normal tissue repair. We then expanded this model by discussing the effects of mechanical stresses, as a satellite cell might experience during the stretching or contraction of myofibers, on skeletal muscle regeneration. Building on an understanding of the biophysical cues required for successful myogenic repair, this chapter also offers to explain how ECM dysregulation could both weaken adult skeletal muscle and limit its capacity for regeneration in several pathological settings. Finally, we summarize advances in bioengineering, from the design of biomimetic scaffolds to the development of novel culture strategies, which paint the notion of "skeletal muscle regenerative medicine" as an increasingly tangible reality.

2. MATRIX REGULATION OF SKELETAL MUSCLE REGENERATION

2.1 Maintenance of Satellite Cell Quiescence

Tissue regeneration is a resource-intensive process which depends on the ability of stem cells to recognize when to enter and exit a readily reversible state of proliferative inactivity, known as quiescence. One mechanism by which potential myogenic cells maintain in a viable but nonproliferative state is the sensing of biomechanical cues generated by neighboring cells, otherwise known as contact inhibition. When adult murine satellite cells are kept in contact with the sarcolemma of an uninjured muscle fiber, these cells remained quiescent even upon administration of a mitogen which would otherwise induce proliferation (Bischoff, 1990). Furthermore, the satellite cells refrained from fusing to the attached myofiber, indicating that they truly were inactive and did not simply begin to undergo myogenic differentiation. Yet, when satellite cells were detached from their native myofiber or grown on a chemically injured fiber, the proliferative response to mitogen was restored. Hence, it can be deduced that the physical architecture of healthy skeletal muscle informs the decision of satellite cells to remain in a quiescent state. Indeed, a recent study demonstrated that culturing prospectively isolated muscle stem cell on top of an artificial muscle fiber with mechanical softness, curvature, and cell surface receptors matching the native muscle fiber niche, it was possible to maintain quiescence in culture for days (Quarta et al., 2016).

On a molecular scale, adhesion proteins such as cadherins enable contact inhibition by sensing mechanical cues that emerge in the presence of neighboring cells. To illustrate, homophilic ligation of E-cadherin molecules on adjacent cells has been associated with activation of Hippo signaling (Kim, Koh, Chen, & Gumbiner, 2011). Further downstream, it has been shown that Hippo signaling prevents the Yap transcriptional coactivator from localizing to the nucleus in response to confluent culture conditions (Ota & Sasaki, 2008). As such, cell contact can inhibit the expression of proteins which promote proliferation and myogenic lineage commitment (Fischer, Rikeit, Knaus, & Coirault, 2016). Altogether, the careful regulation of quiescence by niche biomechanical cues prevents premature differentiation of potential myogenic progenitor cells and is critical to the maintenance of viable satellite cell reserves.

2.2 Induction of Satellite Cell Activation

Once normal tissue architecture is disrupted, satellite cells become exposed to aberrant regenerative cues which prepare them to emerge from a quiescent state. Among the key regulators of this process are again the biophysical properties and structural components of the stem cell niche. In recent years, the view that the ECM only exists to anchor satellite cells in place was superseded by studies which found that individual components of the matrix can act directly on membrane mechanoreceptors and, in doing so, mobilize quiescent satellite cells for myogenic repair. For instance, the network of collagens and laminins which structuralize the basal lamina can also initiate signal transduction by associating with $\alpha7/\beta1$ integrins on the surface of resident satellite cells (Sanes, 2003). Since these integrins subtypes are known to mediate cell adhesion and motility in skeletal muscle (Yao, Ziober, Squillace, & Kramer, 1996; Yao, Ziober, Sutherland, Mendrick, & Kramer, 1996), disruption of the basal lamina may be an initial step toward producing a regenerative phenotype in satellite cells. If not to bind a receptor, components of the basal lamina may alter the presentation of other ligands by binding and concentrating less abundant signaling factors. Due to the porous structure of the ECM, biochemical cues known to trigger satellite cell activation can be sequestered into cavities normally inaccessible to membrane receptors but also kept in close proximity in case of injury. Accordingly, significant levels of hepatocyte growth factor (HGF), a prominent satellite cell activation cue, can even be found in extracts of uninjured rat skeletal muscle, in which no active regeneration would be expected

(Tatsumi, Anderson, Nevoret, Halevy, & Allen, 1998). Evidently, the ECM acts as more than a mere scaffold in synergizing with the activation factors responsible for initiating the muscle repair process.

Following this train of thought, one might hypothesize that biophysical and biochemical cues coalesce during the beginning stages of traumatic injury to "switch on" dormant intracellular repair mechanisms in muscle progenitor cells. Once the sarcolemma and basal lamina are destroyed, for instance, previously sequestered HGF may traverse an unobstructed path toward quiescent satellite cells (Fu, Wang, & Hu, 2015). Binding of HGF to its receptor, transmembrane tyrosine kinase c-Met, then activates downstream pathways which regulate the development of various tissues and organs (Organ & Tsao, 2011). Cellular responses that have been implicated in the activation of the HGF/c-Met axis include proliferation, migration, morphogenesis, and survival—traits which typify cells undergoing both tumor progression and tissue regeneration (Ishikawa et al., 2012; Neuss, Becher, Woltje, Tietze, & Jahnen-Dechent, 2004). Similarly, damage to the blood vessels supplying myofibers can elicit the release of vascular endothelial growth factor and platelet-derived growth factor, which have been shown to directly stimulate satellite cell expansion in vitro (Ceafalan et al., 2014). These factors also help to regulate the angiogenic process which spatiotemporally coordinates with myogenesis, the ultimate goal being to serve the perfusion needs of newly formed myofibers (Christov et al., 2007).

Equally crucial for satellite cell activation is the expression of receptors which recognize ligands unveiled by niche remodeling. Among these receptors are syndecans, a class of proteoglycan which serve as coreceptors for tyrosine kinases. Specifically, syndecans-3 and -4 can be found on the surface of satellite cells and bind a range of activation cues including hepatocyte and fibroblast growth factors (Cornelison, Filla, Stanley, Rapraeger, & Olwin, 2001). These interactions mediate signaling events at the earliest stages of myogenic progression. Indeed, knockout of syndecan-4 impairs the activation of satellite cells, which can also limit the capacity of the cells to subsequently expand, differentiate, and fuse after being explanted onto single myofibers (Cornelison et al., 2004). Meanwhile, knockout of syndecan-3 caused MyoD, a transcription factor required for myogenic commitment, to mislocalize to the cytoplasm of explanted satellite cells. These findings reinforce the idea that several of the necessary components for regeneration are either already expressed by quiescent satellite cells or present in the surrounding niche and may only require matrix remodeling to kickstart the regenerative cascade.

The actual remodeling of the niche can be attributed to the activity of matrix metalloproteinases (MMPs), a family of endopeptidases which may be recruited to degrade laminin, fibronectin, elastin, and various other ECM structural proteins in the event of muscle injury (Carmeli, Moas, Reznick, & Coleman, 2004). In a study examining cardiotoxin-injured and congenitally necrotic muscles, MMP-2 expression and activation were detected specifically at sites of repair where new myofibers were actively being formed (Kherif et al., 1999). Alternately, MMP-9 was found to be expressed by leukocytes and macrophages which infiltrate the extracellular protein network and mediate injury-induced inflammation. The action of these MMPs raises the effective concentration of sequestered biochemical activation factors by releasing them from temporary binding sites within the satellite cell niche (Yamada et al., 2008, 2006).

Given that ECM remolding can assist in the endogenous repair of skeletal muscle, studies have shown that satellite cells themselves can direct regenerative efforts by modulating the activity of MMPs. For example, constitutive expression and secretion of MMP-2 have been detected in satellite cells extracted from human bicep muscles, a muscle group which undergoes frequent tissue reconstruction (Guérin & Holland, 1995). By comparison, the constitutive expression of TIMP-3, an inhibitor of metalloproteinases, becomes downregulated upon satellite cell activation in order to amplify the activity of MMP-2 (Pallafacchina et al., 2010). Thus, through autocrine signaling, already activated satellite cells may strengthen existing biomechanical cues to mobilize neighboring muscle stem cells for repair.

For these reasons, it is thought that the basement membrane is deconstructed to facilitate the early stages of myogenic repair and must later be rebuilt to fully restore muscle architecture. To trace the ECM structural changes which occur during muscle degeneration and regeneration, one study employed an autotransplantation model in which segments of mouse skeletal muscle were removed and then engrafted into the same anatomical location (Gulati, Reddi, & Zalewski, 1982). Within only 2 days of autotransplantation, it was found that fibronectin disappeared from the pericellular basement membrane. The loss of laminin, types IV and V collagen, and matrix proteoglycans ensued, and a new basement membrane only began to emerge as fully committed myoblasts started fusing into myofibers (Gulati, Reddi, & Zalewski, 1983). Ultimately, these results suggest that ECM remodeling, both directly and by altering the presentation of preexisting biochemical cues, introduces an element of spatiotemporal control in guiding naive satellite cells toward muscle regeneration.

2.3 Self-Renewal of Muscle Progenitor Cell Reserves

Once resident satellite cells become cognizant of tissue damage, they rapidly transition into a proliferative state, allowing the activated satellite cells to generate a sizable pool of replicates suitable for later myogenic differentiation. Again, the physical composition of the stem cell niche and other architectural features of skeletal muscle play key roles in the control of satellite cell self-renewal. For example, it has been found that the ECM glycoprotein fibronectin interacts with a coreceptor formed by the complex of Syndecan-4 and Frizzed-7 to stimulate symmetric expansion of satellite cells during muscle regeneration (Bentzinger et al., 2013). Binding of fibronectin coincided with activation of Wnt7a signaling, a pathway that has been implicated in the control of cell proliferation across a variety of developmental lineages. Notably, transient high-level expression of fibronectin has also been detected in newly activated satellite cells, suggesting that satellite cells alone are capable of remolding their niche. These findings reinforce the idea that the ECM is malleable in nature and contributes dynamically to the control of muscle progenitor cell expansion following injury.

Arguably the most distinctive feature of undifferentiated satellite cells is the expression of transcription factor Pax7. Yet, when satellite cells derived from collagen VI-deficient mice were allowed to propagate in vitro, a markedly greater proportion of cells began to express myogenic differentiation markers MyoD and MyoG in place of Pax7 (Urciuolo et al., 2013). Addition of purified collagen VI was able to rescue these defects by both prolonging the survival of the $Pax7^{+}$ $MyoD^{-}$ population and promoting cell proliferation. These results reveal that collagen VI plays a far from trivial role in the regulation of satellite cell self-renewal. To give another example, laminin, a major component of the basal lamina enclosing satellite cells, has been shown to stimulate myoblast proliferation in vitro, and in a dose-dependent manner (Öcalan, Goodman, Kühl, Hauschka, & von der Mark, 1988). When laminin was bound to the surface of a cell-adhesive collagen substrate, the number of actively cycling myoblasts was two- to threefold greater than conditions which employed an equivalent amount of fibronectin. Perhaps most profound is the ability of laminin, but not collagen or fibronectin, to ready proliferating myoblasts by inducing the outgrowth of cell processes which might function as cellular limbs during motile phases of tissue repair.

Even if the composition of the satellite cell niche remains unchanged, matrix stiffness may modulate the decision to differentiate or self-renew. Compared to skeletal muscle stem cells grown on rigid plastic dishes, cells seeded on pliant hydrogel substrates manage to evade terminal differentiation and retain certain "stem-like" features of adult satellite cells, including the expression of Pax7 (Gilbert et al., 2010). Simultaneously, the proportion of cells expressing myogenin, a marker of terminal differentiation, is markedly reduced on softer substrates. Despite delaying differentiation, replicating the stiffness of the muscle stem cell niche in vitro did not change the rate of cell division. In this way, physical properties of the niche retain precise control over the fate of muscle stem cells, ensuring that there are adequate supplies of viable progenitor cells at the various stages of tissue repair.

2.4 Myogenic Commitment and Differentiation

Following the proliferative phase, a subpopulation of nonmyogenic satellite cells returns to quiescence in order to replenish the satellite cell pool. The remainder become designated to continue in the direction of terminal myogenic differentiation. During this process, physical features of the niche provide key information which can alter the fate of progenitor cells not yet committed to a particular lineage. For instance, growing satellite cells on a laminin-coated substrate has yielded a two- to fivefold increase in the number of cells expressing desmin, a marker of myogenic commitment, when compared to the collagen-coated control (Urciuolo et al., 2013). As the rate of division among nonmyogenic satellite cells remained unchanged, this effect could be attributed to a marked increase in proliferation among early myoblasts.

Matrix stiffness offers one additional level of control in the regulation of myoblast differentiation. In one study, it was found that the transverse elastic modulus of mammalian myoblasts increased steadily over the span of days as they progressed toward terminal differentiation (Collinsworth, Zhang, Kraus, & Truskey, 2002). As determined by pharmacological inhibition experiments, factors which contributed to the apparent change in stiffness include the expression of cytoskeletal actin and myosin. Importantly, deviations above or below baseline tissue stiffness may facilitate the transition from self-renewal to myogenic differentiation (Gilbert et al., 2010; Urciuolo et al., 2013). Yet, stiffness in excess has been found to suppress the expression of MyoG and trigger apoptosis in early myoblasts instead (Gilbert et al., 2010). Furthermore, it was found that the sarcomeral striations characteristic of healthy mature myofibers only develop on substrates mimicking typical tissue stiffness of approximately 12 kPa (Engler et al., 2004). It can be concluded that the actin and myosin required for these striations were either not fully expressed or integrated improperly into the cytoskeletal organization. Evidently, tissue stiffness must be regulated to a high degree of specificity in order for muscle progenitor cells to properly differentiate into mature myocytes.

Central to the biophysical regulation of differentiation is the ability of myogenic progenitor cells to detect regenerative cues at the surface membrane and then translate this information into DNA-level changes. Underlying this unique form of mechanosensation are transmembrane cell adhesion molecules, such as cadherins, which can assemble into molecular complexes to stabilize neighboring cells. In its complexed form, cadherins may bind Cdo to trigger further downstream a series of signal intermediaries, activation of transcription factor p38, which directly regulates myoblast differentiation (Lu, 2011). Following this sequence of events, it becomes clear how cell density, and the consequent formation of cadherin-based adhesions, might alter the fate of uncommitted myoblasts. Likewise, it has been shown that the expression of cadherins, made possible by transfection, is critical to the development of sarcomeral myosin in three-dimensional aggregates of transformed fibroblast-like cells (Redfield, Nieman, & Knudsen, 1997). In this model, the recruitment of focal adhesion-associated protein β-catenin and the formation of strong cell–cell adhesions were necessary to induce skeletal muscle differentiation. The crucial role of adhesions in differentiation is not limited to cell–cell interactions either given that the inhibitor TGF-β1, whose action lies in the regulation of extracellular collagen, blocks the upregulation in myogenin that would be expected when myoblasts are switched to differentiation medium (Heino & Massagué, 1990).

Perhaps the most direct demonstration of biophysically induced differentiation is the observation that myogenic precursors begin diving asymmetrically on micropatterned substrates. During this process, cell fate determinants are segregated unequally between daughter cells, giving rise to progeny with dissimilar fates (Rocheteau, Gayraud-Morel, Siegl-Cachedenier, Blasco, & Tajbakhsh, 2012). Asymmetric substrates may therefore promote asymmetric division by creating conditions that spatially favor nonrandom DNA segregation. In fact, the shape of the micropatterns on which mouse skeletal muscle stem cells were grown could be correlated with the frequency of nonrandom chromosome segregation and unequal transcription factor partitioning (Yennek, Burute, Théry, & Tajbakhsh, 2014). Consequently, when grown on asymmetric micropatterns, prospectively isolated $Pax7^+$ satellite cells more frequently gave rise to daughter cells with a differential proliferative phenotype (Fig. 1). However, a symmetric

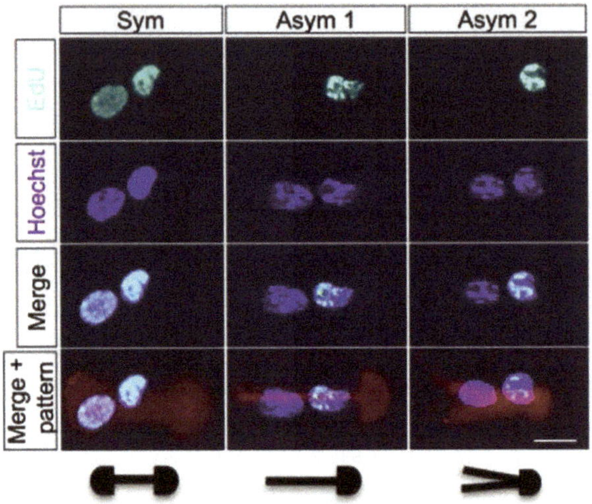

Fig. 1 Satellite cells cultured on asymmetric micropatterns are prone to asymmetric division. Pax7$^+$ satellite cells harvested from mouse *tibialis anterior* muscle and isolated by fluorescence-activated cell sorting. On symmetric micropatterns (Sym), the division of a parent satellite cell frequently gave rise to daughter cells which equally incorporated thymidine analogue EdU, a marker of proliferation. Divisions on asymmetric patterning (Asym 1 and 2) tended to produce one daughter cell that was negative for EdU, indicative of differentially fated progeny. *Reprinted with permission from Creative Commons license: https://creativecommons.org/licenses/by-nc-nd/3.0/legalcode. Yennek, S., Burute, M., Théry, M., Tajbakhsh, S. (2014). Cell adhesion geometry regulates non-random DNA segregation and asymmetric cell fates in mouse skeletal muscle stem cells.* Cell Reports, *7, 961–970. https://doi.org/10.1016/j.celrep.2014.04.016.*

fibronectin/fibrinogen substrate failed to significantly alter the fate of satellite cells that were already committed to asymmetric division. For this reason, regulation of DNA segregation by cell adhesion may be confined to a certain vulnerable stage of cell growth. Nonetheless, myogenic commitment and differentiation are influenced by a plethora of niche biomechanical cues ranging from tissue stiffness to matrix structure and composition.

2.5 Migration of Myogenic Progenitor Cells

In order for muscle precursors to exit their niche and reach the site of repair, necrotic debris must be cleared from the path of migration. Aside from releasing satellite cell activation factors from sequestered sites, MMPs can facilitate the movement of muscle precursors by degrading ECM obstructions, and an autocrine role in this process has been suggested by the expression of select gelatinase MMPs in both human and mouse satellite cells

(Guérin & Holland, 1995; Kherif et al., 1999). In an in vitro model of muscle regeneration, MMP-2 continued to be secreted by myoblasts throughout all stages of repair, and cell migration was severely impaired by antibody-mediated inhibition of MMP-9 (Lewis, Tippett, Sinanan, Morgan, & Hunt, 2000). Moreover, MMP-9 expression was halted once single cells began to fuse, suggesting that matrix degradation could be attenuated following the migratory phase. The notion of temporally graded expression can be extended to TIMPs as well, considering that levels of TIMP-1 mRNA gradually increased as myoblasts progressed toward fiber formation. In tempering the degradative action of MMPs, the inhibitory action of TIMPs could also guide the morphological development of new muscle tissue and assist in the reconstruction of the basal lamina. In mice with cardiotoxin-induced acute muscle injury, upregulation of interstitial collagenase MMP-13 was also implicated in the repair process, and pharmacological inhibition of MMP-13 led to a marked reduction in the migratory capacity of transplanted mouse myoblasts (Lei, Leong, Smith, & Barton, 2013). However, expression of MMP-13 peaked during the postfusion phase, as myoblasts were approaching terminal differentiation. Hence, myoblast migration is likely to be dependent on an intricate interplay between several MMPs coordinating with one another in both space and time.

A more definitive demonstration of the necessity of MMPs was provided by a study on the intramuscular migration of transplanted mouse myoblasts (El Fahime, Torrente, Caron, Bresolin, & Tremblay, 2000). For instance, pretreatment of injured muscle with a synthetic metalloproteinase inhibitor limited the ability of myoblasts to migrate away from their point of origin. Conversely, the distance traveled by transplanted myoblasts in vivo could be increased by inducing overexpression of MMP2 prior to injection. In vitro, it was discovered that myoblasts from the same cell line selected for transplantation expressed both MMP2 and a complementary proteolytic enzyme denoted membrane-type MMP (MT1-MMP). As the expression of MT1-MMP was raised, the activated form of MMP2 became more prevalent as well, suggesting that the migration-stimulating activity of MMP2 was dependent on the presence of another endogenous cofactor. In this way, the activity of factors contributing to matrix degradation can be tightly controlled during the migration of muscle tissue precursors.

In the absence of enzymatic degradation, more subtle changes in ECM composition can also modulate myoblast migration. For instance, a fibronectin substrate has been found to stimulate the development of focal adhesions, stress fibers, and a rigidly anchored actin cytoskeleton in murine myoblasts (Goodman, Risse, & Von Der Mark, 1989). In contrast, actinin and the focal adhesion protein vinculin remain dispersed throughout the cytoplasm on a laminin-coated substrate, generating a more fluid cytoskeletal phenotype. Due to this advantage in structural flexibility, MM14 murine myoblasts grown on laminin possess a unique ability to develop cell processes and locomote five times faster than myoblasts plated on fibronectin (Öcalan et al., 1988). The superiority of laminin as a motile platform can be attributed at least in part to the expression of α7 integrin on the surface of myoblasts and its adhesive interactions with select laminin isoforms (Yao, Ziober, Squillace, et al., 1996; Yao, Ziober, Sutherland, et al., 1996). Namely, locomotion of MM14 and C2C12 myoblasts over laminin-1 and -2/4 was almost completely eliminated when a function-blocking antibody was targeted to α7 integrin. The observed defects coincided with a complete loss of adhesive interactions on laminin-1 and partial inhibition of cell adhesion to laminin-2/4. Yet, perhaps the most telling observation is that a mixed substrate containing basal laminar glycoproteins laminin and nidogen yielded a twofold increase in the speed of myoblast locomotion compared to laminin alone (Goodman et al., 1989). Regardless of the intrinsic ability of any one protein to promote a motile phenotype, the migratory capacity of muscle progenitor cells may prove optimal when culture conditions match the native architecture and composition of the ECM.

One final prerequisite to proper migration is that muscle progenitor cells, in addition to traversing great distances, be capable of sensing the topographical features of their surroundings and orienting themselves in the correct direction toward a site of repair. At a receptor level, syndecan-4 supports directional migration by activating Rac1, a small signaling protein that drives actin polymerization, in response to the formation of cell–matrix engagements (Bass et al., 2007). Activated Rac1 is then recruited to the leading edge of the cell to generate a membrane protrusion which assists with motility. While knockout of syndecan-4 did not impair speed of migration, Rac1 failed to reliably localize to any point of contact between the cell and matrix, causing migration to occur in random directions. The coupling of cell motility to syndecan engagement can also be explained by the observation that Rac1-mediated cell spreading and cytoskeletal polarization are characteristics of a more broadly defined regenerative phenotype. To illustrate, satellite cells bathed in Wnt7a ligand exhibited a greater tendency to become morphologically polarized and migrate further when transplanted into dystrophic muscle (Bentzinger et al., 2014). As the Wnt7a/Fzd7 pathway is known to stimulate proliferation, pretreated satellite cells also conferred hypertrophic behavior to the fibers they fused into. These effects were accompanied by a noticeable improvement in the likelihood of engraftment. Thus, the ability for myogenic cells to morphologically adapt to the topology of the ECM may correlate with not only their capacity for directed migration but also the overall potential to generate ample quantities of replacement tissue following muscle injury.

2.6 Myotube Formation and Reintegration

Once myoblasts are ready to be integrated into functional muscle tissue, biomechanical cues from the ECM again guide the cells to align and fuse into multinucleated myotubes. To do so, prospective myocytes must be able to distinguish firm layers of tightly woven myotubes from softer regions of neuronal tissue enclosing any neuromuscular junctions. When this pattern of alternating elasticity was recapitulated on the surface of a hydrogel substrate, it was found that C2C12 myoblasts consistently localized to stiffer, myogenic surfaces (Choi et al., 2012). Cells that adhered then immediately began to align and, within a matter of days, fused into myotubes expressing proteins specific to skeletal muscle. Importantly, the rate of fusion and number of fusing cells were consistently greater on regions mimicking myogenic stiffness than on gels with entirely uniform surfaces. As such, matrix stiffness may act as a proxy in ensuring that prospective myocytes assemble on and properly integrate into existing muscle tissue.

Similarly, the extent of myoblast fusion may also be influenced by matrix composition. Although ECM components do not appear to mediate fusion by altering the presentation of growth factors bFGF and TGFβ1, the choice of substrate does exert an influence on the size of myotubes formed from cultures of BALB/c mouse myoblasts (Maley, Davies, & Grounds, 1995). On gelatin, laminin, fibronectin, and collagen IV-based substrates, small myotubes containing two to three nuclei predominated over myotubes of any larger size. However, the majority of myotubes spanned four or more nuclei when grown on Matrigel, a substrate that is believed to more closely recapitulate the complexity of the basement membrane matrix. Accompanied with the formation of larger myotubes was a significant increase in the proportion of mononuclear cells expressing desmin, a muscle-specific adhesion molecule. These results suggest that cell adhesions confer crucial mechanosensitive capabilities which enable myoblasts to align and assemble into syntytial structures.

Accordingly, it would not be unexpected that the expression of various adhesion proteins occurs concomitantly with myotube formation following induced muscle regeneration. Of these proteins, α3 integrin was found to form multiprotein complexes with β1 integrin, ADAM12, CD9, and CD81 as myoblasts began to fuse together in vitro (Przewoźniak et al., 2013). Furthermore, downregulation of α3 integrin gene expression resulted in aberrant myoblast fusion, which would suggest that all components contained within adhesion complexes are essential to achieving the mechanosensitive capabilities required for myocyte assembly. By contrast, the loss of M-cadherin, an adhesion protein whose expression pattern would also be indicative of a role in muscle regeneration, seemed to produce no change in satellite cell homing or myoblast fusion in vivo (Hollnagel, Grund, Franke, & Arnold, 2002). That is, necrotic wounds in M-cadherin-null mice healed as efficiently as the wild-type control, a finding that was corroborated by the observation that primary myoblasts derived from these mice fused normally in artificial culture. Therefore, while certain adhesion molecules may be indispensable to myogenic repair, there may exist a baseline level of redundancy which allows select mechanosensitive surface proteins to compensate for the loss of others.

In order for myotubes to mature into functional, responsive myofibers, efferent neurons must be able to locate and innervate newly formed muscle tissue. The expression of adhesion molecules and ECM components on denervated muscle helps axonal outgrowths to recognize appropriate sites for synaptogenesis. One such homing target is the neuronal cell adhesion molecule N-CAM, which can be detected in abundant levels on the surface of myotubes during early embryogenesis, injury, and paralysis, when the muscle is in need of innervation (Sanes, Schachner, & Covault, 1986). In addition to N-CAM, a series of ECM proteins characteristic of the skeletal muscle basement membrane were shown to concentrate in the interstitial regions surrounding denervated muscle, where growing axons can often be found. In this way, restructuring of the skeletal muscle architecture may facilitate myotube innervation.

In summary, matrix biomechanical cues can synergize with available ligands and intrinsic signaling pathways to regulate the present function and fate decisions of myogenic progenitor cells. Knowing this, one might imagine a dynamically changing satellite cell niche, sat atop the sarcolemma of an adult myofiber, that conveys variations in tissue stiffness, interstitial shear stress, tensile strain, as well as general contractile wear-and-tear to the behaviorally malleable stem cell (Fig. 2). As such, the physical features of the ECM, and myofiber bundles at large, have an indispensable role to play in skeletal muscle regeneration.

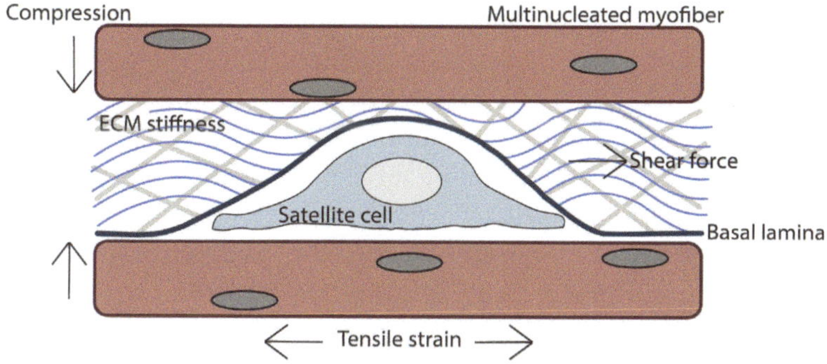

Fig. 2 The satellite cell resides within a mechanically dynamic niche. The adult skeletal muscle stem cell, satellite cell, is positioned between the sarcolemma of an adult myofiber and the basal lamina. Satellite cells reside within a mechanically dynamic niche where they experience compressive and shear forces, as well as tensile strain due to the contraction and relaxation of skeletal muscle.

3. INFLUENCES OF MECHANICAL STIMULATION ON MYOGENESIS

Given that the satellite cell, the basic unit of myogenic repair, is a highly mechanosensitive apparatus which responds to even slight changes in the structure and composition of the ECM, extrinsic mechanical stresses that threaten to interrupt this endogenous balance can tangibly affect the regenerative capacity of skeletal muscle. The mechanical stretching of damaged myofibers is a prominent example of such a stress as it occurs naturally during antagonistic muscle contraction and movements of the body in general. Hence, the effects of stretch on the behavior of satellite cells could inform strategies to enhance muscle growth or promote repair in human disease, injury, as well as aging. At a fundamental level, stretch is known to induce satellite cell activation but occurs through a complex sequence of signaling events that ultimately converges at MMP-mediated ECM remodeling and the consequent delivery of growth factor to the quiescent muscle stem cell (Tatsumi & Allen, 2008). At the very outset of this signaling cascade are mechanosensitive calcium channels which serve to transduce the initial intracellular effects of cyclic myofiber stretch.

3.1 Stretch-Induced Myogenic Responses

3.1.1 Calcium-Dependent Mechanotransduction

Calcium homeostasis has long been known to play a crucial role in skeletal muscle maintenance as an effector of mitogen-activated protein kinase (MAPK) signaling and several other pathways that regulate cell proliferation, differentiation, and cytoskeletal restructuring (Benavides Damm & Egli,

2014). Moreover, myogenic progenitor cells express calcium ion channels which become activated in response to induced membrane deformities (Franco & Lansman, 1990; Guharay & Sachs, 1984). For instance, calcium currents through the transient receptor potential channel 1 (TRPC-1) were amplified when a microscopic pulling force was applied to the plasma membrane of C2C12 satellite cells (Formigli et al., 2009). Correspondingly, TRPC-1 expression was significantly elevated throughout differentiation and other phases of regeneration. In addition, the loss of stretch-activated calcium entry may be a distinguishing feature of aberrant myogenesis. While the vast majority of mechanosensitive ion channels (MS channels) become more permeable in response to stretch, a subpopulation of constitutively open, stretch-inactivated channels appears in myotubes derived from dystrophic *mdx* mice (Lansman, 2007). Critically, satellite cells extracted from this strain of mouse demonstrate a limited capacity to grow new muscle tissue. These results highlight the importance of calcium-dependent signaling to effective myogenic repair.

Additionally, several studies have shown that calcium transit is a necessary first step in stretch-induced satellite cell activation. When a calcium-chelating agent was added to capture free extracellular calcium ions from cultures of primary mouse satellite cells, short repeated intervals of mechanical stretch would no longer stimulate cell proliferation (Hara et al., 2012). A similar effect was observed when either MS channels or long-lasting voltage-gated calcium-ion channels (L-VGC channels) were inhibited by pharmacological antagonists, indicating that the entry of extracellular calcium, and subsequent satellite cell activation, is dependent on the proper functioning and possible cooperative activity of both channel types. Integral to this sequence of events would be the association of intracellular calcium with calmodulin, a ubiquitously expressed but constitutively inactive messenger protein that depends on the availability of calcium to modulate various protein targets (Tatsumi et al., 2009).

Perhaps most notably, the satellite cell activation that was observed in the presence of functioning calcium channels was accompanied by the release of active HGF from temporary extracellular binding sites (Hara et al., 2012). Correspondingly, the inactivation of MS and L-VGC channels limited the presence of soluble HGF from the conditioned medium of cultured satellite cells. Even when an ionophore was introduced to ensure stable influx of calcium, the inhibition of calmodulin produced the same response and also led to a dramatic decrease in the number of cycling satellite cells (Tatsumi et al., 2009). Thus, the accumulation of calcium ions within satellite cells, and subsequent activation of calmodulin, may trigger a cascade of signaling events that serve to, in an autocrine fashion, increase the accessibility of local myogenic cues.

3.1.2 Nitric Oxide Production

Further downstream, it is thought that activated calcium–calmodulin complexes stimulate the activity of nitric oxide synthases (NOS) to generate diffusible messenger molecules that can freely exit the satellite cell and efficiently catalyze niche remodeling. In a mouse model, injection of nitric oxide-donating agents adjacent to the *tibialis anterior* muscle almost immediately elicited satellite cell activation, concomitant with a weakening of the adhesions between satellite cells and their fiber–laminar enclosure (Anderson, 2000). At a physiological level, muscle hypertrophy was observed as well, which would indicate that satellite cells exposed to NO were able to detach from the laminar sheath and succeed to activities related to tissue growth. Based on histology assays, muscles from both *mdx* mice and NOS-1 knockout mice exhibited an approximate 10-fold time delay in the onset of this response. One explanation for the phenotypic similarity between these mouse models is that the abnormal intake of calcium in *mdx* mice prevents the propagation of signaling events necessary to "switch on" nitric oxide synthesis (Tutdibi, Brinkmeier, Rüdel, & Föhr, 1999); these results affirm the sequence of events that places calcium-dependent mechanotransduction upstream of NOS activation. Finally, both transient and prolonged NOS inhibition resulted in stunted myogenic repair, as characterized by unprovoked nuclear apoptosis, deficiencies in satellite cell maneuverability, and other abnormalities related to myofiber morphogenesis (Anderson, 2000). In this regard, disruption of NO production impedes not only satellite cell activation but also the viability of newly formed muscle fibers at the conclusion of myogenesis.

3.1.3 MMP-Motivated HGF Release

Crucial to the eventual shedding of tethered HGF from the ECM is the action of MMPs that are already present in the extracellular network but inactive until exposed to an appropriate messenger molecule. Similar to tissue-resident myogenic cues, MMP-2 transcripts and fully translated proteins can be detected in whole lysate samples of rat satellite cells (Yamada et al., 2008). In the context of stretch-induced satellite cell activation, the messenger molecule that is needed to activate MMPs is gaseous NO, produced from the activity of intracellular NOS. Once sufficient quantities of NO disperse into the ECM, the proform of MMP-2 undergoes proteolytic conversion to a lighter protein with uninhibited enzymatic activity. Cleavage of MMP-2 could be detected in the culture medium of rat satellite cells regardless of whether NO was generated by a synthetic NO donor or a 2-h period of cyclic stretch; however, MMP activation was abolished in the presence of a NOS inhibitor. The identity and importance of MMPs to satellite cell activation were then confirmed by the detection of HGF in cell cultures that constitutively produced NO, as well as the absence of HGF in cultures pretreated with TIMP-1, a MMP inhibitor.

Further studies have formalized the proposed mechanism of stretch-induced satellite cell activation by inserting the release of HGF and its interactions with the c-Met receptor immediately upstream of cell proliferation (Tatsumi, Hattori, Ikeuchi, Anderson, & Allen, 2002; Tatsumi et al., 2006; Tatsumi, Sheehan, Iwasaki, Hattori, & Allen, 2001). Remarkably, unstretched satellite cells remained stably quiescent (Wozniak et al., 2003) but could be prompted to enter cell cycle by mere treatment with the conditioned medium from stretched counterpart cultures (Tatsumi et al., 2001). As well, a more prolonged stretch cycle, spread over 2 days, led to the upregulation of every component of the stretch-activated mechanism described thus far, from intracellular calmodulin and NOS to extracellular MMP-2, HGF, and c-Met (Chen et al., 2013). In aggregate, these studies provide a basic mechanistic explanation for the phenomenon of stretch-induced autocrine satellite cell activation, the precise sequence of events being depicted in Fig. 3. That being said, satellite cell activation in myogenic repair cannot be ascribed to HGF/c-Met signaling alone. Indeed, only a subset of all c-Met^{+} satellite cells detected in single fiber culture entered a proliferative state within a 2-h period of cyclic mechanical stretch (Wozniak et al., 2003). Rather, the theme of stress-gated calcium signaling, nitric oxide production, and MMP-motivated HGF release likely makes up one of many pathways by which extrinsic mechanical stresses can influence the native myogenic cascade.

3.1.4 Gene Regulation

Although not mutually exclusive from one another, there exist several gene-regulating mechanisms that facilitate stretch-motivated satellite cell activation, some of which may act cooperatively or directly downstream of c-Met. For instance, the level of extracellular signal-regulated kinase (ERK) phosphorylation in adult bovine satellite cells was found to increase in response to cyclic strain delivered by a vacuum stretch apparatus

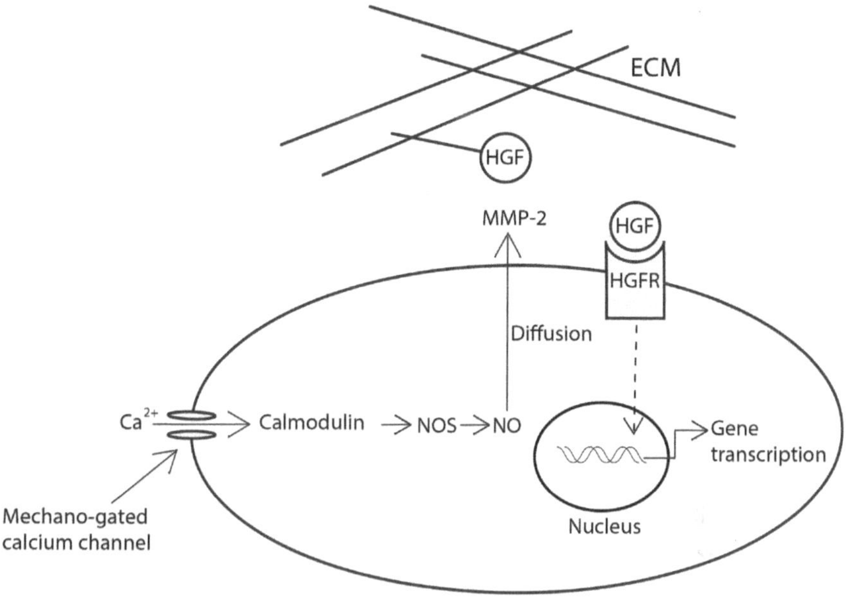

Fig. 3 Myofiber stretch promotes satellite cell activation in an autocrine fashion. Stretching of adjacent muscle fibers promotes activation of mechanosensitive ion channels and consequent influx of calcium into the quiescent satellite cell. Formation of calcium–calmodulin complex subsequently stimulates nitric oxide synthase (NOS) to produce freely diffusible nitric oxide (NO). Further downstream, NO-activated matrix metalloproteinase-2 (MMP-2) to release hepatocyte growth factor (HGF) from its extracellular tethering. Finally, binding of HGF to the c-Met/HGF receptor (HGFR) triggers, in an autocrine fashion, a host of signaling pathways that enact gene-level changes, thereby instructing the satellite cell to reenter cell cycle.

(Kook et al., 2008). The ERK family is a prominent member of the MAPK family of pathways which convey receptor-level signals, such as the activity of HGF-cMet, into gene-level changes with implications for entry into cell cycle. Accordingly, the application of cyclic stretch was found to be associated with the expression of cyclin D1 and proliferating cell nuclear antigen as well as the suppression of p27, acting in concert to maintain the bovine satellite cells in a proliferative state. Simultaneously, cyclic stretching impeded myotube formation and inhibited the expression of myogenic differentiation factors such as myosin heavy chain, reinforcing the idea that mechanical stretch selectively bolsters the beginning stages of myogenesis.

Alternately, stretch has been shown to activate the p38 class of MAPKs, which play a particularly well-characterized role in the regulation of myogenic gene expression and differentiation. From a mechanistic standpoint, both C2C12 and primary rat myoblasts, when subjected to persistent stretch, were found to release a cytokine known as tumor necrosis factor alpha (TNFα) which would amplify the activity of p38 in an autocrine fashion (Zhan, Jin, Chen, Reecy, & Li, 2007). The rate-limiting step in this process is the enzyme responsible for synthesizing TNFα. More recent studies have affirmed that the phosphorylation of a tyrosine residue on this enzyme is what mediates the transduction of mechanical stimulation into p38 MAPK signaling (Niu et al., 2013). Consequently, blocking this step led to a significant reduction in the expression of myogenin and embryonic myosin heavy chain in regenerating mouse myofibers and further impairments in the positioning of myofiber nuclei postfusion.

A final explanation for the myogenic consequences of cyclic stretch lies in the action of inflammatory mediators which infiltrate into the satellite cell niche upon muscle injury. To illustrate, prostaglandin F2α may serve a dual role in regeneration by mobilizing an inflammatory response and also stimulating the production of proteins necessary to support the growth of new tissue (Burd et al., 2010). In this regard, the inducible prostaglandin synthase cyclooxygenase-2 (COX-2) might also be a critical component of stress-regulated myogenesis. Correspondingly, primary myoblasts demonstrated a manifold increase in COX-2 expression immediately after a 5-h stretch cycle (Otis, Burkholder, & Pavlath, 2005). Several hours later, once the myoblasts had sufficient time to assemble molecular precursors for the COX-2 pathway, levels of prostaglandin E2 and F2α rose considerably as well. It is hypothesized that these prostaglandins then act on high-level transcriptional control mechanisms to upregulate muscle-associated protein synthesis and sustain rapid progenitor cell growth. Hence, cyclic stretch would no longer prompt myoblasts to enter cell cycle if COX-2 had been selectively inhibited or knocked out, though the addition of exogenous prostaglandin E2 was able to restore cell proliferation. Ultimately, the sheer number of pathways by which stretch can affect the transcriptome, without even considering the potential for cross talk, reflects the high degree of sensitivity with which muscle progenitor cells respond to mechanical stimulation.

3.1.5 Inhibitory Effects of Stretch

While the majority of the literature converges on the promyogenic effects of cyclic stretch, there is evidence to suggest that excessive mechanical stimulation has deleterious consequences on certain aspects and stages of repair. In one model, myoblasts were plated on a flexible substrate and subjected to varying magnitudes of stretch depending on the degree of surface elongation (Liu, Wang, Yuan, Feng, & Liu, 2010). As the magnitude of applied stretch was increased, a greater proportion of myoblasts were found to undergo apoptosis, and this dependence was abolished upon inhibition of caspase-3, a protease centrally involved in the execution of programmed cell death. However, the seemingly antagonistic responses invoked by cyclic strain might occasionally lead to productive synergies in the context of muscle repair given that apoptosis is crucial to the homeostatic regulation of cell density and tissue morphogenesis. In this regard, low levels of stretch-induced apoptosis may be favorable for myogenesis.

Myoblast differentiation is another aspect of myogenesis that appears to be inhibited in a trade-off for the more conspicuous proliferative effects of mechanical stimulation. For instance, cyclic stretch would override the fact that C2C12 cells had been cultured in differentiation medium, blocking the appearance of typical differentiation markers such as fast myosin heavy chain (Kumar, Murphy, Robinson, Wei, & Boriek, 2004). In this way, physical stretch functions similarly to Notch ligands, which promote the conversion of quiescent satellite cells to proliferative progenitor cells while suppressing differentiation into multinucleated myotubes (Tsivitse, 2010). Uniquely, it was proposed that the induction of proliferation and the suppression of differentiation occur downstream the same cascade of events, beginning with surface-level mechanotransduction by focal adhesion kinase and propagating forward toward the nucleus through the activation of Rac-1 and nuclear factor-kappa B. Alternatively, it is possible that any actin remodeling induced by Rac-1, through changes in cytoskeletal tension, influenced the activity of stretch-gated ion channels and rerouted calcium-dependent signaling pathways governing myoblast maturation (Formigli et al., 2007). Thus, stretch-induced satellite cell activation coincides with a host of complementary inhibitory events which serve to control the pacing of myogenesis.

3.2 Alternate Forms of Mechanical Stress

Thus far, the behavior of muscle progenitor cells has only been examined through the lens of cyclic stretch, but the manner in which a mechanical stimulus is presented may also shape the regenerative response. Alongside ERK and p38, c-Jun NH2-terminal kinase (JNK) belongs to the family of MAPKs and is a potent regulator of cell proliferation. Accordingly, the activity of p38 and JNK rose exponentially when hindlimb muscles from adult rats were subjected to stretch (Boppart, Hirshman, Sakamoto, Fielding, & Goodyear, 2001). However, it is important to note that the pulling force applied to these myofibers was sustained continuously over the span of 10 min to simulate the conditions of isometric exercise. This stands in contrast to the paradigm of alternating stretch and rest cycles employed by previously mentioned experiments. A similar but more moderate increase in the level of JNK and p38 phosphorylation became apparent when muscles were made to contract by way of electrical stimulation. Remarkably, these effects could also be uncoupled at the level of fiber types seeing as static stretch was a more potent effector of MAPK activation in the soleus than the extensor digitorum longus muscle. These findings suggest that the nature of an exercise routine and the muscle groups it targets can generate subtle differences in the ensuing myogenic response.

Although various forms of mechanical stimulation have been linked to hypertrophic responses, the directionality of an applied stress can influence the mechanism by which muscle cells transduce mechanical cues and arrive at the decision to ultimately reenter cell cycle. This "third axis" of myogenic control becomes especially appreciable in the context of muscle groups that are subject to physiological forces other than uniaxial contraction and relaxation. One such muscle is the diaphragm, a skeletal muscle that experiences axial and transverse strain with respiratory-associated motion. In one study, stress in either direction increased the phosphorylation of various MAPK-related signaling proteins in lysate samples of mouse diaphragm muscles, including ERK-1/2, p38, and transcriptional regulator further downstream (Kumar, Chaudhry, Reid, & Boriek, 2002). While the magnitude of the applied force was held constant across conditions, transverse loading of the diaphragm consistently generated a more elevated response in MAPK-related activity. The study then introduced pharmacological antagonists against a host of secondary messengers and observed that some would exclusively block ERK-1/2 activation in response to transverse stretch, and vice versa. As such, an individual myofiber may encode the same essential mechanical stimulus into different signal transduction pathways depending on the way in which that force aligns with the structure and organization of its broader muscle group.

The notion that forces of equal magnitude, presented in different formats, can invoke distinct myogenic signaling mechanisms must also take into account the nature of damage produced by varying forms of mechanical stimulation. For instance, eccentric contractions, whereby the force generated by contraction is insufficient to overcome opposing resistance, tend to elongate muscle and cause more severe tissue injury. Conversely, concentric, shortening contractions often result in more minor damage, leading into the hypothesis that muscle-shortening exercises generate a weaker hypertrophic response. It follows that, in the *tibialis anterior* muscle of adult rats, repeated eccentric contractions produced a more pronounced and immediate increase in the activity of lysosomal enzyme β-glucuronidase, a biochemical marker of muscle damage (Komulainen et al., 2000). These changes were accompanied by a rapid transition to glycolytic fiber types, reflecting a more efficient adaptive response to lengthening contractions. As well, certain characteristics of proliferating myogenic progenitor cells, such as the expression of muscle-specific desmin, could only be detected in fibers subjected to eccentric contractions. Yet, the difference in extent of damage produced by lengthening and shortening contractions had a negligible impact on tissue-level hypertrophic responses. Intriguingly, the cross-sectional area of fibers damaged by either form of contraction was restored to a similar size several days after the initial exercise session, suggesting that some degree of myogenic repair may be possible without the involvement of resident satellite cells. Though speculative, it is possible that a newly identified population of progenitor cells, characterized by the expression of transcription factor Twist2, may in part be responsible for the persistent hypertrophic response (Liu et al., 2017). Nevertheless, there is evidence to suggest that discrepancies in signaling associated with unique forms of mechanical stimulation do not always translate into tangible differences in the progression of myogenesis.

Altogether, mechanical stretch of skeletal myofibers, in models of exercise, injury, or movement at large, can prompt quiescent satellite cells to enter myogenesis via a cascade of autocrine signaling events, starting from receptor-level mechanotransduction to the release of ECM activation cues and, ultimately, the regulation of gene transcription. That being said, the relationship between a physical stimulus and its consequent myogenic response cannot be comprehensive without careful consideration of alternate signaling pathways, possible inhibitory effects, and the exact manner in which the stimulus was presented.

4. PATHOLOGICAL STATES OF MATRIX AND MUSCLE

4.1 Aging

As humans age, the body inevitably undergoes a progressive decline in skeletal muscle mass. However, once age-related muscle degeneration begins to place restrictions on one's mobility, balance, and independence, it can become classified as a disease known as sarcopenia. In diagnostic terms, an unjustified decrease in muscle mass and maximal walking speed are among the most common features used to distinguish aging from pathology (von Haehling, Morley, & Anker, 2010). Yet, a quantitative analysis of quadricep biopsies from young and elderly humans would reveal that the loss of mass cannot fully account for the functional impairments associated with sarcopenia (D'Antona et al., 2003). Rather, because aged muscle contracted at slower speeds and generated less powerful contractions proportional to cross-sectional area, it has been proposed that aging reduces the intrinsic capacity of muscle to propagate force. From a molecular perspective, one contributing factor could be a decrease in the number of actin–myosin sliding interactions responsible for sarcomere contraction. Equally possible is that the individual cross-linking interactions allowing actin and myosin to propel each other forward weaken over time due to demonstrated changes in preference for the various isoforms of myosin heavy chain. In support of these ideas is the fact that the molecular and physiological changes associated with aging could be resolved regardless of whether elderly subjects had been forced into a sedentary lifestyle. Although immobilization increased the severity of age-related muscle degeneration, the two could successfully be uncoupled, indicating that there must exist some mechanism by which aging alone signals for changes in the form or function of healthy myofibers.

One viable candidate for the underlying cause of sarcopenia is the profound transformation in structure and composition of the ECM. Justifying this hypothesis is the finding that single fibers from rat hindlimb muscles in early adulthood (8–12 months) and old age (28–30 months) do not significantly differ in stiffness (Wood et al., 2014). Rather, the two age categories only became discrepant when comparing whole bundles of fiber, where the architecture of skeletal muscle ECM was allowed to play a role. Among the molecular effectors of age-induced ECM remodeling is an increase in collagen deposition, visualized by the upregulation of collagen building blocks and an accumulation of glycation end-products from fibril cross-link formation. Though it remains unclear whether abnormalities in collagen synthesis could be a contributing factor, it has been shown that collagen-degrading enzymes from older animals are less aggressive than their younger counterparts (Kragstrup, Kjaer, & Mackey, 2011).

Aside from promoting collagen deposition, aging inhibits the expression of more compliant ECM components such as elastin and type III collagen, resulting in a stiffer support network for mature myofibers and muscle progenitor cells (Stearns-Reider et al., 2017). There is also evidence to suggest that aging causes fatty tissue to accumulate within the skeletal muscle architecture (Kragstrup et al., 2011); the cytokines and chemokines secreted by invading adipocytes may in turn favor a state of persistent inflammation. For this reason, the ECM of aged skeletal muscle might also succumb to infiltrating immune cells and the edematous leakage of interstitial fluid, as seen in models of wound healing. Evidently, aging is a complex, multifaceted process which imposes gradual changes on the organization of the skeletal muscle ECM, leading to a net increase in bulk tissue stiffness.

The age-transformed ECM may then interfere with known mechanisms of mechanotransduction to alter the behavior of progenitor cell reserves. To illustrate, muscle stem cells seeded on decellularized ECM scaffolds from aged (22–24 months old) mice exhibit a greater tendency to become committed to a fibrogenic lineage, at the expense of decreased myogenicity (Stearns-Reider et al., 2017). Aberrant lineage specification could be attributed to the decrease in collagen tortuosity and stiffening of aged muscle ECM. Consistent with the progressive nature of sarcopenia-related functional decline, any additional deposition due to age-induced fibrogenic conversion would again disrupt the physical properties of the stem cell niche and establish a positive feedback loop which further reduces the regenerative capacity of potential myogenic progenitor cells.

Eventually, aging could even cause progenitor cells to completely lose the property of reversible quiescence. Namely, satellite cells from geriatric (28–32 months old) mice have been shown to descend into a state of permanent dormancy characterized by dysregulation of the Cdkn2a gene (Sousa-Victor et al., 2014). Remarkably, transplantation of geriatric cells into the injured muscle of young mice could not restore baseline levels of myogenic activation and proliferation, the suppression of Cdkn2a gene being the only viable treatment. This stands in contrast to studies suggesting that the loss of "stemness" is not intrinsic to aged cells but, instead, dependent on matrix mechanical cues. In one such study, the age-impaired phenotype could be recapitulated by simply growing myogenic progenitor cells on hydrogels of increasing stiffness (Lacraz et al., 2016). Furthermore, progenitor cells cultivated from aged mice displayed no apparent disability in their ability to expand and fuse into multinucleated myotubes in vitro. Hence, the question remains whether aging inhibits myogenic repair by permanently editing gene expression or by depriving progenitor cells of necessary contact with a niche of recognizable structure and stiffness.

Regardless, a complete understanding of the aging ECM, and its implications on the functional integrity of muscle progenitor cells, is necessary to be able to identify the etiology and progression of sarcopenia. Without even considering the impediments placed on myogenesis, structural changes brought on by aging may muffle the forces generated by muscle contraction. As such, aging is a classic example of how dysregulation of matrix structure and composition can lead to progressive impairments in the form and function of skeletal muscle.

4.2 Muscular Dystrophies

4.2.1 Emery–Dreifuss Muscular Dystrophy

Emery–Dreifuss Muscular Dystrophy (EDMD) is a disease characterized by progressive skeletal muscle wasting and cardiomyopathy. It is caused by mutations in the genes that code for proteins of the LINC complex (linker of nucleoskeleton and cytoskeleton), such as nesprins 1 and 2, or proteins associated with the LINC complex, such as lamins and emerin. The layout of the entire LINC complex, and the interactions between its components, is shown in Fig. 4. While mutations in the *EMD* gene, encoding emerin, an inner nuclear membrane (INM) protein, are responsible for the X-linked form of EDMD, the autosomal forms of this disease are caused by mutations in *LMNA* gene that encodes lamins A and C and by mutations in the *SYNE-1* and *2*, the genes that encode nesprin proteins 1 and 2 (Bonne et al., 1999; Brown et al., 2011; Puckelwartz et al., 2009; Zhang et al., 2007).

The LINC complex spans the INM and outer nuclear membrane (ONM) to bridge actin within the cytoplasm with the nuclear lamina underlying the INM. The LINC complex is composed of nesprins, or KASH (Klarsicht–ANC–Syne homology)-domain proteins, positioned at the ONM, and SUN (Sad1/UNC84) domain proteins found at the INM. The largest mammalian nesprin isoforms Nesprins-1 and 2 (also known as SYNE-1 and 2) are composed of a pair of N-terminal actin-binding calponin homology (CH) domain, a spectrin repeat rich rod domain and a transmembrane KASH domain (Zhang et al., 2001). Within the periplasmic space, the KASH domain of a nesprin protein binds to the SUN domain of a trimer of SUN proteins (Sosa, Rothballer, Kutay, & Schwartz, 2012). It has been demonstrated that the trimerization of the SUN domain proteins is required for binding of the KASH domain protein. Mutations in the coiled-coil domain of SUN proteins impaired trimerization and as such KASH domain proteins were unable to bind (Sosa et al., 2012; Zhou et al., 2012). Furthermore, the localization of nesprin-2 at the ONM appears

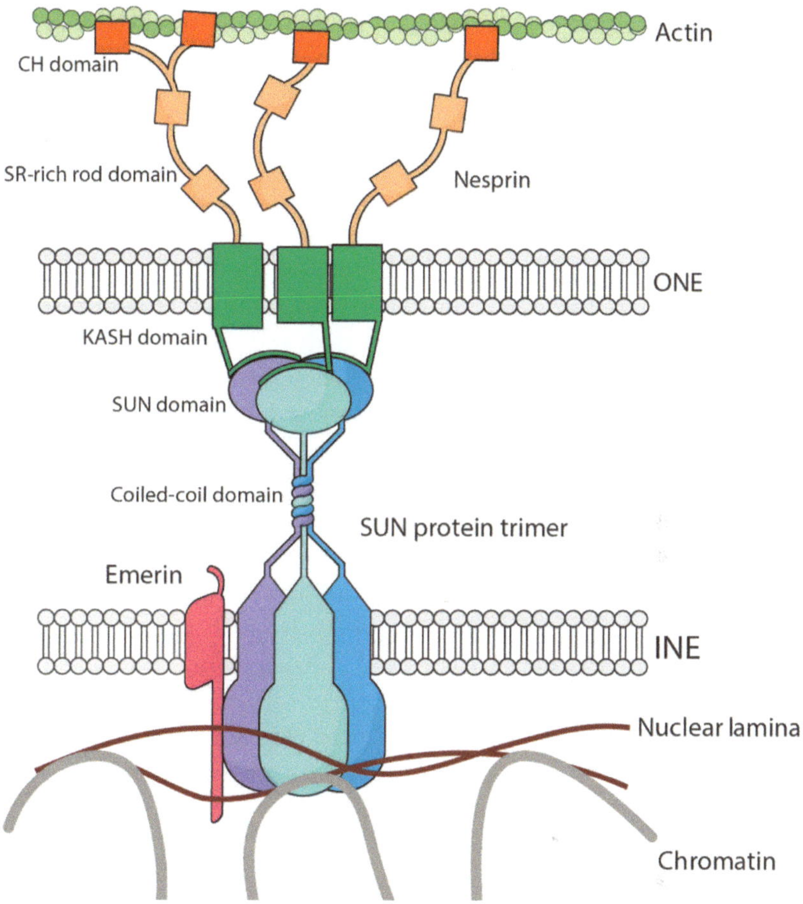

Fig. 4 The LINC complex facilitates mechanotransduction by bridging the cytoskeleton to the nuclear lamina. Nesprin protein calponin homology (CH) domains bind to actin cytoskeleton, while transmembrane KASH domains bind to SUN protein trimer in the periplasmic space. Transmembrane domains of SUN proteins bind to nuclear lamina, which is closely associated with chromatin. Emerin traverses the inner nuclear envelope to bind to nucleoplasmic SUN protein domain and to lamin.

to be dependent on the presence of SUN domain proteins at the INM (Crisp et al., 2006; Padmakumar et al., 2005). To anchor the complex within the nucleus, the nucleoplasmic N-terminal domains of the SUN proteins interact directly A-type lamins (lamins A and C), which result from alternative splicing of the *LMNA* gene (Haque et al., 2006). The nuclear lamina, which is formed by a network of lamin proteins, is extensively associated with chromatin. While nesprins and SUN proteins compose the core of the complex, the LINC complex is associated with the protein emerin. Emerin spans the INM to bind to the nucleoplasmic portion of the SUN domain protein.

One function of the LINC complex is to mechanically couple the cytoskeleton and the nucleus allowing the transduction of mechanical forces imposed upon the cell through to the nucleus (Guilluy et al., 2014; Lombardi et al., 2011). To demonstrate that the LINC complex transduces force into the nucleus directly force was applied to nesprin-1 proteins of isolated nuclei. As a result of the applied force to nesprin-1, the nuclei displayed increased stiffness in order to resist the tension (Guilluy et al., 2014). The roles of this complex within the cell are varied, and disruption of the LINC complex has been demonstrated to induce in defects cellular polarization and migration as well as nuclear positioning (Lombardi et al., 2011). Because the LINC complex binds the nuclear lamina via its SUN domain protein and thus by extension interacts with chromatin, it had been hypothesized that the LINC complex may be involved in the regulation of gene transcription in response to different mechanical cues. Indeed, it has recently been demonstrated that in response to substrate rigidity the LINC complex mediates regulation of transcription for various processes, such as cellular motility, adhesion, and ion transport (Alam et al., 2016). The exact mechanism through which the LINC complex mediates transcriptional changes remains to be elucidated. One possibility is that due to its physical connection to the nuclear lamina and by extension to chromatin, mechanical force applied through to the nucleus by the LINC complex may induces changes in chromatin organization and thus altering the degree of condensation of chromatin and allowing for transcription of otherwise silenced genes.

4.2.2 Duchenne Muscular Dystrophy

Similar to EDMD, Duchenne muscular dystrophy (DMD) is an inherited disorder which presents in early childhood as a progressive atrophy in the skeletal muscle. The progression of DMD is known to be particularly severe as its degenerative effects inevitably "metastasize" from the pelvis and shoulder to skeletal muscles throughout the body, causing a relentless decline in strength and motor function (McDonald, 2012). While EDMD may be caused by loss of function in several of the proteins linking nucleoskeleton and cytoskeleton together, DMD is formally an X-linked disorder resulting from mutations in the dystrophin gene (Blake, Weir, Newey, & Davies, 2002). A critical component of the protein bridge between cytoskeleton and ECM, dystrophin localizes at the inner edge of the sarcolemma, binding to intracellular actin at one end and a chain of transmembrane dystrophin-associated glycoproteins at the other (Campbell, 1995). Moving outward, α-dystroglycan anchors to laminin-α2 chains protruding from the basal lamina. In this way, the dystrophin–glycoprotein complex (DGC) forms a functional link between individual sarcomeres and the broader ECM, allowing forces to be transmitted in either direction. During muscle contraction, for instance, the DGC bears the brunt of the load in order to relieve the sarcolemma of potentially damaging mechanical stress (Jaalouk & Lammerding, 2009). As evidence, diaphragm muscle fibers from the dystrophin-deficient *mdx* mice are far more compliant in response to imposed lengthening and shortening cycles, reaffirming the role of dystrophin in storing force (Kumar, Khandelwal, Malya, Reid, & Boriek, 2004; Kumar, Murphy, et al., 2004).

Equally, the DGC assists biomechanical cues originating from the ECM in entering and initiating signaling inside muscle cells. For example, loss of dystrophin causes an overly exaggerated increase in ERK activity within stretched myofibers, beyond the constraints of what was normal for healthy controls (Kumar, Khandelwal, et al., 2004; Kumar, Murphy, et al., 2004). Moreover, removal of extracellular calcium abolished this response, suggesting that aberrant stretch-induced activation in dystrophin deficiency is mediated by the excessive influx of calcium ions. These results are in general agreement with the theory that a fragile, leaky sarcolemma is the root cause for the pathological effects of DMD (Mokri & Engel, 1975). Further evidence lies in the observations that an increased number of *mdx* tricep fibers incorporate interstitial albumin and that these fibers had not simply began to undergo necrotic membrane permeabilization (Clarke, Khakee, & McNeil, 1993). In addition, a variety of mechanical stresses were found to increase the permeability of *mdx* myofibers to dyes dispersed in the extracellular medium, indicating that the dystrophin-deficient membrane may be more susceptible to damage (Petrof, Shrager, Stedman, Kelly, & Sweeney, 1993). As would be expected with impairments in membrane structural integrity, the accumulation of intracellular calcium in *mdx* fibers also triggers muscle contraction but to the extent that fibers began to deteriorate and clot as a result of sustained activity (Claflin & Brooks, 2008). These results demonstrate that membrane stabilization and mechanotransduction are preconditions for the basic function and viability of skeletal myofibers.

Other characteristic features of DMD muscle cells include a decrease in efficiency of proteolysis, slow oxidative damage, as well as progressive apoptosis and necrosis (Blake et al., 2002). Perhaps most notably, *mdx* hindlimb muscle was found to be significantly stiffer at rest than wild-type controls, an effect that could also complicate any attempts at compensatory tissue regeneration (Engler et al., 2004). Myotubes grown on gels mimicking the stiffness of dystrophic muscle failed to develop the sarcomeric striations typical of mature muscle (Engler et al., 2004) whereas substrates of tissue-like stiffness succeeded, to a remarkable degree, at rescuing proper myotube formation (Serena et al., 2010). Furthermore, it has been shown that dystrophin-deficient satellite cells possess a limited capacity for asymmetric division due to the inability to acquire morphological polarity (Dumont et al., 2015). Considering the crucial role dystrophin plays in progenitor cell mechanosensing, the pathological effects of DMD must not be limited to adult myofibers. Rather, the DGC is an irreplaceable feature of the mechanosensitive apparatus which enables adult myofibers to efficiently propagate force and progenitor cells to retain their capacity for myogenic repair.

4.3 Fibrosis

A characteristic feature of several pathological states of myogenesis is the excessive deposition of connective tissue during attempts at tissue repair, otherwise known as fibrosis (Serrano & Muñoz-Cánoves, 2010). As opposed to other more easily delineable diseases, the causes of fibrosis often overlaps with that of age-related muscle degeneration, inherited muscular dystrophies, and broadly aberrant responses to blunt trauma. In this regard, fibrosis can be viewed as an extension of the norm, whereby the production of ECM components by injury-recruited fibroblastic cells becomes counterproductive to muscle regeneration. Thus, a familiarity with the intended role of fibrotic tissue in myogenic repair is crucial to understanding the mechanisms and implications of its dysregulation.

A simple example of how connective tissue might normally benefit myogenesis is in the development of a structural framework for the expansion and migration of muscle progenitor cells after shearing injury. Once a bundle of skeletal myofibers is severed, muscle contraction no longer elicits movement since any laceration would have disrupted the continuity between the fibers and their associated myotendinous junctions (MTJs). At this point, spontaneous contractions may only widen the gap exposed by the tear (Kääriäinen, Järvinen, Järvinen, Rantanen, & Kalimo, 2000). In an attempt to recoup this loss of connectivity, shear-injured myofibers upregulate α7 integrin to strengthen lateral adhesions with surroundings sheaths of connective tissue (Kääriäinen et al., 2000). However, integrin expression returns to baseline levels as interstitial fibroblasts begin to synthesize new ECM components for reconstruction of the muscle architecture.

Owing in part to the activation of fibroblasts and myogenic progenitor cells, a host of fibrotic ECM components begin to infiltrate the site of injury and nearby areas of inflammatory activity during latter stages of repair. Specifically, when blunt trauma was delivered to the calf muscle of adult rats, type IV collagen, fibronectin, and vinculin would accumulate over areas of necrotic tissue (Kami et al., 1993). As time progressed, these proteins began to pattern into ring-like structures around regenerating myotubes, recapitulating the biomechanical features of the basement membrane in mature skeletal muscle. When the same muscle group was severed by surgical transection, an exaggeration of collagen synthesis, followed by an increase in proteoglycan production, could be visualized in the days following injury (Lehto & Järvinen, 1985). Ultimately, the deposition of connective tissue components contributes to the formation of a cellular scaffold between the ruptured ends of torn muscle, allowing regenerating myotubes to form secondary MTJs at the site of initial injury (Kääriäinen et al., 1998). Consequently, the regulated secretion of fibrotic ECM components promotes the restoration of muscle continuity following blunt trauma.

In pathology, the intrinsic mechanisms controlling ECM production become deregulated. Homeostasis can be even further driven off course by circulating fibroblast-like cells that become drawn to areas of inflammation and then exacerbate connective tissue deposition (Bucala, Spiegel, Chesney, Hogan, & Cerami, 1994). As a result, locally contained deposits of supportive matrix components can readily morph into degradation-resistant scar tissue, leading to morphological abnormalities in the basal lamina and potentially even the substitution of healthy myofibers with fibrotic tissue (Serrano & Muñoz-Cánoves, 2010). In addition to disrupting the optimal biomechanical settings for myogenesis, fibrotic scarring may outlast the initial attempt at repair and render the affected muscle more susceptible to repetitive injury (Stauber, Knack, Miller, & Grimmett, 1996).

One reason why fibrosis follows the onset of numerous other diseases of the muscle is that TGF-β1, a cytokine associated with the injury-induced inflammatory response, signals for connective tissue proliferation as well (Li, Cummins, & Huard, 2001). Correspondingly, patients with dermatomyositis, an inflammatory myopathy which often leads into chronic fibrosis, were found to possess an excess of TGF-β1 in the endomysial and perimysial layers of ECM surrounding their skeletal muscle (Confalonieri, Bernasconi,

Cornelio, & Mantegazza, 1997). However, active TGF-β1 did not stably colocalize with areas populated by infiltrating inflammatory cells, indicating that the release of this cytokine is discontinued beyond the early phases of inflammation. For this reason, it is thought that the ECM functions as a reservoir for biochemical cues associated with both satellite cell activation and fibrogenesis. Likewise, TGF-β1 correlated with the extent of connective tissue overgrowth and was generally overexpressed across all skeletal muscle groups in patients with DMD (Bernasconi et al., 1999). That being said, levels of the cytokine were comparatively lower in another group of congenital muscular dystrophies that are associated with even more severe fibrosis. As such, the amplification of fibroblast proliferation must partner with other mechanisms of connective tissue dysregulation to provide a sufficient explanation for the pathogenesis of fibrosis.

Equally important points of consideration include the overproduction of ECM components by activated fibroblasts and tissue-resident progenitor cells, which is too supported by the role of TGF-β1 (Li et al., 2001). Most notorious among these components are types I and III collagens, structural proteins that become particularly highly expressed during the development of fibrosis (Best & Hunter, 2000). For example, the same collagen subtypes, and their corresponding soluble proforms, would accumulate within the endomysial and perimysial scar tissue of patients with DMD and dermatomyositis/polymyositis (Foidart, Foidart, & Engel, 1981). The lingering presence of procollagens suggests that ECM components may continue to be actively synthesized by interstitial fibroblasts even after scarring has integrated itself into the architecture of skeletal muscle. Given that several functions of muscle regeneration are dependent on an appropriately stiff collagenous substrate, constitutive expression of types I and III collagens could also be traced to satellite cells derived from *mdx* dystrophic mice and wild-type controls (Alexakis, Partridge, & Bou-Gharios, 2007). Remarkably, however, the combination of muscular dystrophy with old age produced a distinct fibrotic phenotype, whereby production of type I collagen was arrested in myogenic cells and instead delegated to a layer of nonmyogenic cells surrounding aged *mdx* fibers. Based on these results, it is possible that, in severely fibrotic settings, muscle precursors may be reprogrammed to a more fibroblast-like cell type that becomes irreversibly committed to collagen production.

Aside from the quantitative increase in collagen content, skeletal muscle burdened with fibrosis is also subject to abnormalities in the posttranslational cross-linking and general organization of collagen fibrils. Indeed, the collagen within *mdx* hindlimb muscle was considerably less densely packed than that of their age-matched controls, reflecting a more loosely structured and disorganized arrangement of ECM components in fibrotic muscle (Smith & Barton, 2014). However, neither collagen content nor packing density was a strong predictor of muscle passive stiffness. As such, there still remains some ambiguity as to whether collagen deposition alone can account for bulk changes in the biomechanical properties of fibrosed muscle. Nonetheless, it is known that excessive connective tissue deposition reduces the contractile strength of skeletal muscle and can act as a physical barrier to myogenic repair. Thus, fibrosis exacerbates the effects of muscle degeneration and is often what defines the changes in structural integrity observed in various muscle pathologies.

Holistically, the pathogenesis of sarcopenia, muscular dystrophy, fibrosis, and various other degenerative diseases almost invariably coincides with alterations in the biomechanical properties of skeletal muscle, and these effects can often be traced to insidious changes in structure and composition of the ECM. Muscle progenitor cells are therefore subject to a milieu of tunable mechanical properties that are held at an optimal constant in health and passivity but varied in the wake of insult, injury, or disease (Table 1). In this way, pathology disrupts the intimate relationship between muscle and matrix, weakening the body's capacity for both movement and tissue regeneration.

5. BIOMIMETIC CULTURE SYSTEMS FOR SKELETAL MUSCLE REGENERATIVE MEDICINE

The goal of skeletal muscle regenerative medicine is to restore form and function to injured, aged, and diseased tissue. Cell transplantation, endogenous repair, gene therapy, and engineered replacement tissue are among the regenerative medicine therapies under development. Understanding the influence of biophysical stresses on skeletal muscle regeneration is expected to provide new therapeutic entry sites for endogenous repair, and this fundamental knowledge has also served to overcome challenges associated with cell transplantation and engineered replacement tissue therapies. Selected examples that demonstrated in vivo relevance with functional assays are reviewed here.

5.1 Culture Substrate Stiffness to Maintain Stemness In Vitro

In the early 1980s, intramuscular transplantation of myogenic progenitors (myoblasts) was explored as a therapeutic approach to restore localized dystrophin expression to dystrophic muscle (Gussoni et al., 1992; Huard et al.,

Table 1 Average Stiffness of Whole Skeletal Muscle and Single *Extensor Digitorum Longus* (EDL) Myofibers in Healthy and Pathological Settings

	Average Stiffness (kPa)	Sample	Measurement Method	References
Whole skeletal muscle	12	C57Bl/6J (*tibialis anterior*)	Shear rheology (Anton Paar Physica MCR 301)	Gilbert et al. (2010)
	18	*mdx* (*extensor digitorum longus*)	Atomic force microscopy	Engler et al. (2004)
	7	Col6a1 −/− (*tibialis anterior*)	Compression testing (Zwixk/Roell Z005)	Urciuolo et al. (2013)
Single *Extensor digitorum longus* (EDL) myofibers	0.5	Young C57Bl/6	Atomic force microscopy	Trensz et al. (2015)
	0.4	Aged C57Bl/6	Atomic force microscopy	Lacraz et al. (2016)
	1.9	Young C57Bl/6		
	2.3	Young C57Bl/6, collapsed		
	10.4	Aged C57Bl/6, collapsed		

1992; Karpati et al., 1993; Law et al., 1992; Tremblay et al., 1993). Although these early clinical trials met with limited success, renewed interest arose upon demonstration that transplanted muscle stem cells provided substantial improvement in muscle strength and dystrophin protein restoration when compared side-by-side with transplanted myoblasts (Cerletti et al., 2008; Collins et al., 2005; Cornelison et al., 2001; Fukada et al., 2004; Kuang, Kuroda, Le Grand, & Rudnicki, 2007; Montarras et al., 2005; Sacco, Doyonnas, Kraft, Vitorovic, & Blau, 2008). However, muscle stem cells constitute a rare subpopulation within muscle necessitating ex vivo expansion prior to use as a cell transplantation therapy. Standard tissue culture plastic substrates in the context of serum containing media supplemented with bFGF2 rapidly diminished the regenerative potential of the therapeutic population, thereby presenting a hurdle to clinical translation (Gilbert et al., 2010; Montarras et al., 2005; Qu-Petersen et al., 2002; Sacco et al., 2008).

Inspired by the first demonstration that stiffness of cell culture substrates can dramatically influence the differentiation of mesenchymal stromal cells

(Engler, Sen, Sweeney, & Discher, 2006), studies began to explore whether stiffness might also control adult stem cell self-renewal. The rationale arose from quantitative measurements of murine skeletal muscle tissue. Atomic force microscopy (Engler et al., 2004) and shear rheometry (Gilbert et al., 2010) measurements showed that homeostatic, uninjured murine *tibialis anterior* muscles are soft (~12 kPa), while tissue culture plastic dishes are orders of magnitude stiffer. Using a synthetic poly(ethyleneglycol)-based hydrogel substrate tuned to match the stiffness of murine skeletal muscle and surface functionalized with the basement membrane protein laminin, it was shown that prospectively isolated murine muscle stem cells maintained stemness outside of the body significantly better than on plastic culture (Gilbert et al., 2010). Specifically, following a week of culture in growth conditions a greater proportion of cells in the population were positive for the stem cell marker Pax7, and fewer cells expressed the differentiation marker Myogenin. Aided by a newly developed computer vision program to track the lineage of individual muscle stem cells cultured on soft hydrogels compared with stiff plastic, it was possible to discern that hydrogel cultured cell populations grew over time in part because there was significantly less death compared to the plastic condition (Gilbert et al., 2010).

Notably, subsequent studies by others also reported the influence of substrate stiffness on murine (Cosgrove et al., 2014; Davoudi & Gilbert, 2017; Quarta et al., 2016; Urciuolo et al., 2013) and human (Monge et al., 2017) muscle stem cell viability and stem cell signatures. Indeed, tuning the propagation substrate stiffness has proven an effective strategy to support Pax7 expression, prevent cell cycle exit and transition to the myocyte fate, as well as to delay both murine and human myogenic progenitor differentiation (Boontheekul, Hill, Kong, & Mooney, 2007; Gribova, Gauthier-Rouvière, Albigès-Rizo, Auzely-Velty, & Picart, 2013; Serena et al., 2010).

Essential to the impact of the afore described study (Gilbert et al., 2010), was the use of in vivo functional assays to support claims that cells meeting the definition of a stem cell (i.e., self-renewal and differentiation potential) were maintained in culture by propagation on soft hydrogel substrates. Following a week of culture on tissue culture plastic or hydrogels softer than or matching the softness of muscle, the putative muscle stem cells were collected and transplanted intramuscularly into hindlimb irradiated immunocompromised mice. Evaluations of satellite cell niche repopulation and contribution to host muscle fibers by transplanted cells revealed that culturing prospectively isolated muscle stem cells on skeletal muscle mimetic hydrogels best maintained their therapeutic potential ex vivo. However, without evidence of division, it was not possible to draw conclusions about the role of substrate stiffness on muscle stem cell self-renewal. To address this experimentally, single muscle stem cells were spatially segregated in small bioengineered "microwells" with soft or rigid culture surfaces and tracked over time. Microwells in which the cell underwent a single division were collected and transplanted intramuscularly. Notably, only doublets harvested from the muscle mimetic hydrogels were able to engraft. This observation lent evidence in support of the notion that soft hydrogel substrates support self-renewal divisions in culture, while plastic culture does not. Since this study evaluated transplant efficiency after the first division, the work also revealed that "stemness" in standard plastic culture conditions is lost within the first 40–72 h of culture.

Interestingly, when compared one-to-one with freshly isolated muscle stem cells, hydrogel cultured cells still were significantly less potent. This, in combination with the rather limited amount of population doubling observed over time ($\sim$2.5-fold over plastic culture) led to a premise that soft hydrogel might support asymmetric divisions and that additional biomechanical or biochemical cues were needed to push symmetric self-renewal divisions. Indeed, subsequent studies found that treating muscle stem cells in culture with a p38MAPK small molecule inhibitor in the context of a muscle mimetic hydrogel substrate supported substantial population expansion that included self-renewal expansion divisions confirmed by in vivo functional assays (Cosgrove et al., 2014). Notably, a follow-up study confirmed that p38MAPK inhibitor treatment in culture also improved human muscle stem cell transplant efficacy (Charville et al., 2015). It is yet to be determined whether a synergy with soft hydrogel culture benefits human muscle stem cells as well. However, given the promising effects of hydrogel culture on human satellite cell (Monge et al., 2017) and myogenic progenitor fate (Boontheekul et al., 2007; Gribova et al., 2013; Serena et al., 2010), it seems likely.

As introduced in Section 2.1 (Quarta et al., 2016), a recent elegant study engineered an artificial muscle fiber to recapitulate certain key biophysical and biochemical features that were found necessary to maintain muscle stem cell quiescence in culture. By culturing prospectively isolated murine or human muscle stem cells on soft, extruded type I collagen nanofibrils matching the softness (1–2 kPa) and curvature of native muscle fibers, and presenting a VCAM receptor in the context of a proprietary culture media allowed for long-term maintenance of the quiescent state in culture (Fig. 5). When the muscle stem cells were then transplanted intramuscularly

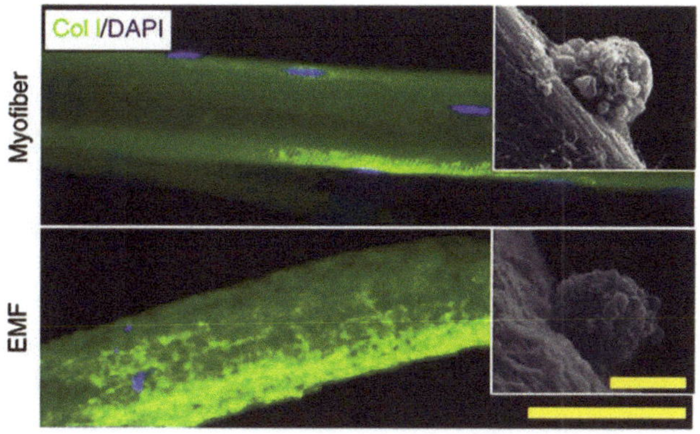

Fig. 5 Comparison of satellite cell localization on a single mouse myofiber and artificial muscle fiber. Immunofluorescent staining for type I collagen (*green*) and myofiber nuclei (*blue*) on a single murine myofiber (*top*) and an artificial muscle fiber designed to mimic the native satellite cell niche (*bottom*). Scanning electron microscopy images (*right*, inset) portray satellite cells localized on their respective muscle fibers. Scale bar, 50 μm. *Reprinted by permission from Macmillan Publishers Ltd.: Nature Biotechnology. Quarta, M., Brett, J.O., DiMarco, R., De Morree, A., Boutet, S.C., Chacon, R., et al. (2016). An artificial niche preserves the quiescence of muscle stem cells and enhances their therapeutic efficacy.* Nature Biotechnology, 34, *752–759. https://doi.org/10.1038/nbt.3576, copyright 2016.*

into the hindlimbs of mice in association with the artificial muscle fiber, they performed on a par with freshly isolated muscle fibers. In addition, by performing genetic manipulation studies in culture, it was shown that culture atop artificial muscle fibers greatly improves the regenerative potential of the genetically modified cell population when compared to standard culture practice. This is a key advance in light of advances in genome editing. Further, since the artificial muscle fibers are strong enough to withstand syringe manipulations, this delivery vehicle inspired by biophysical and biochemical aspects of the native muscle stem cell niche is poised to provide an important skeletal muscle regenerative medicine tool (Quarta et al., 2016).

5.2 Patterning Cell Adhesion to Augment Differentiation

The design of biomaterials with features that support alignment and fusion of myogenic progenitors into parallel rows matching the native structure has been studied extensively to support tissue engineering applications (Qazi, Mooney, Pumberger, Geissler, & Duda, 2015). The rationale is that layers of aligned, nascent muscle fibers might serve as a skeletal muscle "patch" or be layered on top of one another to create a three-dimensional replacement tissue.

A simple strategy to generate aligned rows of muscle fibers in culture is to print the ECM adhesive interface in strips of optimized width and separation directly onto the culture substrate of interest. Using the technique "microcontact printing" (μCP) it is possible to easily and inexpensively generate "islands" of ECM protein in almost any desired shape or size. In μCP, polydimethylsiloxane (PDMS) rubber is cured onto silicone wafers that contain a positive replica of the desired features (e.g., arrays of circles, rectangles, squares, stars, etc.). After hardening, the PDMS is peeled away from the silicone wafer, and the contact surface now presents the arrays of microfeatures. By placing the PDMS rubber feature side down onto a surface containing an ECM protein (e.g., laminin, collagen, fibronectin), the PDMS functions like a "stamp" and adsorbs the proteins only at the top of each feature. By then laying the stamp on a subsequent surface (e.g., tissue culture plastic, a layer of PDMS, etc.) and applying pressure, the ECM is transferred to the new surface effectively creating geometric protein islands. Using this inexpensive and relatively simple technique, or more modern methods that achieve a similar outcome, such as modified inkjet printers to spray the microfeatures, numerous groups have explored the influence of ECM identity and patterned line width and spacing as it relates to myogenic progenitor alignment and differentiation into muscle fibers (Duffy & Feinberg, 2014). In one such example, PMDS stamps with 10, 15, 20, or 30 μm line spacings and 20, 50, 100, or 200 μm line widths were evaluated to create a total of 16-micropattern conditions that were evaluated in the context of fibronectin, laminin, type I or type II collage adhesion (Duffy, Sun, & Feinberg, 2016). For C2C12 murine myoblasts and human skeletal muscle-derived cells, laminin consistently produced the greatest number myotubes that were longer, wider, and contained more nuclei than culture on the other ECMs. Interestingly, while C2C12 myotube maturation and alignment was found to be sensitive to micropattern width and spacing, human-derived skeletal muscle cells performed equally well on all micropatterns revealing potential species specific differences. The study nicely demonstrates a systematic approach to assessing synergies between ECM and substrate patterning.

One strategy to create a skeletal muscle patch is to grow tissues on a scaffold that supports early stages of maturation, but in which later stages of maturation prompt "release" of the skeletal muscle sheet. As an example, by varying PDMS precursor polymer ratio and then plasma oxidizing the material under weak or strong vacuum, a sheet of rubber can be generated that contains high-quality wavy features down the length of each sheet with wavelengths (distance between two adjacent wave peaks) of 3, 6, or 12 μm (Lam, Sim, Zhu, & Takayama, 2006). With laminin added as an adhesive interface, C2C12s plated on the sheets were inclined to adhere to the top of each wave peak such that aligned rows of myogenic progenitors formed; a behavior that was not observed in unpatterned cultures. Transitioning cultures to conditions that support differentiation then revealed that the wavy features support myoblast fusion into multinucleated structures that the myonuclei and actin cytoskeleton aligned in parallel rows with the 6 μm periodicity producing most consistent results. PDMS is not an inherently adhesive material for cells, so as differentiation proceeded, the aligned myotubes released into the culture media as sheets.

Skeletal muscle sheets are thin and difficult to transfer, so biomaterials, which by definition are biocompatible, offer an opportunity to act as a scaffold for building skeletal muscle tissues that can be directly implanted into a patient. Ideally, the scaffold is degraded into a nontoxic product that is cleared from the patient leaving the engineered skeletal muscle behind. As an example, a scaffold created from an engineered protein produced in bacteria (the elastin sequence fused to "RGD" fibronectin adhesion peptides) and designed to present different densities of RGD ligand (930, 4650, 9300 ligands/μm^2) was cross-linked in contact with patterned molds allowing for the generation of grooves with different widths (20–200 μm) (Romano, Sengupta, Chung, & Heilshorn, 2011). C2C12 murine myoblasts and primary human myoblasts each aligned and elongated most optimally within grooves of the smallest widths. Differences between C2C12s and human myoblasts emerged upon evaluating myocyte spreading area and responses to RGD ligand density. While human myoblast cell spreading area increased as groove width and ligand density increased, C2C12s decreased in cell area with increasing groove width and were found to be insensitive to the range of ligand densities tested. Since the scaffold is derived from the elastin ECM molecule, it is degraded over time in vivo, underlying the relevance of this proof-of-concept study.

Recent efforts have focused on demonstrating the in vivo potential of the engineered skeletal muscle "patches" that are built in culture using patterned surfaces. For example, poly(lactide-co-glycolide) (PLGA) is a biocompatible and biodegradable polymer that can be fabricated to present aligned nanogrooves. Consistent with similar studies (Duffy & Feinberg, 2014; Qazi et al., 2015), nanogrooved surfaces coated with a thin layer of gelatin supported murine muscle cell differentiation, fusion, and alignment into rows of parallel multinucleated muscle fibers better than the unpatterned PLGA substrates (Yang et al., 2014). One month after skeletal muscle fiber patches were transplanted parallel to the quadriceps muscle of *mdx* mice, the nanopatterned PLGA patches contained more donor-derived myonuclei and a higher incidence of dystrophin expressing muscle fibers than the unpatterned PLGA patches. This study delivers in vivo validation that nanopatterning is an important design criteria in skeletal muscle regenerative medicine.

5.3 Biophysical Stimulation in Skeletal Muscle Tissue Engineering

A volumetric muscle loss occurs as a result of surgery or traumatic damage and is defined as a loss of skeletal muscle that cannot be resolved by endogenous repair and, therefore, results in functional impairment (Duffy & Feinberg, 2014; Qazi et al., 2015). While biomaterial-based scaffolds alone can reverse cosmetic damage, they offer limited functional improvement, likely due to the inability of cells from remaining tissue to adequately expand and functionally repopulate the scaffold. Indeed, delivering cells within a biomaterial seems to offer greater therapeutic benefit (Fuoco et al., 2015; Levenberg et al., 2005; Rossi et al., 2011). Hence, substantial efforts focus on preseeding 3D scaffolds with cells and preform tissues prior to transplantation. As reviewed in Qazi et al. (2015) and Duffy and Feinberg (2014), the vast majority of former studies focused on engineering three-dimensional skeletal muscle tissues starting with murine myogenic progenitors. Only recently were efforts to engineer human skeletal muscle tissue in a dish that responds to electrical and chemical stimulation met with success as a result of optimizing the scaffold protein composition away from collagen to instead favor a fibrin—basement membrane mix (Bakooshli & Gilbert, 2015; Madden, Juhas, Kraus, Truskey, & Bursac, 2015). Consistent among all such studies was the need to design an element in the culture platform that acts like tends to produce uniaxial tension, a biophysical stress in native skeletal muscle, which then induces muscle fiber alignment in the 3D tissues.

A recent report further extended the field of skeletal muscle tissue engineering by demonstrating the requirement to implement two additional biophysical cues to support the ex vivo maturation and in vivo vascularization of transplanted tissues: shear stress and mechanical stimulation via exercise [39]. In this study, the authors first optimize cellular and scaffold requirements and conclude that murine or human muscle stem cells together with other muscle resident cells in a natural collagen scaffold resulted in the greatest restoration of muscle structure and force generation in response to electrical stimulation. While characterizing their bioconstructs in culture they noted that the muscle stem cells seem to begin dying in the 3D construct as early as 2 h after embedding, but by perfusing the construct in a bioreactor, it was possible to improve muscle stem cell viability and eventual regeneration potential. The introduction of shear flow, a biophysical cue, presumably improves access to nutrients while also removing waste from the immediate vicinity of the cells but may also have a direct impact on muscle stem cell fate through cell surface receptor stimulation. Finally, while the transplanted bioconstructs responded to electrical stimulation with contraction, stimulating the sciatic nerve resulted in no response, suggesting that bioconstruct innervation was incomplete. While waiting additional time to allow for innervation to occur failed to restore functional recovery, encouraging the transplanted animals to exercise (i.e., mechanically stimulating the bioconstruct in vivo) beginning 1 week after transplantation yielded animals with engineered muscle that contracted in response to sciatic nerve stimulation (Quarta et al., 2017). This comprehensive study weighs the translational benefits of numerous stimulatory cues implemented during the process of skeletal muscle tissue engineering and highlights the need to exploring both biochemical and biophysical cues in regenerative medicine.

6. CONCLUSIONS

While it is understood that biomolecules are important regulators of myogenesis, there is an ever-growing body of evidence in support of putting more emphasis on defining the biomechanical nature of the skeletal muscle stem cell niche. In addition to providing a greater understanding of the spatio-temporal cues that guide the process, a holistic view of the muscle stem cell niche is expected to inform future endogenous repair therapies while also overcoming hurdles impeding other skeletal muscle regenerative medicine therapies. For example, bioprocessing methods to expand therapeutic cell populations rather than relying on rooms full of two-dimensional culture dishes; or endogenous repair drugs or gene vectors guided directly to satellite cells within nanoparticles presenting satellite cell specific ligands and tuned to take advantage of the biophysical highway that is unique to the niche; or bioreactors to recapitulate the young muscle environment and support engineered muscle driving the motor activities of soft robots (Duffy & Feinberg, 2014). Arriving at a 4D understanding of the niche biochemical and biophysical milieu in healthy, aged, injured, or diseased muscle will ultimately require intersection with computational scientists to generate models (Huijing, 1999; van Donkelaar et al., 1996; Yucesoy, Koopman, Huijing, & Grootenboer, 2002) that are then validated by wet lab scientists. The next 50 years of satellite cell research is poised to cross multiple disciplines in order to achieve new levels of understanding that drive the future of skeletal muscle regenerative medicine.

REFERENCES

Alam, S. G., Zhang, Q., Prasad, N., Li, Y., Chamala, S., Kuchibhotla, R., et al. (2016). The mammalian LINC complex regulates genome transcriptional responses to substrate rigidity. *Scientific Reports*, *6*, 38063. https://doi.org/10.1038/srep38063.

Alexakis, C., Partridge, T., & Bou-Gharios, G. (2007). Implication of the satellite cell in dystrophic muscle fibrosis: A self-perpetuating mechanism of collagen overproduction. *American Journal of Physiology. Cell Physiology*, *293*, C661–C669. https://doi.org/10.1152/ajpcell.00061.2007.

Anderson, J. E. (2000). A role for nitric oxide in muscle repair: Nitric oxide-mediated activation of muscle satellite cells. *Molecular Biology of the Cell*, *11*, 1859–1874. https://doi.org/10.1091/mbc.11.5.1859.

Bakooshli, M. A., & Gilbert, P. M. (2015). Muscling in on the third dimension. *eLife*, *4*, https://doi.org/10.7554/eLife.06430.

Bass, M. D., Roach, K. A., Morgan, M. R., Mostafavi-Pour, Z., Schoen, T., Muramatsu, T., et al. (2007). Syndecan-4-dependent Rac1 regulation determines directional migration in response to the extracellular matrix. *The Journal of Cell Biology*, *177*, 527–538. https://doi.org/10.1083/jcb.200610076.

Benavides Damm, T., & Egli, M. (2014). Calcium's role in mechanotransduction during muscle development. *Cellular Physiology and Biochemistry*, *33*, 249–272. https://doi.org/10.1159/000356667.

Bentzinger, C. F., von Maltzahn, J., Dumont, N. A., Stark, D. A., Wang, Y. X., Nhan, K., et al. (2014). Wnt7a stimulates myogenic stem cell motility and engraftment resulting in improved muscle strength. *The Journal of Cell Biology*, *205*, 97–111. https://doi.org/10.1083/jcb.201310035.

Bentzinger, C. F., Wang, Y. X., Von Maltzahn, J., Soleimani, V. D., Yin, H., & Rudnicki, M. A. (2013). Fibronectin regulates Wnt7a signaling and satellite cell expansion. *Cell Stem Cell*, *12*, 75–87. https://doi.org/10.1016/j.stem.2012.09.015.

Bernasconi, P., Di Blasi, C., Mora, M., Morandi, L., Galbiati, S., Confalonieri, P., et al. (1999). Transforming growth factor-beta1 and fibrosis in congenital muscular dystrophies. *Neuromuscular Disorders*, *9*, 28–33. https://doi.org/10.1016/S0960-8966(98)00093-5.

Best, T. M., & Hunter, K. D. (2000). Muscle injury and repair. *Physical Medicine and Rehabilitation Clinics of North America*, *11*, 251–266.

Bischoff, R. (1990). Interaction between satellite cells and skeletal muscle fibers. *Development*, *109*, 943–952.

Blake, D. J., Weir, A., Newey, S. E., & Davies, K. E. (2002). Function and genetics of dystrophin and dystrophin-related proteins in muscle. *Physiological Reviews*, *82*, 291–329. https://doi.org/10.1152/physrev.00028.2001.

Bonne, G., Di Barletta, M. R., Varnous, S., Bécane, H. M., Hammouda, E. H., Merlini, L., et al. (1999). Mutations in the gene encoding lamin A/C cause autosomal dominant

Emery-Dreifuss muscular dystrophy. *Nature Genetics, 21*, 285–288. https://doi.org/10.1038/6799.
Boontheekul, T., Hill, E. E., Kong, H.-J., & Mooney, D. J. (2007). Regulating myoblast phenotype through controlled gel stiffness and degradation. *Tissue Engineering, 13*, 1431–1442. https://doi.org/10.1089/ten.2006.0356.
Boppart, M. D., Hirshman, M. F., Sakamoto, K., Fielding, R. A., & Goodyear, L. J. (2001). Static stretch increases c-Jun NH2-terminal kinase activity and p38 phosphorylation in rat skeletal muscle. *American Journal of Physiology Cell Physiology, 280*, C352–358.
Brown, C. A., Scharner, J., Felice, K., Meriggioli, M. N., Tarnopolsky, M., Bower, M., et al. (2011). Novel and recurrent EMD mutations in patients with Emery-Dreifuss muscular dystrophy, identify exon 2 as a mutation hot spot. *Journal of Human Genetics, 56*, 589–594. https://doi.org/10.1038/jhg.2011.65.
Bucala, R., Spiegel, L. A., Chesney, J., Hogan, M., & Cerami, A. (1994). Circulating fibrocytes define a new leukocyte subpopulation that mediates tissue repair. *Molecular Medicine, 1*, 71–81.
Burd, N. A., Dickinson, J. M., Lemoine, J. K., Carroll, C. C., Sullivan, B. E., Haus, J. M., et al. (2010). Effect of a cyclooxygenase-2 inhibitor on postexercise muscle protein synthesis in humans. *American Journal of Physiology. Endocrinology and Metabolism, 298*, E354–E361. https://doi.org/10.1152/ajpendo.00423.2009.
Campbell, K. P. (1995). Three muscular dystrophies: Loss of cytoskeleton-extracellular matrix linkage. *Cell, 80*, 675–679. https://doi.org/10.1016/0092-8674(95)90344-5.
Carmeli, E., Moas, M., Reznick, A. Z., & Coleman, R. (2004). Matrix metalloproteinases and skeletal muscle: A brief review. *Muscle & Nerve, 29*, 191–197. https://doi.org/10.1002/mus.10529.
Ceafalan, L. C., Popescu, B. O., & Hinescu, M. E. (2014). Cellular players in skeletal muscle regeneration. *BioMed Research International, 2014*, https://doi.org/10.1155/2014/957014.
Cerletti, M., Jurga, S., Witczak, C. A., Hirshman, M. F., Shadrach, J. L., Goodyear, L. J., et al. (2008). Highly efficient, functional engraftment of skeletal muscle stem cells in dystrophic muscles. *Cell, 134*, 37–47. https://doi.org/10.1016/j.cell.2008.05.049.
Charville, G. W., Cheung, T. H., Yoo, B., Santos, P. J., Lee, G. K., Shrager, J. B., et al. (2015). Ex vivo expansion and in vivo self-renewal of human muscle stem cells. *Stem Cell Reports, 5*, 621–632. https://doi.org/10.1016/j.stemcr.2015.08.004.
Chen, R., Feng, L., Ruan, M., Liu, X., Adriouch, S., & Liao, H. (2013). Mechanical-stretch of C2C12 myoblasts inhibits expression of toll-like receptor 3 (TLR3) and of autoantigens associated with inflammatory myopathies. *PloS One, 8*, https://doi.org/10.1371/journal.pone.0079930.
Choi, Y. S., Vincent, L. G., Lee, A. R., Kretchmer, K. C., Chirasatitsin, S., Dobke, M. K., et al. (2012). The alignment and fusion assembly of adipose-derived stem cells on mechanically patterned matrices. *Biomaterials, 33*, 6943–6951. https://doi.org/10.1016/j.biomaterials.2012.06.057.
Christov, C., Chrétien, F., Abou-Khalil, R., Bassez, G., Vallet, G., Authier, F.-J., et al. (2007). Muscle satellite cells and endothelial cells: Close neighbors and privileged partners. *Molecular Biology of the Cell, 18*, 1397–1409. https://doi.org/10.1091/mbc.E06-08-0693.
Claflin, D. R., & Brooks, S. V. (2008). Direct observation of failing fibers in muscles of dystrophic mice provides mechanistic insight into muscular dystrophy. *American Journal of Physiology. Cell Physiology, 294*, C651–C658. https://doi.org/10.1152/ajpcell.00244.2007.
Clarke, M. S., Khakee, R., & McNeil, P. L. (1993). Loss of cytoplasmic basic fibroblast growth factor from physiologically wounded myofibers of normal and dystrophic muscle. *Journal of Cell Science, 106*(Pt. 1), 121–133.

Collins, C. A., Olsen, I., Zammit, P. S., Heslop, L., Petrie, A., Partridge, T. A., et al. (2005). Stem cell function, self-renewal, and behavioral heterogeneity of cells from the adult muscle satellite cell niche. *Cell, 122*, 289–301. https://doi.org/10.1016/j.cell.2005.05.010.

Collinsworth, A. M., Zhang, S., Kraus, W. E., & Truskey, G. A. (2002). Apparent elastic modulus and hysteresis of skeletal muscle cells throughout differentiation. *American Journal of Physiology. Cell Physiology, 283*, C1219–C1227. https://doi.org/10.1152/ajpcell.00502.2001.

Confalonieri, P., Bernasconi, P., Cornelio, F., & Mantegazza, R. (1997). Transforming growth factor-beta 1 in polymyositis and dermatomyositis correlates with fibrosis but not with mononuclear cell infiltrate. *Journal of Neuropathology and Experimental Neurology, 56*, 479–484. https://doi.org/10.1097/00005072-199705000-00003.

Cornelison, D. D., Filla, M. S., Stanley, H. M., Rapraeger, A. C., & Olwin, B. B. (2001). Syndecan-3 and syndecan-4 specifically mark skeletal muscle satellite cells and are implicated in satellite cell maintenance and muscle regeneration. *Developmental Biology, 239*, 79–94. https://doi.org/10.1006/dbio.2001.0416.

Cornelison, D. D. W., Wilcox-Adelman, S. A., Goetinck, P. F., Rauvala, H., Rapraeger, A. C., & Olwin, B. B. (2004). Essential and separable roles for Syndecan-3 and Syndecan-4 in skeletal muscle development and regeneration. *Genes & Development, 18*, 2231–2236. https://doi.org/10.1101/gad.1214204.

Cosgrove, B. D., Gilbert, P. M., Porpiglia, E., Mourkioti, F., Lee, S. P., Corbel, S. Y., et al. (2014). Rejuvenation of the muscle stem cell population restores strength to injured aged muscles. *Nature Medicine, 20*, 255–264. https://doi.org/10.1038/nm.3464.

Crisp, M., Liu, Q., Roux, K., Rattner, J. B., Shanahan, C., Burke, B., et al. (2006). Coupling of the nucleus and cytoplasm: Role of the LINC complex. *The Journal of Cell Biology, 172*, 41–53. https://doi.org/10.1083/jcb.200509124.

D'Antona, G., Pellegrino, M. A., Adami, R., Rossi, R., Carlizzi, C. N., Canepari, M., et al. (2003). The effect of ageing and immobilization on structure and function of human skeletal muscle fibres. *The Journal of Physiology, 552*, 499–511. https://doi.org/10.1113/jphysiol.2003.046276.

Davoudi, S., & Gilbert, P. M. (2017). Optimization of satellite cell culture through biomaterials. *Methods in Molecular Biology (Clifton, NJ), 1556*, 329–341. https://doi.org/10.1007/978-1-4939-6771-1_18.

Duffy, R. M., & Feinberg, A. W. (2014). Engineered skeletal muscle tissue for soft robotics: Fabrication strategies, current applications, and future challenges. *Wiley Interdisciplinary Reviews. Nanomedicine and Nanobiotechnology, 6*, 178–195. https://doi.org/10.1002/wnan.1254.

Duffy, R. M., Sun, Y., & Feinberg, A. W. (2016). Understanding the role of ECM protein composition and geometric micropatterning for engineering human skeletal muscle. *Annals of Biomedical Engineering, 44*, 2076–2089. https://doi.org/10.1007/s10439-016-1592-8.

Dumont, N. A., Wang, Y. X., von Maltzahn, J., Pasut, A., Bentzinger, C. F., Brun, C. E., et al. (2015). Dystrophin expression in muscle stem cells regulates their polarity and asymmetric division. *Nature Medicine, 21*, 1455–1463. https://doi.org/10.1038/nm.3990.

El Fahime, E., Torrente, Y., Caron, N. J., Bresolin, M. D., & Tremblay, J. P. (2000). In vivo migration of transplanted myoblasts requires matrix metalloproteinase activity. *Experimental Cell Research, 258*, 279–287. https://doi.org/10.1006/excr.2000.4962.

Engler, A. J., Griffin, M. A., Sen, S., Bönnemann, C. G., Sweeney, H. L., & Discher, D. E. (2004). Myotubes differentiate optimally on substrates with tissue-like stiffness: Pathological implications for soft or stiff microenvironments. *The Journal of Cell Biology, 166*, 877–887. https://doi.org/10.1083/jcb.200405004.

Engler, A. J., Sen, S., Sweeney, H. L., & Discher, D. E. (2006). Matrix elasticity directs stem cell lineage specification. *Cell, 126*, 677–689. https://doi.org/10.1016/j.cell.2006.06.044.

Fischer, M., Rikeit, P., Knaus, P., & Coirault, C. (2016). YAP-mediated mechanotransduction in skeletal muscle. *Frontiers in Physiology, 7*. https://doi.org/10.3389/fphys.2016.00041.

Foidart, M., Foidart, J. M., & Engel, W. K. (1981). Collagen localization in normal and fibrotic human skeletal muscle. *Archives of Neurology, 38*, 152–157. https://doi.org/10.1001/archneur.1981.00510030046006.

Formigli, L., Meacci, E., Sassoli, C., Squecco, R., Nosi, D., Chellini, F., et al. (2007). Cytoskeleton/stretch-activated ion channel interaction regulates myogenic differentiation of skeletal myoblasts. *Journal of Cellular Physiology, 211*, 296–306. https://doi.org/10.1002/jcp.20936.

Formigli, L., Sassoli, C., Squecco, R., Bini, F., Martinesi, M., Chellini, F., et al. (2009). Regulation of transient receptor potential canonical channel 1 (TRPC1) by sphingosine 1-phosphate in C2C12 myoblasts and its relevance for a role of mechanotransduction in skeletal muscle differentiation. *Journal of Cell Science, 122*, 1322–1333. https://doi.org/10.1242/jcs.035402.

Franco, A., & Lansman, J. B. (1990). Stretch-sensitive channels in developing muscle cells from a mouse cell line. *The Journal of Physiology, 427*, 361–380.

Fu, X., Wang, H., & Hu, P. (2015). Stem cell activation in skeletal muscle regeneration. *Cellular and Molecular Life Sciences, 72*, 1663–1677. https://doi.org/10.1007/s00018-014-1819-5.

Fukada, S. I., Higuchi, S., Segawa, M., Koda, K. I., Yamamoto, Y., Tsujikawa, K., et al. (2004). Purification and cell-surface marker characterization of quiescent satellite cells from murine skeletal muscle by a novel monoclonal antibody. *Experimental Cell Research, 296*, 245–255. https://doi.org/10.1016/j.yexcr.2004.02.018.

Fuoco, C., Rizzi, R., Biondo, A., Longa, E., Mascaro, A., Shapira-Schweitzer, K., et al. (2015). In vivo generation of a mature and functional artificial skeletal muscle. *EMBO Molecular Medicine, 7*, 411–422. https://doi.org/10.15252/emmm.201404062.

Gilbert, P., Havenstrite, K., Magnusson, K., Sacco, A., Leonardi, N., Kraft, N., et al. (2010). Substrate elasticity regulates skeletal muscle stem cell self-renewal in culture. *Science (New York, N.Y.), 329*, 1078–1081. https://doi.org/10.1126/science.1191035. Substrate.

Gillies, A. R., & Lieber, R. L. (2011). Structure and function of the skeletal muscle extracellular matrix. *Muscle & Nerve, 44*, 318–331. https://doi.org/10.1002/mus.22094.

Goodman, S. L., Risse, G., & Von Der Mark, K. (1989). The E8 subfragment of laminin promotes locomotion of myoblasts over extracellular matrix. *The Journal of Cell Biology, 109*, 799–809. https://doi.org/10.1083/jcb.109.2.799.

Gribova, V., Gauthier-Rouvière, C., Albigès-Rizo, C., Auzely-Velty, R., & Picart, C. (2013). Effect of RGD functionalization and stiffness modulation of polyelectrolyte multilayer films on muscle cell differentiation. *Acta Biomaterialia, 9*, 6468–6480. https://doi.org/10.1016/j.actbio.2012.12.015.

Guérin, C. W., & Holland, P. C. (1995). Synthesis and secretion of matrix-degrading metalloproteases by human skeletal muscle satellite cells. *Developmental Dynamics, 202*, 91–99. https://doi.org/10.1002/aja.1002020109.

Guharay, F., & Sachs, F. (1984). Stretch-activated single ion channel currents in tissue-cultured embryonic chick skeletal muscle. *The Journal of Physiology, 352*, 685–701. https://doi.org/10.1113/jphysiol.1984.sp015317.

Guilluy, C., Osborne, L. D., Van, L. L., Sharek, L., Superfine, R., Garcia-Mata, R., et al. (2014). Isolated nuclei adapt to force and reveal a mechanotransduction pathway in the nucleus. *Nature Cell Biology, 16*, 376–381. https://doi.org/10.1038/ncb2927.

Gulati, A. K., Reddi, A. H., & Zalewski, A. A. (1982). Distribution of fibronectin in normal and regenerating skeletal muscle. *The Anatomical Record*, *204*, 175–183. https://doi.org/10.1002/ar.1092040302.

Gulati, A. K., Reddi, A. H., & Zalewski, A. A. (1983). Changes in the basement membrane zone components during skeletal muscle fiber degeneration and regeneration. *The Journal of Cell Biology*, *97*, 957–962. https://doi.org/10.1083/jcb.97.4.957.

Gussoni, E., Pavlath, G. K., Lanctot, A. M., Sharma, K. R., Miller, R. G., Steinman, L., et al. (1992). Normal dystrophin transcripts detected in Duchenne muscular dystrophy patients after myoblast transplantation. *Nature*, *356*, 435–438. https://doi.org/10.1038/356435a0.

Haque, F., Lloyd, D. J., Smallwood, D. T., Dent, C. L., Shanahan, C. M., Fry, A. M., et al. (2006). SUN1 interacts with nuclear Lamin A and cytoplasmic Nesprins to provide a physical connection between the nuclear lamina and the cytoskeleton. *Molecular and Cellular Biology*, *26*, 3738–3751. https://doi.org/10.1128/MCB.26.10.3738-3751.2006.

Hara, M., Tabata, K., Suzuki, T., Do, M.-K. Q., Mizunoya, W., Nakamura, M., et al. (2012). Calcium influx through a possible coupling of cation channels impacts skeletal muscle satellite cell activation in response to mechanical stretch. *American Journal of Physiology. Cell Physiology*, *302*, C1741–C1750. https://doi.org/10.1152/ajpcell.00068.2012.

Heino, J., & Massagué, J. (1990). Cell adhesion to collagen and decreased myogenic gene expression implicated in the control of myogenesis by transforming growth factor β. *The Journal of Biological Chemistry*, *265*, 10181–10184.

Hollnagel, A., Grund, C., Franke, W. W., & Arnold, H.-H. (2002). The cell adhesion molecule M-cadherin is not essential for muscle development and regeneration. *Molecular and Cellular Biology*, *22*, 4760–4770. https://doi.org/10.1128/MCB.22.13.4760–4770.2002.

Huard, J., Bouchard, J. P., Roy, R., Malouin, F., Dansereau, G., Labrecque, C., et al. (1992). Human myoblast transplantation: Preliminary results of 4 cases. *Muscle & Nerve*, *15*, 550–560. https://doi.org/10.1002/mus.880150504.

Huijing, P. A. (1999). Muscle as a collagen fiber reinforced composite: A review of force transmission in muscle and whole limb. *Journal of Biomechanics*, *32*, 329–345. https://doi.org/10.1016/S0021-9290(98)00186-9.

Ishikawa, T., Factor, V. M., Marquardt, J. U., Raggi, C., Seo, D., Kitade, M., et al. (2012). Hepatocyte growth factor/c-met signaling is required for stem-cell-mediated liver regeneration in mice. *Hepatology*, *55*, 1215–1226. https://doi.org/10.1002/hep.24796.

Jaalouk, D. E., & Lammerding, J. (2009). Mechanotransduction gone awry. *Nature Reviews. Molecular Cell Biology*, *10*, 63–73. https://doi.org/10.1038/nrm2597.

Kääriäinen, M., Järvinen, T., Järvinen, M., Rantanen, J., & Kalimo, H. (2000). Relation between myofibers and connective tissue during muscle injury repair. *Scandinavian Journal of Medicine & Science in Sports*, *10*, 332–337. https://doi.org/10.1034/j.1600-0838.2000.010006332.x.

Kääriäinen, M., Kääriäinen, J., Järvinen, T. L. N., Sievanen, H., Kalimo, H., & Järvinen, M. (1998). Correlation between biomechanical and structural changes during the regeneration of skeletal muscle after laceration injury. *Journal of Orthopaedic Research*, *16*, 197–206. https://doi.org/10.1002/jor.1100160207.

Kami, K., Masuhara, M., Kashiba, H., Kawai, Y., Noguchi, K., & Senba, E. (1993). Changes of vinculin and extracellular matrix components following blunt trauma to rat skeletal muscle. *Medicine and Science in Sports and Exercise*, *25*, 832–840.

Karpati, G., Ajdukovic, D., Arnold, D., Gledhill, R. B., Guttmann, R., Holland, P., et al. (1993). Myoblast transfer in Duchenne muscular dystrophy. *Annals of Neurology*, *34*, 8–17. https://doi.org/10.1002/ana.410340105.

Kherif, S., Lafuma, C., Dehaupas, M., Lachkar, S., Fournier, J. G., Verdière-Sahuqué, M., et al. (1999). Expression of matrix metalloproteinases 2 and 9 in regenerating skeletal

muscle: A study in experimentally injured and mdx muscles. *Developmental Biology, 205*, 158–170. https://doi.org/10.1006/dbio.1998.9107.
Kim, N.-G., Koh, E., Chen, X., & Gumbiner, B. M. (2011). E-cadherin mediates contact inhibition of proliferation through hippo signaling-pathway components. *Proceedings of the National Academy of Sciences of the United States of America, 108*, 11930–11935. https://doi.org/10.1073/pnas.1103345108.
Komulainen, J., Kalliokoski, R., Koskinen, S. O. A., Drost, M. R., Kuipers, H., & Hesselink, M. K. C. (2000). Controlled lengthening or shortening contraction-induced damage is followed by fiber hypertrophy in rat skeletal muscle. *International Journal of Sports Medicine, 21*, 107–112. https://doi.org/10.1055/s-2000-8869.
Kook, S.-H., Son, Y.-O., Choi, K.-C., Lee, H.-J., Chung, W.-T., Hwang, I.-H., et al. (2008). Cyclic mechanical stress suppresses myogenic differentiation of adult bovine satellite cells through activation of extracellular signal-regulated kinase. *Molecular and Cellular Biochemistry, 309*, 133–141. https://doi.org/10.1007/s11010-007-9651-y.
Kragstrup, T. W., Kjaer, M., & Mackey, A. L. (2011). Structural, biochemical, cellular, and functional changes in skeletal muscle extracellular matrix with aging. *Scandinavian Journal of Medicine & Science in Sports, 21*, 749–757. https://doi.org/10.1111/j.1600-0838.2011.01377.x.
Kuang, S., Gillespie, M. A., & Rudnicki, M. A. (2008). Niche regulation of muscle satellite cell self-renewal and differentiation. *Cell Stem Cell, 2*, 22–31. https://doi.org/10.1016/j.stem.2007.12.012.
Kuang, S., Kuroda, K., Le Grand, F., & Rudnicki, M. A. (2007). Asymmetric self-renewal and commitment of satellite stem cells in muscle. *Cell, 129*, 999–1010. https://doi.org/10.1016/j.cell.2007.03.044.
Kumar, A., Chaudhry, I., Reid, M. B., & Boriek, A. M. (2002). Distinct signaling pathways are activated in response to mechanical stress applied axially and transversely to skeletal muscle fibers. *The Journal of Biological Chemistry, 277*, 46493–46503. https://doi.org/10.1074/jbc.M203654200.
Kumar, A., Khandelwal, N., Malya, R., Reid, M. B., & Boriek, A. M. (2004). Loss of dystrophin causes aberrant mechanotransduction in skeletal muscle fibers. *The FASEB Journal, 18*, 102–113. https://doi.org/10.1096/fj.03-0453com.
Kumar, A., Murphy, R., Robinson, P., Wei, L., & Boriek, A. M. (2004). Cyclic mechanical strain inhibits skeletal myogenesis through activation of focal adhesion kinase, Rac-1 GTPase, and NF-kappaB transcription factor. *The FASEB Journal, 18*, 1524–1535. https://doi.org/10.1096/fj.04-2414com.
Lacraz, G., Rouleau, A. J., Couture, V., Söllrald, T., Drouin, G., Veillette, N., et al. (2016). Increased stiffness in aged skeletal muscle impairs muscle progenitor cell proliferative activity. *PloS One, 11*, https://doi.org/10.1371/journal.pone.0136217.
Lam, M. T., Sim, S., Zhu, X., & Takayama, S. (2006). The effect of continuous wavy micropatterns on silicone substrates on the alignment of skeletal muscle myoblasts and myotubes. *Biomaterials, 27*, 4340–4347. https://doi.org/10.1016/j.biomaterials.2006.04.012.
Lansman, J. B. (2007). Mechanosensitive ion channels in dystrophic muscle. *Current Topics in Membranes, 59*, 467–484. https://doi.org/10.1016/S1063-5823(06)59017-9.
Law, P. K., Goodwin, T. G., Fang, Q., Duggirala, V., Larkin, C., Florendo, J. A., et al. (1992). Feasibility, safety, and efficacy of myoblast transfer therapy on Duchenne muscular dystrophy boys. *Cell Transplantation, 1*, 235–244.
Lehto, M., & Järvinen, M. (1985). Collagen and glycosaminoglycan synthesis of injured gastrocnemius muscle in rat. *European Surgical Research, 17*, 179–185. https://doi.org/10.1159/000128465.
Lei, H., Leong, D., Smith, L. R., & Barton, E. R. (2013). Matrix metalloproteinase 13 is a new contributor to skeletal muscle regeneration and critical for myoblast migration.

American Journal of Physiology. Cell Physiology, 305, C529–C538. https://doi.org/10.1152/ajpcell.00051.2013.

Levenberg, S., Rouwkema, J., Macdonald, M., Garfein, E. S., Kohane, D. S., Darland, D. C., et al. (2005). Engineering vascularized skeletal muscle tissue. *Nature Biotechnology, 23*, 879–884. https://doi.org/10.1038/nbt1109.

Lewis, M. P., Tippett, H. L., Sinanan, A. C. M., Morgan, M. J., & Hunt, N. P. (2000). Gelatinase-B (matrix metalloproteinase-9; MMP-9) secretion is involved in the migratory phase of human and murine muscle cell cultures. *Journal of Muscle Research and Cell Motility, 21*, 223–233. https://doi.org/10.1023/A:1005670507906.

Li, Y., Cummins, J., & Huard, J. (2001). Muscle injury and repair. *Current Opinion in Orthopaedics, 12*, 409–415. https://doi.org/10.1097/00001433-200110000-00008.

Liu, N., Garry, G. A., Li, S., Bezprozvannaya, S., Sanchez-Ortiz, E., Chen, B., et al. (2017). A Twist2-dependent progenitor cell contributes to adult skeletal muscle. *Nature Cell Biology, 19*, 202–213. https://doi.org/10.1038/ncb3477.

Liu, J., Wang, Y., Yuan, X., Feng, Y., & Liu, H. (2010). Cyclic-stretch induces the apoptosis of myoblast by activation of Caspase-3 protease in a magnitude-dependent manner. *The International Journal of Biochemistry & Cell Biology, 42*, 2004–2011. https://doi.org/10.1016/j.biocel.2010.08.013.

Lombardi, M. L., Jaalouk, D. E., Shanahan, C. M., Burke, B., Roux, K. J., & Lammerding, J. (2011). The interaction between nesprins and sun proteins at the nuclear envelope is critical for force transmission between the nucleus and cytoskeleton. *The Journal of Biological Chemistry, 286*, 26743–26753. https://doi.org/10.1074/jbc.M111.233700.

Lu, M. (2011). *Regulation of myoblast differentiation by cell-cell contact and adhesion*. ProQuest Dissertations & Theses Global.

Madden, L., Juhas, M., Kraus, W. E., Truskey, G. A., & Bursac, N. (2015). Bioengineered human myobundles mimic clinical responses of skeletal muscle to drugs. *eLife, 4*, e04885. https://doi.org/10.7554/eLife.04885.

Maley, M. A., Davies, M. J., & Grounds, M. D. (1995). Extracellular matrix, growth factors, genetics: Their influence on cell proliferation and myotube formation in primary cultures of adult mouse skeletal muscle. *Experimental Cell Research, 219*, 169–179. https://doi.org/10.1006/excr.1995.1217.

McDonald, C. M. (2012). Clinical approach to the diagnostic evaluation of hereditary and acquired neuromuscular diseases. *Physical Medicine and Rehabilitation Clinics of North America, 23*, 495–563. https://doi.org/10.1016/j.pmr.2012.06.011.

Mokri, B., & Engel, A. G. (1975). Duchenne dystrophy: Electron microscopic findings pointing to a basic or early abnormality in the plasma membrane of the muscle fiber. *Neurology, 25*, 1111–1120. https://doi.org/10.1212/WNL.51.1.1-a.

Monge, C., DiStasio, N., Rossi, T., Sébastien, M., Sakai, H., Kalman, B., et al. (2017). Quiescence of human muscle stem cells is favored by culture on natural biopolymeric films. *Stem Cell Research & Therapy, 8*, 104. https://doi.org/10.1186/s13287-017-0556-8.

Montarras, D., Morgan, J., Collins, C., Relaix, F., Zaffran, S., Cumano, A., et al. (2005). Direct isolation of satellite cells for skeletal muscle regeneration. *Science (New York, N.Y.), 309*, 2064–2067. https://doi.org/10.1126/science.1114758.

Morrissey, J. B., Cheng, R. Y., Davoudi, S., & Gilbert, P. M. (2015). Biomechanical origins of muscle stem cell signal transduction. *Journal of Molecular Biology, 428*, 1441–1454. https://doi.org/10.1016/j.jmb.2015.05.004.

Neuss, S., Becher, E., Woltje, M., Tietze, L., & Jahnen-Dechent, W. (2004). Functional expression of HGF and HGF receptor/c-met in adult human mesenchymal stem cells suggests a role in cell mobilization, tissue repair, and wound healing. *Stem Cells, 22*, 405–414. https://doi.org/10.1634/stemcells.22-3-405.

Niu, A., Wen, Y., Liu, H., Zhan, M., Jin, B., & Li, Y.-P. (2013). Src mediates the mechanical activation of myogenesis by activating TNFα-converting enzyme. *Journal of Cell Science, 126*, 4349–4357. https://doi.org/10.1242/jcs.125328.

Öcalan, M., Goodman, S. L., Kühl, U., Hauschka, S. D., & von der Mark, K. (1988). Laminin alters cell shape and stimulates motility and proliferation of murine skeletal myoblasts. *Developmental Biology, 125*, 158–167. https://doi.org/10.1016/0012-1606(88)90068-1.

Organ, S. L., & Tsao, M.-S. (2011). An overview of the c-MET signaling pathway. *Therapeutic Advances in Medical Oncology, 3*, S7–S19. https://doi.org/10.1177/1758834011422556.

Osses, N., & Brandan, E. (2002). ECM is required for skeletal muscle differentiation independently of muscle regulatory factor expression. *American Journal of Physiology. Cell Physiology, 282*, C383–C394. https://doi.org/10.1152/ajpcell.00322.2001.

Ota, M., & Sasaki, H. (2008). Mammalian Tead proteins regulate cell proliferation and contact inhibition as transcriptional mediators of hippo signaling. *Development, 135*, 4059–4069. https://doi.org/10.1242/dev.027151.

Otis, J. S., Burkholder, T. J., & Pavlath, G. K. (2005). Stretch-induced myoblast proliferation is dependent on the COX2 pathway. *Experimental Cell Research, 310*, 417–425. https://doi.org/10.1016/j.yexcr.2005.08.009.

Padmakumar, V. C., Libotte, T., Lu, W., Zaim, H., Abraham, S., Noegel, A. A., et al. (2005). The inner nuclear membrane protein Sun1 mediates the anchorage of Nesprin-2 to the nuclear envelope. *Journal of Cell Science, 118*, 3419–3430. https://doi.org/10.1242/jcs.02471.

Pallafacchina, G., François, S., Regnault, B., Czarny, B., Dive, V., Cumano, A., et al. (2010). An adult tissue-specific stem cell in its niche: A gene profiling analysis of in vivo quiescent and activated muscle satellite cells. *Stem Cell Research, 4*, 77–91. https://doi.org/10.1016/j.scr.2009.10.003.

Petrof, B. J., Shrager, J. B., Stedman, H. H., Kelly, A. M., & Sweeney, H. L. (1993). Dystrophin protects the sarcolemma from stresses developed during muscle contraction. *Proceedings of the National Academy of Sciences of the United States of America, 90*, 3710–3714. https://doi.org/10.1073/pnas.90.8.3710.

Przewoźniak, M., Czaplicka, I., Czerwińska, A. M., Markowska-Zagrajek, A., Moraczewski, J., Stremińska, W., et al. (2013). Adhesion proteins—An impact on skeletal myoblast differentiation. *PloS One, 8*, e61760. https://doi.org/10.1371/journal.pone.0061760.

Puckelwartz, M. J., Kessler, E., Zhang, Y., Hodzic, D., Randles, K. N., Morris, G., et al. (2009). Disruption of nesprin-1 produces an Emery-Dreifuss muscular dystrophy-like phenotype in mice. *Human Molecular Genetics, 18*, 607–620. https://doi.org/10.1093/hmg/ddn386.

Qazi, T. H., Mooney, D. J., Pumberger, M., Geissler, S., & Duda, G. N. (2015). Biomaterials based strategies for skeletal muscle tissue engineering: Existing technologies and future trends. *Biomaterials, 53*, 502–521. https://doi.org/10.1016/j.biomaterials.2015.02.110.

Quarta, M., Brett, J. O., DiMarco, R., De Morree, A., Boutet, S. C., Chacon, R., et al. (2016). An artificial niche preserves the quiescence of muscle stem cells and enhances their therapeutic efficacy. *Nature Biotechnology, 34*, 752–759. https://doi.org/10.1038/nbt.3576.

Quarta, M., Cromie, M., Chacon, R., Blonigan, J., Garcia, V., Akimenko, I., et al. (2017). Bioengineered constructs combined with exercise enhance stem cell-mediated treatment of volumetric muscle loss. *Nature Communications, 8*, 15613. https://doi.org/10.1038/ncomms15613.

Qu-Petersen, Z., Deasy, B., Jankowski, R., Ikezawa, M., Cummins, J., Pruchnic, R., et al. (2002). Identification of a novel population of muscle stem cells in mice: Potential for muscle regeneration. *The Journal of Cell Biology, 157*, 851–864. https://doi.org/10.1083/jcb.200108150.

Redfield, A., Nieman, M. T., & Knudsen, K. A. (1997). Cadherins promote skeletal muscle differentiation in three-dimensional cultures. *The Journal of Cell Biology, 138*, 1323–1331. https://doi.org/10.1083/jcb.138.6.1323.

Rocheteau, P., Gayraud-Morel, B., Siegl-Cachedenier, I., Blasco, M. A., & Tajbakhsh, S. (2012). A subpopulation of adult skeletal muscle stem cells retains all template DNA strands after cell division. *Cell, 148*, 112–125. https://doi.org/10.1016/j.cell.2011.11.049.

Romano, N. H., Sengupta, D., Chung, C., & Heilshorn, S. C. (2011). Protein-engineered biomaterials: Nanoscale mimics of the extracellular matrix. *Biochimica et Biophysica Acta, 1810*, 339–349. https://doi.org/10.1016/j.bbagen.2010.07.005.

Rossi, C. A., Flaibani, M., Blaauw, B., Pozzobon, M., Figallo, E., Reggiani, C., et al. (2011). In vivo tissue engineering of functional skeletal muscle by freshly isolated satellite cells embedded in a photopolymerizable hydrogel. *The FASEB Journal, 25*, 1–9. https://doi.org/10.1096/fj.10-174755.

Sacco, A., Doyonnas, R., Kraft, P., Vitorovic, S., & Blau, H. M. (2008). Self-renewal and expansion of single transplanted muscle stem cells. *Nature, 456*, 502–506. https://doi.org/10.1038/nature07384.

Sambasivan, R., Yao, R., Kissenpfennig, A., Van Wittenberghe, L., Paldi, A., Gayraud-Morel, B., et al. (2011). Pax7-expressing satellite cells are indispensable for adult skeletal muscle regeneration. *Development, 138*, 3647–3656. https://doi.org/10.1242/dev.067587.

Sanes, J. R. (2003). The basement membrane/basal lamina of skeletal muscle. *The Journal of Biological Chemistry, 278*, 12601–12604. https://doi.org/10.1074/jbc.R200027200.

Sanes, J. R., Schachner, M., & Covault, J. (1986). Expression of several adhesive macromolecules (N-CAM, L1, J1, NILE, uvomorulin, laminin, fibronectin, and a heparan sulfate proteoglycan) in embryonic, adult, and denervated adult skeletal muscle. *The Journal of Cell Biology, 102*, 420–431. https://doi.org/10.1083/jcb.102.2.420.

Seale, P., Sabourin, L. A., Girgis-Gabardo, A., Mansouri, A., Gruss, P., & Rudnicki, M. A. (2000). Pax7 is required for the specification of myogenic satellite cells. *Cell, 102*, 777–786. https://doi.org/10.1016/S0092-8674(00)00066-0.

Serena, E., Zatti, S., Reghelin, E., Pasut, A., Cimetta, E., & Elvassore, N. (2010). Soft substrates drive optimal differentiation of human healthy and dystrophic myotubes. *Integrative Biology: Quantitative Biosciences from Nano to Macro, 2*, 193–201. https://doi.org/10.1039/b921401a.

Serrano, A. L., & Muñoz-Cánoves, P. (2010). Regulation and dysregulation of fibrosis in skeletal muscle. *Experimental Cell Research, 316*, 3050–3058. https://doi.org/10.1016/j.yexcr.2010.05.035.

Smith, L. R., & Barton, E. R. (2014). Collagen content does not alter the passive mechanical properties of fibrotic skeletal muscle in mdx mice. *American Journal of Physiology. Cell Physiology, 306*, C889–C898. https://doi.org/10.1152/ajpcell.00383.2013.

Sosa, B. A., Rothballer, A., Kutay, U., & Schwartz, T. U. (2012). LINC complexes form by binding of three KASH peptides to domain interfaces of trimeric SUN proteins. *Cell, 149*, 1035–1047. https://doi.org/10.1016/j.cell.2012.03.046.

Sousa-Victor, P., Gutarra, S., García-Prat, L., Rodriguez-Ubreva, J., Ortet, L., Ruiz-Bonilla, V., et al. (2014). Geriatric muscle stem cells switch reversible quiescence into senescence. *Nature, 506*, 316–321. https://doi.org/10.1038/nature13013.

Stauber, W. T., Knack, K. K., Miller, G. R., & Grimmett, J. G. (1996). Fibrosis and intercellular collagen connections from four weeks of muscle strains. *Muscle and Nerve, 19*, 423–430. https://doi.org/10.1002/mus.880190402.

Stearns-Reider, K. M., D'Amore, A., Beezhold, K., Rothrauff, B., Cavalli, L., Wagner, W. R., et al. (2017). Aging of the skeletal muscle extracellular matrix drives a stem cell fibrogenic conversion. *Aging Cell, 16*, 518–528. https://doi.org/10.1111/acel.12578.

Stern, M. M., Myers, R. L., Hammam, N., Stern, K. A., Eberli, D., Kritchevsky, S. B., et al. (2009). The influence of extracellular matrix derived from skeletal muscle tissue on the proliferation and differentiation of myogenic progenitor cells ex vivo. *Biomaterials, 30*, 2393–2399. https://doi.org/10.1016/j.biomaterials.2008.12.069.

Stilhano, R. S., Martins, L., Ingham, S. J. M., Pesquero, J. B., & Huard, J. (2015). Gene and cell therapy for muscle regeneration. *Current Reviews in Musculoskeletal Medicine, 8*, 182–187. https://doi.org/10.1007/s12178-015-9268-9.

Tatsumi, R., & Allen, R. E. (2008). Mechano-biology of resident myogenic stem cells: Molecular mechanism of stretch-induced activation of satellite cells. *Animal Science Journal, 79*, 279–290. https://doi.org/10.1111/j.1740-0929.2008.00528.x.

Tatsumi, R., Anderson, J. E., Nevoret, C. J., Halevy, O., & Allen, R. E. (1998). HGF/SF is present in normal adult skeletal muscle and is capable of activating satellite cells. *Developmental Biology, 194*, 114–128. https://doi.org/10.1006/dbio.1997.8803.

Tatsumi, R., Hattori, A., Ikeuchi, Y., Anderson, J. E., & Allen, R. E. (2002). Release of hepatocyte growth factor from mechanically stretched skeletal muscle satellite cells and role of pH and nitric oxide. *Molecular Biology of the Cell, 13*, 2909–2918. https://doi.org/10.1091/mbc.E02.

Tatsumi, R., Liu, X., Pulido, A., Morales, M., Sakata, T., Dial, S., et al. (2006). Satellite cell activation in stretched skeletal muscle and the role of nitric oxide and hepatocyte growth factor. *American Journal of Physiology. Cell Physiology, 290*, C1487–C1494. https://doi.org/10.1152/ajpcell.00513.2005.

Tatsumi, R., Sheehan, S. M., Iwasaki, H., Hattori, A., & Allen, R. E. (2001). Mechanical stretch induces activation of skeletal muscle satellite cells in vitro. *Experimental Cell Research, 267*, 107–114. https://doi.org/10.1006/excr.2001.5252.

Tatsumi, R., Wuollet, A. L., Tabata, K., Nishimura, S., Tabata, S., Mizunoya, W., et al. (2009). A role for calcium-calmodulin in regulating nitric oxide production during skeletal muscle satellite cell activation. *American Journal of Physiology. Cell Physiology, 296*, C922–C929. https://doi.org/10.1152/ajpcell.00471.2008.

Tedesco, F. S., Dellavalle, A., Diaz-Manera, J., Messina, G., & Cossu, G. (2010). Repairing skeletal muscle: Regenerative potential of skeletal muscle stem cells. *The Journal of Clinical Investigation, 120*, 11–19. https://doi.org/10.1172/JCI40373.

Tremblay, J. P., Malouin, F., Roy, R., Huard, J., Bouchard, J. P., Satoh, A., et al. (1993). Results of a triple blind clinical study of myoblast transplantations without immunosuppressive treatment in young boys with Duchenne muscular dystrophy. *Cell Transplantation, 2*, 99–112.

Trensz, F., Lucien, F., Couture, V., Söllrald, T., Drouin, G., Rouleau, A.-J., et al. (2015). Increased microenvironment siffness in damaged myofibers promotes myogenic progenitor cell proliferation. *Skeletal Muscle, 5*, 5.

Tsivitse, S. (2010). Notch and Wnt signaling, physiological stimuli and postnatal myogenesis. *International Journal of Biological Sciences, 6*, 268–281.

Tutdibi, O., Brinkmeier, H., Rüdel, R., & Föhr, K. J. (1999). Increased calcium entry into dystrophin-deficient muscle fibres of MDX and ADR-MDX mice is reduced by ion channel blockers. *The Journal of Physiology, 515*(Pt. 3), 859–868. https://doi.org/10.1111/j.1469-7793.1999.859ab.x.

Urciuolo, A., Quarta, M., Morbidoni, V., Gattazzo, F., Molon, S., Grumati, P., et al. (2013). Collagen VI regulates satellite cell self-renewal and muscle regeneration. *Nature Communications, 4*, 1964. https://doi.org/10.1038/ncomms2964.

van Donkelaar, C. C., Drost, M. R., van Mameren, H., Tuinenburg, C. F., Janssen, J. D., & Huson, A. (1996). Three-dimensional reconstruction of the rat triceps surae muscle and finite element mesh generation of the gastrocnemius medialis muscle. *European Journal of Morphology, 34*, 31–37. https://doi.org/10.1076/ejom.34.1.31.13161.

von Haehling, S., Morley, J. E., & Anker, S. D. (2010). An overview of sarcopenia: Facts and numbers on prevalence and clinical impact. *Journal of Cachexia, Sarcopenia and Muscle, 1*, 129–133. https://doi.org/10.1007/s13539-010-0014-2.

Wood, L. K., Kayupov, E., Gumucio, J. P., Mendias, C. L., Claflin, D. R., & Brooks, S. V. (2014). Intrinsic stiffness of extracellular matrix increases with age in skeletal muscles of mice. *Journal of Applied Physiology*, *117*, 363–369. https://doi.org/10.1152/japplphysiol.00256.2014.

Wozniak, A. C., Pilipowicz, O., Yablonka-Reuveni, Z., Greenway, S., Craven, S., Scott, E., et al. (2003). C-Met expression and mechanical activation of satellite cells on cultured muscle fibers. *The Journal of Histochemistry and Cytochemistry*, *51*, 1437–1445. https://doi.org/10.1177/002215540305101104.

Yamada, M., Sankoda, Y., Tatsumi, R., Mizunoya, W., Ikeuchi, Y., Sunagawa, K., et al. (2008). Matrix metalloproteinase-2 mediates stretch-induced activation of skeletal muscle satellite cells in a nitric oxide-dependent manner. *The International Journal of Biochemistry & Cell Biology*, *40*, 2183–2191. https://doi.org/10.1016/j.biocel.2008.02.017.

Yamada, M., Tatsumi, R., Kikuiri, T., Okamoto, S., Nonoshita, S., Mizunoya, W., et al. (2006). Matrix metalloproteinases are involved in mechanical stretch-induced activation of skeletal muscle satellite cells. *Muscle and Nerve*, *34*, 313–319. https://doi.org/10.1002/mus.20601.

Yang, H. S., Ieronimakis, N., Tsui, J. H., Kim, H. N., Suh, K. Y., Reyes, M., et al. (2014). Nanopatterned muscle cell patches for enhanced myogenesis and dystrophin expression in a mouse model of muscular dystrophy. *Biomaterials*, *35*, 1478–1486. https://doi.org/10.1016/j.biomaterials.2013.10.067.

Yao, C. C., Ziober, B. L., Squillace, R. M., & Kramer, R. H. (1996). α7 integrin mediates cell adhesion and migration on specific laminin isoforms. *The Journal of Biological Chemistry*, *271*, 25598–25603. https://doi.org/10.1074/jbc.271.41.25598.

Yao, C. C., Ziober, B. L., Sutherland, A. E., Mendrick, D. L., & Kramer, R. H. (1996). Laminins promote the locomotion of skeletal myoblasts via the alpha 7 integrin receptor. *Journal of Cell Science*, *109*, 3139–3150.

Yennek, S., Burute, M., Théry, M., & Tajbakhsh, S. (2014). Cell adhesion geometry regulates non-random DNA segregation and asymmetric cell fates in mouse skeletal muscle stem cells. *Cell Reports*, 7, 961–970. https://doi.org/10.1016/j.celrep.2014.04.016.

Yucesoy, C. A., Koopman, B. H. F. J. M., Huijing, P. A., & Grootenboer, H. J. (2002). Three-dimensional finite element modeling of skeletal muscle using a two-domain approach: Linked fiber-matrix mesh model. *Journal of Biomechanics*, *35*, 1253–1262. https://doi.org/10.1016/S0021-9290(02)00069-6.

Zhan, M., Jin, B., Chen, S.-E., Reecy, J. M., & Li, Y.-P. (2007). TACE release of TNF-alpha mediates mechanotransduction-induced activation of p38 MAPK and myogenesis. *Journal of Cell Science*, *120*, 692–701. https://doi.org/10.1242/jcs.03372.

Zhang, Q., Bethmann, C., Worth, N. F., Davies, J. D., Wasner, C., Feuer, A., et al. (2007). Nesprin-1 and -2 are involved in the pathogenesis of Emery-Dreifuss muscular dystrophy and are critical for nuclear envelope integrity. *Human Molecular Genetics*, *16*, 2816–2833. https://doi.org/10.1093/hmg/ddm238.

Zhang, Q., Skepper, J. N., Yang, F., Davies, J. D., Hegyi, L., Roberts, R. G., et al. (2001). Nesprins: A novel family of spectrin-repeat-containing proteins that localize to the nuclear membrane in multiple tissues. *Journal of Cell Science*, *114*, 4485–4498. https://doi.org/10.1074/jbc.274.50.35375.

Zhou, Z., Du, X., Cai, Z., Song, X., Zhang, H., Mizuno, T., et al. (2012). Structure of Sad1-UNC84 homology (SUN) domain defines features of molecular bridge in nuclear envelope. *The Journal of Biological Chemistry*, *287*, 5317–5326. https://doi.org/10.1074/jbc.M111.304543.

CHAPTER SIX

Satellite Cell Self-Renewal

Lorenzo Giordani*, Alice Parisi†, Fabien Le Grand*,[1]
*Sorbonne Universités, UPMC Univ Paris 06, INSERM UMRS974, Center for Research in Myology, Paris, France
†Nestlé Institute of Health Sciences (NIHS), Campus École Polytechnique Fédérale de Lausanne, École Polytechnique Fédérale de Lausanne Innovation Park, Lausanne, Switzerland
[1]Corresponding author: e-mail address: fabienlegrand@gmail.com

Contents

Abstract

Adult skeletal muscle is endowed with regenerative potential through partially recapitulating the embryonic developmental program. Upon acute injury or in pathological conditions, quiescent muscle-resident stem cells, called satellite cells, become activated and give rise to myogenic progenitors that massively proliferate, differentiate, and fuse to form new myofibers and restore tissue functionality. In addition, a proportion of activated cells returns back to quiescence and replenish the pool of satellite cells in order to maintain the ability of skeletal muscle tissue to repair. Self-renewal is the process by which stem cells divide to make more stem cells to maintain the stem cell population throughout life. This process is controlled by cell-intrinsic transcription factors regulated by cell-extrinsic signals from the niche and the microenvironment. This chapter provides an overview about the general aspects of satellite cell biology and focuses on the

Current Topics in Developmental Biology, Volume 126
ISSN 0070-2153
https://doi.org/10.1016/bs.ctdb.2017.08.001
© 2018 Elsevier Inc.
All rights reserved.

cellular and molecular aspects of satellite cell self-renewal. To date, we are still far from understanding how a very small proportion of the satellite cell progeny maintain their stem cell identity when most of their siblings progress through the myogenic program to construct myofibers.

1. INTRODUCTION

Somatic stem cells, or adult stem cells, are essential for tissue maintenance. In addition to normal homeostasis, they are also involved in tissue repair in case of injury. Skeletal muscles are one of the two sorts of striated muscles of the vertebrate body, the other being the cardiac muscle. They are fixed to the bones by the tendons and have the function to move the body in space under the control of the central nervous system. Skeletal muscle tissue is constituted of muscle fibers (myofibers), cylindrical multinucleated cells containing contracting myofibrils. Throughout the whole adult life, skeletal muscle relies primarily on muscle stem cells called satellite cells to achieve regeneration and preserves their structure in response to injury. Satellite cells express the transcription factor paired box 7 (Pax7), necessary for their specification during development and their maintenance in postnatal life. Their ability to return to quiescence, thus maintaining the homeostasis of the stem cell pool, is crucial to ensure functional integrity in response to recurrent external insults. Perturbation of this fine-tuned equilibrium results in aberrant regeneration ultimately leading to morbidity and chronic pathological conditions. Understanding the mechanisms controlling satellite cell fate decision and self-renewal is therefore of fundamental biological importance and therapeutic relevance.

2. MUSCLE SATELLITE CELLS

Satellite cells have been first identified in 1961 during the characterization of frog myofibers by electron microscopy (Mauro, 1961). In his brief communication, Mauro reports "[...] the presence of certain cells, intimately associated with the muscle fibre, [...] which we have chosen to call satellite cells [...]." The implications and relevance of the unique positioning of satellite cells: in a niche between the host myofiber and the basal lamina (see Fig. 1) have yet to be explored but recent data suggest the niche is required for biophysical cues such as controlled stiffness (Gilbert et al., 2010) and contains sensor molecules that maintain satellite cell quiescence

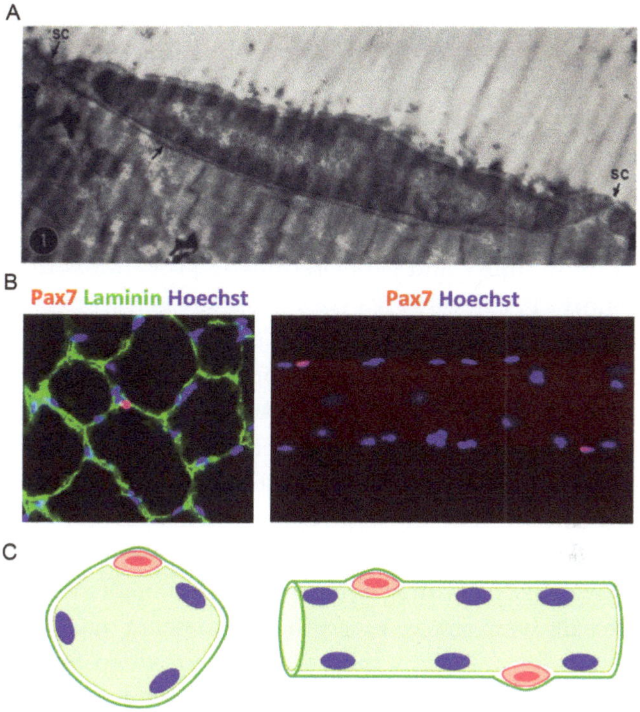

Fig. 1 The satellite cell. (A) Electron microscopy photo showing a longitudinal view of a satellite cell in close contact with the host myofiber (Mauro, 1961). (B) *Left panel*: Cross-section of a portion of tibialis anterior muscle immunostained with antibodies against Pax7 (*red*), one of the most widely used marker for adult satellite cells, and Laminin (*green*), a component of the myofibers basal lamina. *Right panel*: Longitudinal view of a portion of a single myofibers isolated from extensor digitorum longus (EDL) muscle and immunostained for Pax7 (*red*). Nuclei are stained with Hoechst (*blue*). (C) Schematic representation of a cross-sectional view (*left panel*) and a longitudinal view (*right panel*) of a myofiber with satellite cells in their niche (*pink*) and myonuclei (*blue*). The basal lamina is represented as an external *green thick layer*, and the sarcoplasm in *light green*, delimited by the sarcolemma, visible as a thin *light green line*.

(Rozo, Li, & Fan, 2016). Owing to their close association with to the myofibers (see Fig. 1), satellite cells soon became good candidates for the solution of what Mauro defined "the vexing problem of skeletal muscle regeneration." Indeed, it was already well known that skeletal muscle regenerates after injury, but the cellular origin of new adult myofibers was still obscure.

The first evidence that satellite cells are not myonuclei and participate to myofiber regeneration came from ex vivo experiments in which single myofibers were isolated from adult rats and cultured for several days. In this setup, continuous observation of the isolated myofibers revealed that satellite cells start to divide synchronously 22 h after isolation and give rise to a progeny that fuse and form multinucleated myotubes within the basement membrane of the isolated fiber 5 days after isolation (Bischoff, 1975). Furthermore, in vivo experiments based on radioactive labeling and cell tracing during skeletal muscle regeneration confirmed that satellite cells are activated upon injury and generate muscle progenitors that build new myofibers (Snow, 1977).

The physiological role of the satellite cell as an adult muscle stem cell remained controversial for decades until better characterizations were performed. As such, it has been matter of debate if satellite cells possessed the ability to self-renew and therefore could been considered "true stem cell." It was only with the publication of a series of seminal works in the 2000s that satellite cells finally acquired the status of adult stem cells (Collins et al., 2005; Zammit et al., 2004). Zammit and colleagues used a single myofiber approach to elegantly show that, while in culture, a small subfraction of activated satellite cells were able to revert to the quiescent stage (Zammit et al., 2004). One year later, Collins and colleagues provided further evidences transplanting single myofibers into radiation-ablated muscles. In fact, satellite cells associated with the transplanted myofiber were not only able to generate new myofibers but also to expand and regenerate the host satellite cell pool. In addition, when the transplanted muscle was challenged with injury, the engrafted satellite cells would proliferate and regenerate the muscle tissue (Collins et al., 2005).

In the same year, GFP-expressing satellite cells were isolated by flow cytometry (FACS) and grafted into muscles of adult dystrophic mice. The transplanted satellite cells contributed both to myofiber repair and to the muscle satellite cell compartment. Interestingly, this study demonstrated that expansion of satellite cells in culture before engraftment reduced their regenerative capacity (Montarras et al., 2005). Lastly, in 2008, Sacco and colleagues were able to transplant a single muscle stem cell and recapitulate both the regenerative and self-renewal properties already shown with the myofiber transplantation (Sacco, Doyonnas, Kraft, Vitorovic, & Blau, 2008). These experiments proved that satellite cells were archetypal stem cell self-sufficient to sustain muscle regeneration in the adult tissue.

Over the years, an extremely large body of evidence accumulated proving that satellite cells were the main actors of muscle tissue repair. The definitive proof that satellite cells are necessary for muscle regeneration came from three independent research teams in 2011. Using genetic engineering in mice, the three groups showed that conditional depletion of Pax7-expressing satellite cells completely blocks adult myogenesis. Following acute injury, skeletal muscles devoid of satellite cells failed to regenerate. Interestingly, the loss of satellite cells was not compensated for endogenously by other cell types (Lepper, Partridge, & Fan, 2011; Murphy, Lawson, Mathew, Hutcheson, & Kardon, 2011; Sambasivan et al., 2011). In line with these studies, we recently found that targeted disruption of the Adenomatous polyposis coli (APC) gene in satellite cells results in complete abrogation of skeletal muscle regeneration. Loss of APC in satellite cells renders them unable to enter in the cell cycle and results in program cell death (Parisi et al., 2015).

As currently understood, quiescent satellite cells are activated following damage to the myofibers and give rise to a population of transient amplifying myoblasts. Most of the myoblasts will then permanently exit the cell cycle and fuse to form new myofibers and regenerate the tissue, while a subfraction will repopulate the satellite cell niche (see Fig. 2). In this process,

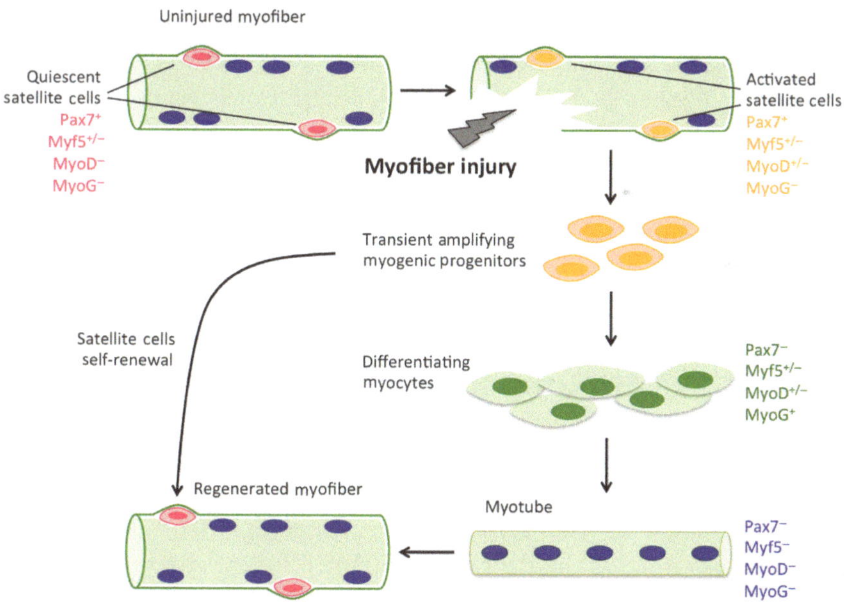

Fig. 2 Schematic representation of skeletal muscle regeneration. Skeletal muscle tissue repair is orchestrated by a genetic cascade that involves Pax7 and the MRFs and drives each step of satellite cell activation, generation of transient amplifying progenitors, differentiation and formation of new myofibers. Satellite cell self-renewal replenishes the stem cell pool of the tissue after regeneration.

extrinsic signaling cascades as well as intrinsic transcription factors instruct cycling satellite cell descendants and control their fate to properly orchestrate the creation of new myofibers and self-renewal of the stem cell pool.

3. SATELLITE CELL HETEROGENEITY

For a long time, satellite cells have been considered a homogeneous population of muscle stem cells. Careful molecular characterization revealed a certain level of heterogeneity in terms of expression of satellite cells specific markers and in the degree of commitment.

Myogenic regulatory factors (MRFs) are the master regulators of skeletal myogenesis (Buckingham et al., 2003). Among the MRFs, myogenic factor 5 (Myf5) is the earliest to be expressed during embryogenesis. Lineage tracing experiments used Myf5-Cre*Rosa-YFP mice, in which whenever Myf5 locus is transcribed, the Cre recombinase irreversibly activates the expression of YFP, thereby labeling cells that expressed Myf5 at some moment of their developmental history. This approach identified two satellite cell subpopulations in intact adult muscle: a committed subgroup that expresses both Pax7 and YFP, and a more "stem" subgroup that represents only 10% of satellite cells and expresses only Pax7 but not YFP (Kuang, Kuroda, Le Grand, & Rudnicki, 2007). Interestingly, transplantation of FACS-isolated Pax7+/YFP− cells into satellite cell-devoid muscles efficiently supported their regeneration, as demonstrated by the presence of YFP-positive fibers and of Pax7-expressing cells located in the sublaminar position. Conversely, Pax7+/YFP+ were able to differentiate and form new myofibers, but were inefficient in generating satellite cells. Furthermore, upon division Pax7+/YFP− satellite cell can give rise to both Pax7+/YFP− and Pax7+/YFP+ daughters, whereas Pax7+/YFP+ only generate Pax7+/YFP+ cells, suggesting that Pax7+/YFP− have a broader lineage potential. These data suggest that uniquely Pax7+/YFP− satellite cells can truly be defined stem cells.

In several tissues, it has been shown that during asymmetric stem cell division, the daughter cell that retains stem properties inherits older template DNA strands, whereas the daughter cell that differentiates receives newly synthetized DNA strands (Yennek & Tajbakhsh, 2013). This mechanism is referred to as nonrandom template DNA segregation and is thought to prevent accumulation of genetic mutations in stem cells and preserves their genome integrity. Interestingly, two studies reported that only a subpopulation of satellite cells performs nonrandom template DNA segregation during mitosis (Conboy, Karasov, & Rando, 2007; Shinin, Gayraud-Morel, Gomès, & Tajbakhsh, 2006). Recently, this feature has been correlated to Pax7 expression levels. Indeed, two subpopulations of quiescent satellite cells have been identified by FACS analysis, and it has been proposed that high levels of Pax7 (Pax7Hi) correspond to a less committed state (Rocheteau, Gayraud-Morel, Siegl-Cachedenier, Blasco, & Tajbakhsh, 2012). When Pax7Hi cells divide, they display a higher frequency of asymmetric template DNA segregation compared to cells expressing lower levels of Pax7 (Pax7Lo). However, the relevance of these observations on satellite cell self-renewal is arguable, since the two subpopulations seem to have comparable efficiency in participating to regeneration and satellite cell repopulation when transplanted into injured muscles. Moreover, it is not clear how Pax7Hi satellite cells could represent a less committed state since the authors report that they express higher levels of Myf5, the key muscle commitment gene.

Satellite cells are heterogeneous also regarding their cell cycle progression. An early examination of satellite cell-mediated postnatal muscle growth showed that 80% of satellite cells have a fast cell cycle progression, whereas the remaining 20% divide more slowly and are referred to as "reserve cells" (Schultz, Chamberlain, McCormick, & Mozdziak, 2006). More recently, FACS isolation of fast- and slow-cycling satellite cells from adult muscle followed by analysis of their myogenic and self-renewal potential revealed that only the slow-dividing subpopulation can give rise to a progeny with long-term self-renewal ability in vitro (Ono et al., 2012). Furthermore, slow-dividing but not fast-dividing satellite cells efficiently contributed to regeneration when transplanted into injured muscles. A recent study showed that slow-dividing satellite cells express higher levels of the cyclin-dependent kinase inhibitor 1B ($p27^{Kip1}$; Cdkn1b), which is required for their long-term maintenance in vivo (Chakkalakal et al., 2014).

In summary, multiple studies highlighted their heterogeneity in terms of lineage commitment, ability to self-renew and repopulate the niche, and amplification potential ex vivo. However, in the present state of our knowledge, no clear and dogmatic hierarchy is established within the satellite cell lineage.

4. INTRINSIC REGULATION OF SATELLITE CELL SELF-RENEWAL

While being beyond the scope of this review to describe in full detail the cascade of events that lead from satellite cell activation to myogenic specification and ultimately differentiation and fusion it is important to mention briefly the core transcription factors involved in the process. In particular the MRFs: Myf5, myogenic differentiation 1 (Myod1) and myogenic factor 6 (Herculin, Mrf4) that control in large part the acquisition of the myogenic identity by progenitor cells (Tajbakhsh, 2009) and myogenin (Myog), the essential muscle differentiation factor (Hasty et al., 1993). Moreover upstream to the MRF, we find Pax7 and its paralog Pax3 (Relaix, 2015) and sineoculis homeobox homolog 1 (Six1) and 4 (Six4). Differential expression of one or more of these transcription factors identifies the different stages of the myogenic progression. In particular, Pax7 and Myod1 protein expression discriminate between quiescent (Pax7+/Myod1−) activated (Pax7+/Myod1+) and committed to differentiation (Pax7−/Myod1+) states on isolated single myofibers (Zammit et al., 2004). Quiescent adult satellite cells express high levels of Pax7 proteins and transcripts (Seale et al., 2000), Myf5 transcripts and low levels of Myf5 proteins (Crist, Montarras, & Buckingham, 2012). Upon injury, satellite cells activate, start to express Myod1 (Cornelison & Wold, 1997), and upregulate Myf5 protein expression. The population of activated satellite cells will transiently amplify and rapidly increase in numbers. A fraction of these myogenic progenitors will completely turn off Pax7 expression and progress to the following steps of the differentiation program: cell cycle withdrawal, activation of Myog expression, and ultimately fusion and new fibers formation (Abmayr & Pavlath, 2012). Another fraction will instead turn off Myod1 while maintaining Pax7 expression to return to quiescence thus replenishing the stem cell pool.

4.1 Pax7

Pax7 role in satellite cells has been widely studied using different approaches including constitutive (Kuang, Chargé, Seale, Huh, & Rudnicki, 2006; Oustanina, Hause, & Braun, 2004; Seale et al., 2000) or conditional genetic ablation in mice (Günther et al., 2013; Lepper, Conway, & Fan, 2009; von Maltzahn, Jones, Parks, & Rudnicki, 2013). Pax7-null mice have craniofacial defects and display on average a 50%–70% decrease in body mass as compared to their wild-type littermates (Mansouri, Stoykova, Torres, & Gruss, 1996). In the C57/B6 background, only 5%–10% of the mice survive the 2 weeks of age. At juvenile age (P11), Pax7-null mice have smaller myofibers but the overall organization of the muscle is not affected (Oustanina et al., 2004; Seale et al., 2000). Postnatal growth is, however, extremely defective due to inadequate myonuclear accretion into myofibers. In the sv129 background, Pax7-null mice can survive to adulthood but display kyphosis and a marked decreased in myofibers numbers mostly due to the loss of fast 2B fibers compared to wild-type mice (Kuang et al., 2006). Muscle regeneration is also markedly compromised in Pax7-null mice; exposure to cardiotoxin (CTX) or mechanical (crush) injuries result in extensive calcium deposition and fibrosis (Kuang et al., 2006; Oustanina et al., 2004). Interestingly, satellite cell large proliferative capacity has been highlighted by the differential behavior of the Pax7-CreER*Pax7-flox conditional-null mice under acute or continuous tamoxifen (to activate the Cre recombinase activity) administration (Lepper et al., 2009; von Maltzahn et al., 2013). Lepper and colleagues showed that at the adult stage, after a series of five tamoxifen injection (acute treatment) Pax7 conditional-null mice were able to undergo multiple rounds of regeneration with no obvious defect (Lepper et al., 2009). On the other hand, following studies using the same model but with a continuous Tamoxifen administration obtained a marked defect in muscle tissue regeneration that was aggravated by additional rounds of injuries and loss of the satellite cell population (Günther et al., 2013; von Maltzahn et al., 2013). Furthermore, in vitro studies have demonstrated that Pax7-deficient cells are incapable of self-renewal and undergo growth arrest and precocious differentiation. The discrepancies are due to Pax7+ escapers cells that albeit low in number can expand and repopulate the satellite cell niche if the Cre recombinase is not continuously activated (Günther et al., 2013, von Maltzahn et al., 2013).

While little is known on the regulation of Pax7 expression, a recent study identified Caspase 3 cleavage inactivation of Pax7 as a crucial step for terminating the self-renewal process. Dick and colleagues demonstrated that inhibition of Caspase3 activity results in elevated Pax7 protein and satellite cell self-renewal in vitro, and that conversely caspase activation leads to Pax7 cleavage and initiation of the differentiation program. Importantly, ex vivo inhibition of Caspase3 activity in satellite cells before transplantation resulted in an accumulation of grafted satellite cells in the host muscle (Dick et al., 2015).

Pax7 binding regions have been recently mapped across the genome (Soleimani et al., 2012). Despite a partial overlap with Pax3, Pax7 controls a distinct nonredundant myogenic subprogram that regulates promotion of proliferation and inhibition of differentiation. In particular, two Pax7 unique binding sites are located upstream of Myod1. Moreover, Myf5 regulatory regions harbor a bona fide Pax7-dependent enhancer that if deleted abrogates Myf5 expression specifically in adult quiescent satellite cells.

Pax3/7 binding sites have also been identified near to genes related to myogenic inhibition, signaling, and adhesion. Additionally, Pax7 binds to the promoter of the Id3 gene, a well-characterized inhibitor of skeletal myogenesis, in quiescent satellite cells, thus preventing premature differentiation (Kumar, Shadrach, Wagers, & Lassar, 2009). From a molecular standpoint, Pax7 regulates satellite cell commitment into the myogenic lineage through the recruitment of a histone H3K4 methyltransferase complex to the Myf5 regulatory sequences and the establishment of epigenetic modification that stimulates transcriptional activation (McKinnell et al., 2008). This interaction between Pax7 and the histone remodeling complex is made possible by the coactivator-associated arginine methyltransferase 1 (Carm1; Prmt4)-mediated methylation of multiple arginines to the N-terminus of Pax7 (Kawabe, Wang, McKinnell, Bedford, & Rudnicki, 2012).

Considering the progressively degenerative phenotype of germline Pax7-null mice, the inability of satellite cell with targeted Pax7 gene disruption to expand and differentiate and the transcriptional control of myogenic genes by Pax7 we can conclude that this transcription factor is absolutely indispensable for satellite cell self-renewal in the adult.

4.2 Six1/4

The SIX transcription factors have been first identified in Drosophila, where together with their cofactors of the Eyes Absent (Eya) and Dachshund (Dach) families, they play a crucial role in eye development (Kawakami, Sato, Ozaki, & Ikeda, 2000). In mouse, Six1 and Six4 control different stages of embryonic myogenesis by binding through their homeodomain to consensus DNA sequences called MEF3, present in the promoter of several muscle-specific genes (Giordani et al., 2007; Spitz et al., 1998). During development, loss of Six1 results in extensive muscle hypoplasia affecting most of the body muscles and in particular certain limb muscles (Laclef et al., 2003) and Six1/Six4 double-null embryos show an aggravation of the phenotype with a complete absence of skeletal muscle in the limbs (Grifone et al., 2005).

We demonstrated that Six1 is expressed by quiescent and activated satellite cells and also plays an important role in skeletal muscle regeneration. Targeted Six1 gene recombination in adult satellite cells results in delayed myogenic cell differentiation, impaired myofiber repair, and perturbed homeostasis of the satellite cell niche during skeletal muscle regeneration. We observed that Six1 directly activates expression of Myod1 and Myog to allow for satellite cell progeny transition from proliferation to differentiation (Le Grand et al., 2012). Moreover, Six1 negatively regulates satellite cell self-renewal by tuning ERK signaling (see Section 5.1). Mechanistically, it has been revealed that Six1 binds to and regulate the chromatin structure of the core enhancer region of Myod1 (Liu et al., 2013).

Six4 is also expressed by adult satellite cells and its silencing in vitro has been shown to inhibit primary myoblast differentiation (Yajima et al., 2010). Six4, in cooperation with Myod1, recruits the histone demethylase lysine-specific demethylase 6A (KDM6A, also known as UTX) to the regulatory region of muscle genes to promote the removal of the repressive H3K27me3 epigenetic marks (Chakroun et al., 2015; Seenundun et al., 2010). Interestingly, Six4 and Myod1 coordinate the expression of several genes critical for proper in vivo muscle regeneration in myogenic cells in vitro. As such, future in vivo evaluation of the role of Six4 in satellite cell function through targeted gene disruption is of importance for our comprehension of adult myogenesis and satellite cell self-renewal.

4.3 Myod1

Myod1 is a master regulator of myogenesis and its role is prominent during the amplification of muscle progenitor cells and the onset of myogenic differentiation (Blais et al., 2005; Davis, Cheng, Lassar, & Weintraub, 1990). Surprisingly, mice lacking Myod1 display no major physiological or morphological defect in muscle formation. This is due to a substantial redundancy with Myf5 that can compensate for the loss of Myod1 expression (Rudnicki, Braun, Hinuma, & Jaenisch, 1992). However, Myod1-null mice exhibit impairment in adult muscle tissue regeneration. Although these mice display an increase in satellite cell numbers, the proliferation of myogenic cells following acute injury is dramatically reduced (Cornelison, Olwin, Rudnicki, & Wold, 2000; Megeney, Kablar, Garrett, Anderson, & Rudnicki, 1996). This is possibly caused by an imbalance in cell fate determination that promotes self-renewal at the expenses of commitment. Supporting this idea, more recent studies have shown some peculiarities at the satellite cell level providing additional insight on the role of Myod1. Transplantation experiments into regenerating muscle highlighted a higher survival and engraft rate of Myod1-null myoblasts. In this setup, not only Myod1-null cells fused with myofibers but also gave rise to satellite cells. Moreover, Myod1-null myoblasts are resistant to apoptosis and this explains their improved engraftment capabilities (Asakura et al., 2007). At molecular this is mediated via the downregulation of the microRNAs mir-1 and mir-206 (the latter is a direct Myod1 target) (Sweetman et al., 2008). This leads to upregulation of the Pax7 paralog Pax3 expression (Hirai et al., 2010; Li, Sarver, Alamgir, & Subramanian, 2012). Pax3 serve as survival factor through upregulating antiapoptotic genes.

Myod1 also regulates Pax7 expression via mir-1, mir-206, and mir-486 (Chen et al., 2010; Dey, Gagan, & Dutta, 2011). Transcription of both these micro-RNAs is directly activated by Myod1 and target the Pax7 3′UTR. Pax7, on the other hand, promotes Id2 expression and leads to repression of Myod1 activity resulting in downregulation of mir-206 and mir-486 expressions. This bistable switch network can exist in both Pax7-positive and Myod1-positive states reinforcing satellite cell fate decision. Although is still not fully clear which are the dynamics that trigger Myod1 down-regulation in vivo it has been shown in C2C12 that Calpain3 is responsible for Myod1 degradation. This, in turn, promotes the formation of the "reserve pool"; a phenomena that could be considered the in vitro equivalent to the return to quiescence (Stuelsatz et al., 2010). In conclusion, without Myod1, myogenic progression is delayed and inefficient whereas return to quiescence is favored. Myod1 downregulation is therefore a key step to enable self-renewal and the balance between Pax7 and Myod1 is crucial in regulating satellite cell fate choice.

4.4 Myf5

Myf5 transcript is heterogeneously expressed in adult satellite cells at the quiescent stage (Beauchamp et al., 2000; Pallafacchina et al., 2010) however protein levels are extremely low. This discrepancy is due to the fact that its mRNA is accumulated in messenger ribonucleoprotein (mRNP) granules and its translation is repressed by mir-31. Myf5 transcript is then released upon activation stimuli by dephosphorylation of fragile X mental retardation 1 (FMR1, also known as FMRP; a mRNP granules component protein) (Crist et al., 2012). This mechanism prevents the unscheduled initiation of the myogenic program during quiescence while allowing for the rapid appearance of Myf5 protein shortly after satellite cell activation.

Yet, Myf5 itself was shown to have a subtle role in adult myogenesis. Myf5-null mice exhibit perturbed muscle regeneration and Myf5-null satellite cell proliferation is primarily impaired (Gayraud-Morel et al., 2007; Ustanina, Carvajal, Rigby, & Braun, 2007). Hence, despite a large body of work the role of Myf5 in self-renewal is however, still to be fully elucidated; while being on one side a marker of myogenic lineage commitment its genetic deregulation leads to contrasting results suggesting that additional regulatory layers could be involved. The study of MyoD/Myf5 double-null satellite cells should shed some light in the exact roles of these redundant MRFs function in satellite cell function and self-renewal.

5. EXTRINSIC REGULATION OF SATELLITE CELL SELF-RENEWAL

In addition to the aforementioned transcription factors that orchestrate the myogenic program cell autonomously, extrinsic mechanisms governing muscle satellite cell self-renewal have been described in the recent years. Diffusible signals and cell-to-cell contacts allow the synchronization of the regeneration process at the population level ensuring the homogenous transition between different stages of the myogenic program.

5.1 ERK Signaling

The first description of an extracellular signaling pathway promoting muscle satellite cell withdrawal from cell cycle to enter into quiescence during muscle regeneration came from Abou-Khalil and colleagues. They found that quiescent satellite cells express high levels of the endothelial-specific receptor tyrosine kinase (Tek, also known as Tie2) receptor and that its ligand angiopoietin 1 (Ang1) was secreted by cells located in the satellite cell neighborhood such as smooth muscle cells and fibroblasts. Ang1/Tie-2 signaling activate extracellular signal-regulated kinase (ERK1/2) signaling in satellite cells to control upregulation of markers associated with the reserve cell phenotype and downregulation of markers associated with myogenic differentiation. In vivo assays performed during muscle regeneration showed that blocking Tie-2 promotes satellite cells proliferation, while conversely, Ang1 overexpression stimulates satellite cells return to quiescence (Abou-Khalil et al., 2009).

More recently, the adaptor protein TNF receptor-associated factor 6 (Traf6) was shown to be required for the activation of ERK1/2 and c-Jun N-terminal kinases (JNK1/2) that in turn augments Pax7 expression in muscle progenitor cells and satellite cells. Selective deletion of Traf6 in satellite cells of adult mice led to profound muscle regeneration defects due to loss of Pax7 expression and a strong impairment of satellite cell self-renewal (Hindi & Kumar, 2016). In line with these studies, a negative feedback regulator of ERK signaling (Hanafusa, Torii, Yasunaga, & Nishida, 2002),

Sprouty1 (Spry1) was further shown to be directly linked to satellite cells self-renewal. Spry1 is expressed by quiescent satellite cells and its level rapidly decline during amplification and terminal differentiation. Conditional genetic disruption of Spry1 in satellite cells has no effect on the satellite cell pool in resting muscle. In contrast, Spry1-null muscle display reduced numbers of satellite cells following muscle injury. This reduction is not aggravated after a second round of injury strongly suggesting that Spry1 is required for the self-renewal of a specific subset of satellite cells (Shea et al., 2010). ERK signaling is also elevated in Six1-null satellite cells that are endowed with high self-renewal potential. In mutant satellite cells, the increase in ERK signaling has been linked to the absence of the Six1 target gene dual specificity phosphatase 6 (Dusp6, also called Mkp3) expression in proliferating satellite cells. Dusp6 is a selective inhibitor of ERK signaling and while muscles lacking Dusp6 were able to regenerate properly they showed a striking increase in satellite cell numbers after regeneration (Le Grand et al., 2012).

Altogether these studies identified ERK signaling as a crucial regulator of satellite cell self-renewal but also proposed divergent roles for known inhibitors of this signaling pathway. Return to the niche is positively regulated by Sprty1, but limited by Dusp6. As such, the different expression profiles of these two inhibitors highlight the timely requirement for inhibition/activation of ERK signaling during satellite cell self-renewal.

5.2 Noncanonical Wnt7a Signaling

The Wnt signaling cascades are critical regulators of stem cells biology in many adult tissues. Our previous work highlighted that many Wnt proteins are secreted during muscle regeneration. We further demonstrated that satellite cells express the noncanonical Wnt receptor Frizzled-7 (Fzd7) and that its ligand Wnt7a is markedly upregulated in newly formed myofibers during regenerative myogenesis. The Fzd7 receptor is necessary for Wnt7a binding at the surface of myogenic cells and signaling through the noncanonical Wnt7a/Fzd7 planar cell polarity (PCP) pathway drives Pax7+/Myf5− satellite cell symmetric cell divisions and self-renewal (Le Grand, Jones, Seale, Scimè, & Rudnicki, 2009).

During symmetric planar divisions, Wnt7a stimulation induces the polarized localization of the core PCP component Vang-like 2 (Vangl2) at opposite poles in pairs of dividing MuSCs in a manner consistent with Wnt7a activating PCP signaling, thus leading both daughters cells to maintain contact with their niche. We proposed this "niche connection" to be instrumental in regulating satellite cell self-renewal. Furthermore, Wnt7a over-expression in injured muscle resulted in an impressive enhancement of the regenerative process. Treated animals displayed a marked expansion of the satellite cell pool and a higher number of myofibers of bigger caliber, independently of an effect on myoblast proliferation (Le Grand et al., 2009).

A following study highlighted the role of extracellular matrix (ECM) remodeling in Wnt7a/Fzd7 signaling. Bentzinger and colleagues showed that satellite cells remodel their niche via transient high-level expression of fibronectin (Fn1) during muscle regeneration. Fn1 binding to a coreceptor complex of the proteoglycan Syndecan-4 (Sdc4) and Fzd7 stimulated the ability of Wnt7a to induce the symmetric expansion of Pax7+ /Myf5− satellite cells. Silencing of Fn1 in satellite cells severely impaired their ability to repopulate the satellite cell niche following transplantation while in vivo overexpression of Fn1 with Wnt7a dramatically stimulated satellite cell self-renewal (Bentzinger et al., 2013).

Taken together these studies demonstrated the synergistic effect of extracellular signals and ECM composition in the regulation of stem cell division orientation and self-renewal.

5.3 Calcitonin Receptor Signaling

In 2007, a genome-wide gene expression profiling of satellite cell purified by FACS revealed numerous genes upregulated in quiescent satellite cells, among them the calcitonin receptor (Calcr). Calcr is a G protein-coupled receptor that binds the peptide hormone calcitonin and is involved in intracellular calcium homeostasis (Pondel, 2000). Calcr expression marks both resting and newly self-renewed satellite cells but not proliferating or differentiating myoblasts. Interestingly, ex vivo calcitonin stimulation delayed the activation of quiescent satellite cells (Fukada et al., 2007). Targeted Calcr gene disruption in satellite cells lead to a marked decrease in the satellite cell pool with time (35%–80% 16 weeks after depletion). This reduction in satellite cell numbers resulted in a moderate impairment of muscle regeneration (Yamaguchi et al., 2015). Furthermore, using the single myofiber model Yamaguchi and colleagues showed that Calcr exert a dual role on satellite cells. Calcr signaling prevents the exit from quiescence via the cyclic adenosine monophosphate (cAMP)–protein kinase A (PKA) pathway and, on the other hand, blocks the emergence from the niche by both cAMP–PKA and the cAMP–exchange factor directly activated by cAMP 1 (Epac) pathway. This regulatory circuit is moreover reinforced by miRNA contribution.

Mir-489, which is encoded in intron 4 of calcitonin receptor gene, promotes the persistence of the state of quiescence. This is achieved via the direct downregulation of the oncogene Dek that is asymmetrically expressed in the satellite cell fraction that eventually will undergo amplification (Cheung et al., 2012).

5.4 p38 MAPK Signaling

The p38 mitogen-activated protein kinase (MAPK) pathway, initially described as a transducer of the response to environmental stress conditions, plays an important role in skeletal muscle differentiation. As such, p38α-null mice exhibit defective myogenesis and neonatal muscle deficient in p38α shows cellular hyperproliferation and delayed maturation mainly caused by delayed cell-cycle exit of myogenic progenitors (Perdiguero et al., 2007). Muscle regeneration after cardiotoxin-induced injury is delayed in the absence of p38α and the satellite cell pool is increased following regeneration (Brien, Pugazhendhi, Woodhouse, Oxley, & Pell, 2013). Mechanistically, p38 MAPK activity targets chromatin remodeling complex to the loci of myogenic genes (Rampalli et al., 2007; Simone et al., 2004).

Many upstream signals can result in p38 MAPK activation. In muscle cells, the promyogenic cell surface protein cell adhesion molecule-related/downregulated by oncogenes (Cdon, also known as Cdo) has been shown to interact with JNK-associated leucine zipper protein (JLP), a scaffold protein for the p38α/β MAPK pathway through its intracellular region (Takaesu et al., 2006). Cdo activation at the cell membrane stimulates the small GTPase cell division control protein 42 homolog (Cdc42) activity in the cytoplasm, which in turn promotes p38 α/β activity and muscle cell differentiation (Kang et al., 2008).

Also, in pathological conditions (e.g., muscle dystrophy), tumor necrosis factor (TNF)α extracellular signaling activates p38α to promote the interaction between YY1 and the chromatin remodeling complex PRC2 (polycomb repressive complex 2) and lead to the formation of repressive chromatin on Pax7 promoter (Palacios et al., 2010).

p38α/β MAPK are required for the activation of quiescent satellite cells and the induction of Myod1 expression (Jones et al., 2005). During the first satellite cell division, p38α/β MAPK are asymmetrically activated in only one daughter cell, in which Myod1 is induced, thus generating a committed daughter. In contrast, the daughter cell without p38α/β MAPK signaling will not activate Myod1 expression and is postulated to self-renew. Importantly, during satellite cells divisions, p38α/β colocalize on one side of the dividing cell with the Par-complex proteins partitioning defective 3 homolog (Pard3) and atypical protein kinase C (PKC). These proteins are components of a polarity complex essential for asymmetric division in many cell types. As such, preventing the physical interaction of phosphorylated-p38α and Pard3 on one side of dividing satellite cells by Pard3 silencing results in abrogation of asymmetric divisions (Troy et al., 2012).

Unexpectedly, a recent study demonstrated that Dystrophin, the large protein that connects the myofiber cytoskeleton to the ECM is also expressed in activated satellite cells. Dystrophin-null satellite cells were further shown to be impaired in their capacity to undergo asymmetric division, and this phenotype is attributable to a lack of polarization of Pard3. This defect in polarity establishment results in a higher proportion of abnormal divisions, and impaired muscle regeneration. Therefore, Dystrophin integrate positional cues in satellite cell to limit self-renewal and stimulate the generation of committed satellite cell daughters (Dumont et al., 2015).

5.5 Notch Signaling

Early work showed that Notch signaling could repress myogenesis in different model systems. In the frog embryo, Notch intracellular domain blocks myogenesis via the suppression of Myod1 transactivation ability without interfering with its DNA binding (Kopan, Nye, & Weintraub, 1994). Similar results were further obtained in the chick embryo (Delfini, Hirsinger, Pourquié, & Duprez, 2000), and in C2C12 myoblasts (Kuroda et al., 1999). Supporting these evidences, interference with Notch signaling through mutation of the Notch ligand delta-like protein 1 (Dll1) (Schuster-Gossler, Cordes, & Gossler, 2007) or of the Notch transcriptional effector recombining binding protein suppressor of hairless (Rbpj) (Vasyutina et al., 2007) results in premature depletion of the myogenic progenitor pool in the mouse embryo. Moreover, Notch signaling controls the homing of muscle progenitors to the satellite cell niche during development. During fetal development, Notch signals stimulate the expression of basal lamina components and adhesion molecules by Pax7-positive cells, inducing them to contribute to their own microenvironment and to adhere to myofibers (Bröhl et al., 2012).

In the adult, Notch signaling is active in quiescent satellite cells, declines after activation before the first mitosis and increases again during proliferation. During muscle regeneration, Notch signaling promotes the expansion of myogenic progenitors from satellite cells while Numb-mediated inhibition of Notch promotes satellite cell descendants differentiation (Conboy & Rando, 2002). Numb is asymmetrically inherited in expanding myogenic progenitors and its depletion leads to impaired satellite cell proliferation thus advocating for a unique role of Numb in the amplification step (George et al., 2013).

Targeted genetic deletion of Rbpj in satellite cells corroborated the critical role of Notch in muscle stem cell maintenance. Conditional loss of Notch signaling resulted in a sharp decline of satellite cells number and in marked regeneration impairment in response to injury. This satellite cell pool reduction is due to spontaneous activation followed by terminal differentiation (Bjornson et al., 2012; Mourikis et al., 2012). In line with these findings, the Notch target genes Hairy/enhancer-of-split related with YRPW motif protein 1 (Hey1, also called Hers1) and hairy/enhancer-of-split related with YRPW motif-like (Heyl, also called Hers3) are highly expressed in quiescent cells. While either Hey1-null mice and Heyl-null mice are generally normal, the double-null animals show decreased myofibers numbers and sizes. Animals devoid of Hey1 and Heyl have a reduced satellite cell pool and a regeneration deficit. Furthermore Hey1/Heyl double-null satellite cells are unable to enter in the undifferentiated quiescent state and fail to produce reserve cells in vitro (Fukada et al., 2011).

Notch signaling also acts on satellite cell self-renewal through Pax7 regulation. Overexpression of a constitutively activated form of Notch1 intracellular domain (NICD) upregulates Pax7 expression in primary myoblast. This activation is mediated by the interaction of NCID with Rbpj that binds upstream of the Pax7 gene. Consistently, fibers derived from mice with a constitutively active NICD ($NICD^{CA}$) in satellite cells show higher self-renewal in vivo (Wen et al., 2012). Furthermore, the expression of $NICD^{CA}$ at later stage of myocite differentiation forces the cells to revert to the quiescence state (Bi et al., 2016). Interestingly simultaneous deletion of Pax7 and activation of NICD in satellite cells prevents their depletion. However, these cells fail to upregulate Myod1 expression and undergo brown adipogenic conversion (Pasut et al., 2016).

Paradoxically in the adult Notch3 deficiency leads to an increase in the number of satellite cells in vivo and ex vivo on isolated myofibers. This data would suggest that Notch3 is a negative regulator of self-renewal. To explain the opposing effects of Notch1 and Notch3, it has been speculated that the latter could trigger a regulatory negative feedback loop and act as an inhibitor of Notch1 (Kitamoto & Hanaoka, 2010).

Recently, sex hormones were shown to induce Mind bomb 1 (Mib1) expression in myofibers at puberty, which activates Notch signaling in cycling juvenile satellite cells and promotes entry into the quiescent state. Myofibers lacking Mib1 fail to send Notch signals to satellite cells, leading to their depletion. Consequently, the hypothalamic–pituitary–gonadal axis controls both establishment of the satellite cell niche and self-renewal of quiescent satellite cell populations following injury (Kim et al., 2016).

All of the aforementioned genetic model that, to different extents, abrogates Notch signaling in satellite cells display similar characteristics. In conclusion, Notch signaling acts at many different levels to direct self-renewal in satellite cells both by promoting their return to the niche and by controlling the expression of myogenic differentiation factors.

6. SATELLITE CELL METABOLISM

In recent years, an increasing body of work showed that changes in energy balance and metabolic status are critical for adult stem cell function. Altogether, many research teams are investigating how metabolic flux in stem cells is linked to cell fate changes. The typical quiescent state of satellite cell suggests important changes during activation and self-renewal, since quiescence is defined as a metabolically unique state characterized by suppressed catabolism.

A recent study demonstrated that two distinct phases exist within satellite cell quiescence in vivo, G0 and GAlert. Indeed, in a muscle contralateral to an injured muscle satellite cells are in a particular status that slightly differs from the normal widely described G0 quiescence. In this status, named GAlert, cells display what the authors called an alert phenotype, which is characterized by a small but significant increase in metabolic and transcriptional activity, a more pronounced propensity to enter cell cycle and higher level of the central metabolic regulator mammalian target of rapamycin (mTOR) activation. Therefore, GAlert corresponds to a more responsive state in which cells can rapidly react to muscle injury (Rodgers et al., 2014).

Following exit from quiescence, muscle satellite cells experience a metabolic switch from fatty acid oxidation to glycolysis. This metabolic reprogramming is translated in a global changes in gene expression through the activation of the nicotinamide adenine dinucleotide (NAD)-dependent histone deacetylase Sirtuin1 (Sirt1), and activation of muscle gene transcription (Ryall et al., 2015). Mechanistically, Sirt1 is necessary for satellite cell proliferation in vitro (Rathbone, Booth, & Lees, 2009) and targeted Sirt1 gene deletion in muscle progenitor cells in vivo lead to an impairment of muscle tissue regeneration, and lower levels of Pax7 expression in the tissue (Ryall et al., 2015). Analysis of the role of Sirt1 specifically in satellite cell self-renewal has not been conducted so far, but one can speculate that return to quiescence would be associated with another metabolic reprogramming in satellite cells (possibly a return to an oxidative metabolism) and that Sirt1 activity would have to be quenched for satellite cell to self-renew and repopulate the niche.

The AMP-activated protein kinase (AMPK) is a master regulator of metabolic homeostasis. In almost all cell types, AMPK regulates the balance between catabolic and anabolic processes, particularly in skeletal muscle cells in which it controls myofiber growth and size (Lantier et al., 2010; Mounier et al., 2009). Conditional genetic analyses clearly indicated that AMPK and its upstream kinase, liver kinase b1 (Lkb1), control satellite cell activation and self-renewal. As such, Lkb1-null satellite cells fail to maintain quiescence in noninjured resting muscles and exhibit accelerated proliferation but reduced differentiation kinetics (Shan et al., 2014). In contrast, Ampkα1-null satellite cells display a high self-renewal rate which impairs muscle regeneration. Loss of Ampkα1 switched satellite cell to a higher glycolysis and increased the enzyme lactate dehydrogenase (LDH) expression which stimulated satellite cell return to quiescence (Theret et al., 2017).

Altogether, these studies emphasize the importance of metabolism in satellite cells and demonstrate that the balance between oxidative phosphorylation and aerobic glycolysis can be shifted depending on the their cell cycle. Integration of changes in the metabolic flux is thus an important driver of satellite reversible quiescence and cell self-renewal.

7. CONCLUDING REMARKS

Since Alexander Mauro seminal discovery, many groups contributed to build a large body of evidences dissecting the role of satellite cells in skeletal muscle tissue repair. Historically, histology and morphometrics have been the founding pillars of the muscle field, gradually defining overtime a precise compartmentalization of the different subtypes of muscle resident cells in vivo. The advent of flow cytometry and of cell lineage tracing in mice within the last decade has pushed forward our knowledge, highlighting the roles of specific satellite cell subpopulations and contributing decisively in the definition of the identity of these muscle stem cells. Moreover, the development of novel tools enabling single-cell analyses will soon become widely available (e.g., single-cell RNA-seq and single-cell proteomics) to further comprehend satellite cells heterogeneity and hierarchy within the niche.

Despite more than 50 years of efforts, our comprehension of satellite cell self-renewal remain elusive. Taking as an example the regeneration of a single fiber from the EDL muscle. Following injury, the resident 6 satellite cells will be able to both generate 250 muscle progenitor cells that will fuse to make a new syncytia and self-renew to birth 6 satellite cells that will populate the reconstructed niche. This whole process happens within 5–7 doublings and monitoring the elusive divisions that give rise to stem cell self-renewal within the bulk of myoblast amplification still remains technically harsh. And while the single floating myofiber system is still powerful and widely used, it only allows for studying the first divisions. Furthermore, satellite cell fate is directed by the myofiber culture conditions and freed from the architectural constraints of the in vivo niche. Future work may thus focus on developing new techniques for intravital imaging to track satellite cell fate in situ and/or on reconstructing a three-dimensional niche in vitro to mimic satellite cell niche.

As presented in this chapter, satellite cell self-renewal is dependent, at least, on three distinct phases: activation from quiescence, self-renewing division (asymmetric or symmetric), and return to dormancy. At the tissue level, self-renewal of the muscle stem cell pool can be hindered by genetic mutations or disease-secreted factors acting on any steps and leading to inefficient skeletal muscle repair. As currently understood, the distinct cellular process leading to satellite cell pool self-renewal is controlled by extrinsic signaling pathways that collaborate or diverge to control the activity of intrinsic regulators of cell polarity and/or transcriptional regulators. The more "nodes" are added to the regeneration circuitry the more is evident that each signals acts in synchrony with the others. The whole system acts coherently, constantly regulating and adjusting each component to achieve harmonic progression trough the different stages of muscle regeneration. Up until now we have isolated and studied each component of the regeneration machinery singularly, the next challenge would be able to elucidate how all these signals are integrated together.

REFERENCES

Abmayr, S. M., & Pavlath, G. K. (2012). Myoblast fusion: Lessons from flies and mice. *Development (Cambridge, England)*, *139*, 641–656.

Abou-Khalil, R., Le Grand, F., Pallafacchina, G., Valable, S., Authier, F.-J., Rudnicki, M. A., et al. (2009). Autocrine and paracrine angiopoietin 1/Tie-2 signaling promotes muscle satellite cell self-renewal. *Cell Stem Cell*, *5*, 298–309.

Asakura, A., Hirai, H., Kablar, B., Morita, S., Ishibashi, J., Piras, B. A., et al. (2007). Increased survival of muscle stem cells lacking the MyoD gene after transplantation into regenerating skeletal muscle. *Proceedings of the National Academy of Sciences of the United States of America, 104*, 16552–16557.
Beauchamp, J. R., Heslop, L., Yu, D. S., Tajbakhsh, S., Kelly, R. G., Wernig, A., et al. (2000). Expression of CD34 and Myf5 defines the majority of quiescent adult skeletal muscle satellite cells. *The Journal of Cell Biology, 151*, 1221–1234.
Bentzinger, C. F., Wang, Y. X., von Maltzahn, J., Soleimani, V. D., Yin, H., & Rudnicki, M. A. (2013). Fibronectin regulates Wnt7a signaling and satellite cell expansion. *Cell Stem Cell, 12*, 75–87.
Bi, P., Yue, F., Sato, Y., Wirbisky, S., Liu, W., Shan, T., et al. (2016). Stage-specific effects of Notch activation during skeletal myogenesis. *eLife, 5*.
Bischoff, R. (1975). Regeneration of single skeletal muscle fibres in vitro. *The Anatomical Record, 182*, 215–235.
Bjornson, C. R. R., Cheung, T. H., Liu, L., Tripathi, P. V., Steeper, K. M., & Rando, T. A. (2012). Notch signaling is necessary to maintain quiescence in adult muscle stem cells. *Stem Cells, 30*, 232–242.
Blais, A., Tsikitis, M., Acosta-Alvear, D., Sharan, R., Kluger, Y., & Dynlacht, B. D. (2005). An initial blueprint for myogenic differentiation. *Genes & Development, 19*, 553–569.
Brien, P., Pugazhendhi, D., Woodhouse, S., Oxley, D., & Pell, J. M. (2013). p38α MAPK regulates adult muscle stem cell fate by restricting progenitor proliferation during postnatal growth and repair. *Stem Cells, 31*, 1597–1610.
Bröhl, D., Vasyutina, E., Czajkowski, M. T., Griger, J., Rassek, C., Rahn, H.-P., et al. (2012). Colonization of the satellite cell niche by skeletal muscle progenitor cells depends on Notch signals. *Developmental Cell, 23*, 469–481.
Buckingham, M., Bajard, L., Chang, T., Daubas, P., Hadchouel, J., Meilhac, S., et al. (2003). The formation of skeletal muscle: From somite to limb. *Journal of Anatomy, 202*, 59–68.
Chakkalakal, J. V., Christensen, J., Xiang, W., Tierney, M. T., Boscolo, F. S., Sacco, A., et al. (2014). Early forming label-retaining muscle stem cells require p27kip1 for maintenance of the primitive state. *Development (Cambridge, England), 141*, 1649–1659.
Chakroun, I., Yang, D., Girgis, J., Gunasekharan, A., Phenix, H., Kærn, M., et al. (2015). Genome-wide association between Six4, MyoD, and the histone demethylase Utx during myogenesis. *The FASEB Journal, 29*, 4738–4755.
Chen, J.-F., Tao, Y., Li, J., Deng, Z., Yan, Z., Xiao, X., et al. (2010). microRNA-1 and microRNA-206 regulate skeletal muscle satellite cell proliferation and differentiation by repressing Pax7. *The Journal of Cell Biology, 190*, 867–879.
Cheung, T. H., Quach, N. L., Charville, G. W., Liu, L., Park, L., Edalati, A., et al. (2012). Maintenance of muscle stem-cell quiescence by microRNA-489. *Nature, 482*, 524–528.
Collins, C. A., Olsen, I., Zammit, P. S., Heslop, L., Petrie, A., Partridge, T. A., et al. (2005). Stem cell function, self-renewal, and behavioral heterogeneity of cells from the adult muscle satellite cell niche. *Cell, 122*, 289–301.
Conboy, M. J., Karasov, A. O., & Rando, T. A. (2007). High incidence of non-random template strand segregation and asymmetric fate determination in dividing stem cells and their progeny. *PLoS Biology, 5*, e102.
Conboy, I. M., & Rando, T. A. (2002). The regulation of Notch signaling controls satellite cell activation and cell fate determination in postnatal myogenesis. *Developmental Cell, 3*, 397–409.
Cornelison, D. D., Olwin, B. B., Rudnicki, M. A., & Wold, B. J. (2000). MyoD(-/-) satellite cells in single-fibre culture are differentiation defective and MRF4 deficient. *Developmental Biology, 224*, 122–137.
Cornelison, D. D., & Wold, B. J. (1997). Single-cell analysis of regulatory gene expression in quiescent and activated mouse skeletal muscle satellite cells. *Developmental Biology, 191*, 270–283.

Crist, C. G., Montarras, D., & Buckingham, M. (2012). Muscle satellite cells are primed for myogenesis but maintain quiescence with sequestration of Myf5 mRNA targeted by microRNA-31 in mRNP granules. *Cell Stem Cell*, *11*, 118–126.

Davis, R. L., Cheng, P. F., Lassar, A. B., & Weintraub, H. (1990). The MyoD DNA binding domain contains a recognition code for muscle-specific gene activation. *Cell*, *60*, 733–746.

Delfini, M. C., Hirsinger, E., Pourquié, O., & Duprez, D. (2000). Delta 1-activated notch inhibits muscle differentiation without affecting Myf5 and Pax3 expression in chick limb myogenesis. *Development (Cambridge, England)*, *127*, 5213–5224.

Dey, B. K., Gagan, J., & Dutta, A. (2011). miR-206 and -486 induce myoblast differentiation by downregulating Pax7. *Molecular and Cellular Biology*, *31*, 203–214.

Dick, S. A., Chang, N. C., Dumont, N. A., Bell, R. A. V., Putinski, C., Kawabe, Y., et al. (2015). Caspase 3 cleavage of Pax7 inhibits self-renewal of satellite cells. *Proceedings of the National Academy of Sciences of the United States of America*, *112*, E5246–5252.

Dumont, N. A., Wang, Y. X., von Maltzahn, J., Pasut, A., Bentzinger, C. F., Brun, C. E., et al. (2015). Dystrophin expression in muscle stem cells regulates their polarity and asymmetric division. *Nature Medicine*, *21*, 1455–1463.

Fukada, S., Uezumi, A., Ikemoto, M., Masuda, S., Segawa, M., Tanimura, N., et al. (2007). Molecular signature of quiescent satellite cells in adult skeletal muscle. *Stem Cells*, *25*, 2448–2459.

Fukada, S., Yamaguchi, M., Kokubo, H., Ogawa, R., Uezumi, A., Yoneda, T., et al. (2011). Hesr1 and Hesr3 are essential to generate undifferentiated quiescent satellite cells and to maintain satellite cell numbers. *Development (Cambridge, England)*, *138*, 4609–4619.

Gayraud-Morel, B., Chrétien, F., Flamant, P., Gomès, D., Zammit, P. S., & Tajbakhsh, S. (2007). A role for the myogenic determination gene Myf5 in adult regenerative myogenesis. *Developmental Biology*, *312*, 13–28.

George, R. M., Biressi, S., Beres, B. J., Rogers, E., Mulia, A. K., Allen, R. E., et al. (2013). Numb-deficient satellite cells have regeneration and proliferation defects. *Proceedings of the National Academy of Sciences of the United States of America*, *110*, 18549–18554.

Gilbert, P. M., Havenstrite, K. L., Magnusson, K. E. G., Sacco, A., Leonardi, N. A., Kraft, P., et al. (2010). Substrate elasticity regulates skeletal muscle stem cell self-renewal in culture. *Science*, *329*, 1078–1081.

Giordani, J., Bajard, L., Demignon, J., Daubas, P., Buckingham, M., & Maire, P. (2007). Six proteins regulate the activation of Myf5 expression in embryonic mouse limbs. *Proceedings of the National Academy of Sciences of the United States of America*, *104*, 11310–11315.

Grifone, R., Demignon, J., Houbron, C., Souil, E., Niro, C., Seller, M. J., et al. (2005). Six1 and Six4 homeoproteins are required for Pax3 and Mrf expression during myogenesis in the mouse embryo. *Development (Cambridge, England)*, *132*, 2235–2249.

Günther, S., Kim, J., Kostin, S., Lepper, C., Fan, C.-M., & Braun, T. (2013). Myf5-positive satellite cells contribute to Pax7-dependent long-term maintenance of adult muscle stem cells. *Cell Stem Cell*, 1–12.

Hanafusa, H., Torii, S., Yasunaga, T., & Nishida, E. (2002). Sprouty1 and Sprouty2 provide a control mechanism for the Ras/MAPK signalling pathway. *Nature Cell Biology*, *4*, 850–858.

Hasty, P., Bradley, A., Morris, J. H., Edmondson, D. G., Venuti, J. M., Olson, E. N., et al. (1993). Muscle deficiency and neonatal death in mice with a targeted mutation in the myogenin gene. *Nature*, *364*, 501–506.

Hindi, S. M., & Kumar, A. (2016). TRAF6 regulates satellite stem cell self-renewal and function during regenerative myogenesis. *The Journal of Clinical Investigation*, *126*, 151–168.

Hirai, H., Verma, M., Watanabe, S., Tastad, C., Asakura, Y., & Asakura, A. (2010). MyoD regulates apoptosis of myoblasts through microRNA-mediated down-regulation of Pax3. *The Journal of Cell Biology*, *191*, 347–365.

Jones, N. C., Tyner, K. J., Nibarger, L., Stanley, H. M., Cornelison, D. D. W., Fedorov, Y. V., et al. (2005). The p38alpha/beta MAPK functions as a molecular switch to activate the quiescent satellite cell. *The Journal of Cell Biology*, *169*, 105–116.

Kang, J.-S., Bae, G.-U., Yi, M.-J., Yang, Y.-J., Oh, J.-E., Takaesu, G., et al. (2008). A Cdo-Bnip-2-Cdc42 signaling pathway regulates p38alpha/beta MAPK activity and myogenic differentiation. *The Journal of Cell Biology*, *182*, 497–507.

Kawabe, Y.-I., Wang, Y. X., McKinnell, I. W., Bedford, M. T., & Rudnicki, M. A. (2012). Carm1 regulates Pax7 transcriptional activity through MLL1/2 recruitment during asymmetric satellite stem cell divisions. *Cell Stem Cell*, *11*, 333–345.

Kawakami, K., Sato, S., Ozaki, H., & Ikeda, K. (2000). Six family genes—Structure and function as transcription factors and their roles in development. *BioEssays*, *22*, 616–626.

Kim, J.-H., Han, G.-C., Seo, J.-Y., Park, I., Park, W., Jeong, H.-W., et al. (2016). Sex hormones establish a reserve pool of adult muscle stem cells. *Nature Cell Biology*, *18*, 930–940.

Kitamoto, T., & Hanaoka, K. (2010). Notch3 null mutation in mice causes muscle hyperplasia by repetitive muscle regeneration. *Stem Cells*, *28*, 2205–2216.

Kopan, R., Nye, J. S., & Weintraub, H. (1994). The intracellular domain of mouse Notch: A constitutively activated repressor of myogenesis directed at the basic helix-loop-helix region of MyoD. *Development (Cambridge, England)*, *120*, 2385–2396.

Kuang, S., Chargé, S. B., Seale, P., Huh, M., & Rudnicki, M. A. (2006). Distinct roles for Pax7 and Pax3 in adult regenerative myogenesis. *The Journal of Cell Biology*, *172*, 103–113.

Kuang, S., Kuroda, K., Le Grand, F., & Rudnicki, M. A. (2007). Asymmetric self-renewal and commitment of satellite stem cells in muscle. *Cell*, *129*, 999–1010.

Kumar, D., Shadrach, J. L., Wagers, A. J., & Lassar, A. B. (2009). Id3 is a direct transcriptional target of Pax7 in quiescent satellite cells. *Molecular Biology of the Cell*, *20*, 3170–3177.

Kuroda, K., Tani, S., Tamura, K., Minoguchi, S., Kurooka, H., & Honjo, T. (1999). Delta-induced Notch signaling mediated by RBP-J inhibits MyoD expression and myogenesis. *The Journal of Biological Chemistry*, *274*, 7238–7244.

Laclef, C., Hamard, G., Demignon, J., Souil, E., Houbron, C., & Maire, P. (2003). Altered myogenesis in Six1-deficient mice. *Development (Cambridge, England)*, *130*, 2239–2252.

Lantier, L., Mounier, R., Leclerc, J., Pende, M., Foretz, M., & Viollet, B. (2010). Coordinated maintenance of muscle cell size control by AMP-activated protein kinase. *The FASEB Journal*, *24*, 3555–3561.

Le Grand, F., Grifone, R., Mourikis, P., Houbron, C., Gigaud, C., Pujol, J., et al. (2012). Six1 regulates stem cell repair potential and self-renewal during skeletal muscle regeneration. *The Journal of Cell Biology*, *198*, 815–832.

Le Grand, F., Jones, A. E., Seale, V., Scimè, A., & Rudnicki, M. A. (2009). Wnt7a activates the planar cell polarity pathway to drive the symmetric expansion of satellite stem cells. *Cell Stem Cell*, *4*, 535–547.

Lepper, C., Conway, S. J., & Fan, C.-M. (2009). Adult satellite cells and embryonic muscle progenitors have distinct genetic requirements. *Nature*, *460*, 627–631.

Lepper, C., Partridge, T. A., & Fan, C.-M. (2011). An absolute requirement for Pax7-positive satellite cells in acute injury-induced skeletal muscle regeneration. *Development (Cambridge, England)*, *138*, 3639–3646.

Li, L., Sarver, A. L., Alamgir, S., & Subramanian, S. (2012). Downregulation of microRNAs miR-1, -206 and -29 stabilizes PAX3 and CCND2 expression in rhabdomyosarcoma. *Laboratory Investigation*, *92*, 571–583.

Liu, Y., Chakroun, I., Yang, D., Horner, E., Liang, J., Aziz, A., et al. (2013). Six1 regulates MyoD expression in adult muscle progenitor cells. *PloS One*, *8*, e67762.

Mansouri, A., Stoykova, A., Torres, M., & Gruss, P. (1996). Dysgenesis of cephalic neural crest derivatives in Pax7 −/− mutant mice. *Development (Cambridge, England)*, *122*, 831–838.

Mauro, A. (1961). Satellite cell of skeletal muscle fibres. *The Journal of Biophysical and Biochemical Cytology, 9*, 493–495.
McKinnell, I. W., Ishibashi, J., Le Grand, F., Punch, V. G. J., Addicks, G. C., Greenblatt, J. F., et al. (2008). Pax7 activates myogenic genes by recruitment of a histone methyltransferase complex. *Nature Cell Biology, 10*, 77–84.
Megeney, L. A., Kablar, B., Garrett, K., Anderson, J. E., & Rudnicki, M. A. (1996). MyoD is required for myogenic stem cell function in adult skeletal muscle. *Genes & Development, 10*, 1173–1183.
Montarras, D., Morgan, J., Collins, C., Relaix, F., Zaffran, S., Cumano, A., et al. (2005). Direct isolation of satellite cells for skeletal muscle regeneration. *Science, 309*, 2064–2067.
Mounier, R., Lantier, L., Leclerc, J., Sotiropoulos, A., Pende, M., Daegelen, D., et al. (2009). Important role for AMPKalpha1 in limiting skeletal muscle cell hypertrophy. *The FASEB Journal, 23*, 2264–2273.
Mourikis, P., Sambasivan, R., Castel, D., Rocheteau, P., Bizzarro, V., & Tajbakhsh, S. (2012). A critical requirement for notch signaling in maintenance of the quiescent skeletal muscle stem cell state. *Stem Cells, 30*, 243–252.
Murphy, M. M., Lawson, J. A., Mathew, S. J., Hutcheson, D. A., & Kardon, G. (2011). Satellite cells, connective tissue fibroblasts and their interactions are crucial for muscle regeneration. *Development (Cambridge, England), 138*, 3625–3637.
Ono, Y., Masuda, S., Nam, H., Benezra, R., Miyagoe-Suzuki, Y., & Takeda, S. (2012). Slow-dividing satellite cells retain long-term self-renewal ability in adult muscle. *Journal of Cell Science, 125*, 1309–1317.
Oustanina, S., Hause, G., & Braun, T. (2004). Pax7 directs postnatal renewal and propagation of myogenic satellite cells but not their specification. *The EMBO Journal, 23*, 3430–3439.
Palacios, D., Mozzetta, C., Consalvi, S., Caretti, G., Saccone, V., Proserpio, V., et al. (2010). TNF/p38α/polycomb signaling to Pax7 locus in satellite cells links inflammation to the epigenetic control of muscle regeneration. *Cell Stem Cell, 7*, 455–469.
Pallafacchina, G., François, S., Regnault, B., Czarny, B., Dive, V., Cumano, A., et al. (2010). An adult tissue-specific stem cell in its niche: A gene profiling analysis of in vivo quiescent and activated muscle satellite cells. *Stem Cell Research, 4*, 77–91.
Parisi, A., Lacour, F., Giordani, L., Colnot, S., Maire, P., & Le Grand, F. (2015). APC is required for muscle stem cell proliferation and skeletal muscle tissue repair. *The Journal of Cell Biology, 210*, 717–726.
Pasut, A., Chang, N. C., Gurriaran-Rodriguez, U., Faulkes, S., Yin, H., Lacaria, M., et al. (2016). Notch signaling rescues loss of satellite cells lacking Pax7 and promotes brown adipogenic differentiation. *Cell Reports, 16*, 333–343.
Perdiguero, E., Ruiz-Bonilla, V., Gresh, L., Hui, L., Ballestar, E., Sousa-Victor, P., et al. (2007). Genetic analysis of p38 MAP kinases in myogenesis: Fundamental role of p38alpha in abrogating myoblast proliferation. *The EMBO Journal, 26*, 1245–1256.
Pondel, M. (2000). Calcitonin and calcitonin receptors: Bone and beyond. *International Journal of Experimental Pathology, 81*, 405–422.
Rampalli, S., Li, L., Mak, E., Ge, K., Brand, M., Tapscott, S. J., et al. (2007). p38 MAPK signaling regulates recruitment of Ash2L-containing methyltransferase complexes to specific genes during differentiation. *Nature Structural & Molecular Biology, 14*, 1150–1156.
Rathbone, C. R., Booth, F. W., & Lees, S. J. (2009). Sirt1 increases skeletal muscle precursor cell proliferation. *European Journal of Cell Biology, 88*, 35–44.
Relaix, F. (2015). Pax genes: Master regulators of development and tissue homeostasis. *Seminars in Cell & Developmental Biology, 44*, 62–63.
Rocheteau, P., Gayraud-Morel, B., Siegl-Cachedenier, I., Blasco, M. A., & Tajbakhsh, S. (2012). A subpopulation of adult skeletal muscle stem cells retains all template DNA strands after cell division. *Cell, 148*, 112–125.

Rodgers, J. T., King, K. Y., Brett, J. O., Cromie, M. J., Charville, G. W., Maguire, K. K., et al. (2014). mTORC1 controls the adaptive transition of quiescent stem cells from G0 to G(Alert). *Nature, 510*, 393–396.

Rozo, M., Li, L., & Fan, C.-M. (2016). Targeting β1-integrin signaling enhances regeneration in aged and dystrophic muscle in mice. *Nature Medicine, 22*, 889–896.

Rudnicki, M. A., Braun, T., Hinuma, S., & Jaenisch, R. (1992). Inactivation of MyoD in mice leads to up-regulation of the myogenic HLH gene Myf-5 and results in apparently normal muscle development. *Cell, 71*, 383–390.

Ryall, J. G., Dell'Orso, S., Derfoul, A., Juan, A., Zare, H., Feng, X., et al. (2015). The NAD (+)-dependent SIRT1 deacetylase translates a metabolic switch into regulatory epigenetics in skeletal muscle stem cells. *Cell Stem Cell, 16*, 171–183.

Sacco, A., Doyonnas, R., Kraft, P., Vitorovic, S., & Blau, H. M. (2008). Self-renewal and expansion of single transplanted muscle stem cells. *Nature, 456*, 502–506.

Sambasivan, R., Yao, R., Kissenpfennig, A., Van Wittenberghe, L., Paldi, A., Gayraud-Morel, B., et al. (2011). Pax7-expressing satellite cells are indispensable for adult skeletal muscle regeneration. *Development (Cambridge, England), 138*, 3647–3656.

Schultz, E., Chamberlain, C., McCormick, K. M., & Mozdziak, P. E. (2006). Satellite cells express distinct patterns of myogenic proteins in immature skeletal muscle. *Developmental Dynamics: An Official Publication of the American Association of the Anatomists, 235*, 3230–3239.

Schuster-Gossler, K., Cordes, R., & Gossler, A. (2007). Premature myogenic differentiation and depletion of progenitor cells cause severe muscle hypotrophy in Delta1 mutants. *Proceedings of the National Academy of Sciences of the United States of America, 104*, 537–542.

Seale, P., Sabourin, L. A., Girgis-Gabardo, A., Mansouri, A., Gruss, P., & Rudnicki, M. A. (2000). Pax7 is required for the specification of myogenic satellite cells. *Cell, 102*, 777–786.

Seenundun, S., Rampalli, S., Liu, Q.-C., Aziz, A., Palii, C., Hong, S., et al. (2010). UTX mediates demethylation of H3K27me3 at muscle-specific genes during myogenesis. *The EMBO Journal, 29*, 1401–1411.

Shan, T., Zhang, P., Liang, X., Bi, P., Yue, F., & Kuang, S. (2014). Lkb1 is indispensable for skeletal muscle development, regeneration, and satellite cell homeostasis. *Stem Cells, 32*, 2893–2907.

Shea, K. L., Xiang, W., LaPorta, V. S., Licht, J. D., Keller, C., Basson, M. A., et al. (2010). Sprouty1 regulates reversible quiescence of a self-renewing adult muscle stem cell pool during regeneration. *Cell Stem Cell, 6*, 117–129.

Shinin, V., Gayraud-Morel, B., Gomès, D., & Tajbakhsh, S. (2006). Asymmetric division and cosegregation of template DNA strands in adult muscle satellite cells. *Nature Cell Biology, 8*, 677–682.

Simone, C., Forcales, S. V., Hill, D. A., Imbalzano, A. N., Latella, L., & Puri, P. L. (2004). p38 pathway targets SWI-SNF chromatin-remodeling complex to muscle-specific loci. *Nature Genetics, 36*, 738–743.

Snow, M. H. (1977). Myogenic cell formation in regenerating rat skeletal muscle injured by mincing. II. An autoradiographic study. *The Anatomical Record, 188*, 201–217.

Soleimani, V. D., Punch, V. G., Kawabe, Y., Jones, A. E., Palidwor, G. A., Porter, C. J., et al. (2012). Transcriptional dominance of Pax7 in adult myogenesis is due to high-affinity recognition of homeodomain motifs. *Developmental Cell, 22*, 1208–1220.

Spitz, F., Demignon, J., Porteu, A., Kahn, A., Concordet, J. P., Daegelen, D., et al. (1998). Expression of myogenin during embryogenesis is controlled by six/sine oculis homeoproteins through a conserved MEF3 binding site. *Proceedings of the National Academy of Sciences of the United States of America, 95*, 14220–14225.

Stuelsatz, P., Pouzoulet, F., Lamarre, Y., Dargelos, E., Poussard, S., Leibovitch, S., et al. (2010). Down-regulation of MyoD by calpain 3 promotes generation of reserve cells in C2C12 myoblasts. *The Journal of Biological Chemistry, 285*, 12670–12683.

Sweetman, D., Goljanek, K., Rathjen, T., Oustanina, S., Braun, T., Dalmay, T., et al. (2008). Specific requirements of MRFs for the expression of muscle specific micro-RNAs, miR-1, miR-206 and miR-133. *Developmental Biology, 321*, 491–499.

Tajbakhsh, S. (2009). Skeletal muscle stem cells in developmental versus regenerative myogenesis. *Journal of Internal Medicine, 266*, 372–389.

Takaesu, G., Kang, J.-S., Bae, G.-U., Yi, M.-J., Lee, C. M., Reddy, E. P., et al. (2006). Activation of p38alpha/beta MAPK in myogenesis via binding of the scaffold protein JLP to the cell surface protein Cdo. *The Journal of Cell Biology, 175*, 383–388.

Theret, M., Gsaier, L., Schaffer, B., Juban, G., Ben Larbi, S., Weiss-Gayet, M., et al. (2017). AMPKα1-LDH pathway regulates muscle stem cell self-renewal by controlling metabolic homeostasis. *The EMBO Journal, 36*, 1946–1962.

Troy, A., Cadwallader, A. B., Fedorov, Y., Tyner, K., Tanaka, K. K., & Olwin, B. B. (2012). Coordination of satellite cell activation and self-renewal by par-complex-dependent asymmetric activation of p38α/β MAPK. *Cell Stem Cell, 11*, 541–553.

Ustanina, S., Carvajal, J., Rigby, P., & Braun, T. (2007). The myogenic factor Myf5 supports efficient skeletal muscle regeneration by enabling transient myoblast amplification. *Stem Cells, 25*, 2006–2016.

Vasyutina, E., Lenhard, D. C., Wende, H., Erdmann, B., Epstein, J. A., & Birchmeier, C. (2007). RBP-J (Rbpsuh) is essential to maintain muscle progenitor cells and to generate satellite cells. *Proceedings of the National Academy of Sciences of the United States of America, 104*, 4443–4448.

von Maltzahn, J., Jones, A. E., Parks, R. J., & Rudnicki, M. A. (2013). Pax7 is critical for the normal function of satellite cells in adult skeletal muscle. *Proceedings of the National Academy of Sciences of the United States of America, 110*, 16474–16479.

Wen, Y., Bi, P., Liu, W., Asakura, A., Keller, C., & Kuang, S. (2012). Constitutive Notch activation upregulates Pax7 and promotes the self-renewal of skeletal muscle satellite cells. *Molecular and Cellular Biology, 32*, 2300–2311.

Yajima, H., Motohashi, N., Ono, Y., Sato, S., Ikeda, K., Masuda, S., et al. (2010). Six family genes control the proliferation and differentiation of muscle satellite cells. *Experimental Cell Research, 316*, 2932–2944.

Yamaguchi, M., Watanabe, Y., Ohtani, T., Uezumi, A., Mikami, N., Nakamura, M., et al. (2015). Calcitonin receptor signaling inhibits muscle stem cells from escaping the quiescent state and the niche. *Cell Reports, 13*, 302–314.

Yennek, S., & Tajbakhsh, S. (2013). DNA asymmetry and cell fate regulation in stem cells. *Seminars in Cell & Developmental Biology, 24*, 627–642.

Zammit, P. S., Golding, J. P., Nagata, Y., Hudon, V., Partridge, T. A., & Beauchamp, J. R. (2004). Muscle satellite cells adopt divergent fates: A mechanism for self-renewal? *The Journal of Cell Biology, 166*, 347–357.

CHAPTER SEVEN

"Known Unknowns": Current Questions in Muscle Satellite Cell Biology

DDW Cornelison[1]
University of Missouri, Columbia, MO, United States
Christopher S. Bond Life Sciences Center, University of Missouri, Columbia, MO, United States
[1]Corresponding author: e-mail address: cornelisond@missouri.edu

Contents

Abstract

Our understanding of satellite cells, now known to be the obligate stem cells of skeletal muscle, has increased dramatically in recent years due to the introduction of new molecular, genetic, and technical resources. In addition to their role in acute repair of damaged muscle, satellite cells are of interest in the fields of aging, exercise, neuromuscular disease, and stem cell therapy, and all of these applications have driven a dramatic increase in our understanding of the activity and potential of satellite cells. However, many fundamental questions of satellite cell biology remain to be answered, including their emergence as a specific lineage, the degree and significance of heterogeneity within the satellite cell population, the roles of their interactions with other resident and infiltrating cell types during homeostasis and regeneration, and the relative roles of intrinsic vs extrinsic factors that may contribute to satellite cell dysfunction in the context of aging or disease. This review will address the current state of these open questions in satellite cell biology.

Current Topics in Developmental Biology, Volume 126
ISSN 0070-2153
https://doi.org/10.1016/bs.ctdb.2017.08.006
© 2018 Elsevier Inc.
All rights reserved.

1. INTRODUCTION

The muscle satellite cell was first described in 1961 by Alexander Mauro and by Bernard Katz as a small, lamina-bound cell adjacent to the sarcolemma of a mature frog extraocular muscle fiber (Katz, 1961; Mauro, 1961). Over the following 50 years, it has emerged as the obligate tissue-specific stem cell of skeletal muscle, with roles in postnatal growth, hypertrophy, homeostasis, and repair of this critical tissue. With the discovery and development of new molecular and genetic tools, we now understand a great deal more about the developmental origins, molecular programs, and physiological capabilities of satellite cells, yet in these and other key areas significant unanswered questions remain.

2. EMERGING DEFINITION

Skeletal muscle is a highly adaptive and resilient tissue, comprising greater than one-third of an adult human's body mass. Even after full mature growth is achieved, muscle cells will grow (hypertrophy) in response to increased physiological loading; it is also one of the most highly regenerative tissues in the body. Although satellite cells were first described over 50 years ago, it is only recently that they have been definitively shown to be necessary for skeletal muscle regeneration. Until the 1990s, satellite cells could be definitively identified in situ only in electron micrographs, on the basis of their position between the sarcolemma (cell membrane) and the external lamina of a mature myofiber, their sparse cytoplasm, and their heavily heterochomatinized nucleus (Katz, 1961; Mauro, 1961). However, this method is time-consuming, is necessarily limited to very few cells, and does not permit longitudinal or population-level analysis. Over the last two decades, identification of a growing suite of molecular markers including lineage determination factors (primarily transcriptional activators) and cell surface molecules has led to significant advances in prospective identification of satellite cells (reviewed in Yin, Price, & Rudnicki, 2013). The most useful marker for satellite cells to date has been the paired-box containing transcription factor Pax7 (Seale et al., 2000), which is ubiquitously expressed by both the quiescent satellite cell population and activated satellite cells that retain stem cell potential (Zammit et al., 2004).

While both Mauro and Katz originally hypothesized that satellite cells could potentially act to repair and regenerate damaged muscle, there was considerable doubt surrounding this idea. When satellite cells were noted in electron microscopy studies, they were considered to be simply myoblasts that had not incorporated into the muscle (Allbrook, 1962; Price, Howes, & Blumberg, 1964)—at that time, regeneration of muscle was thought to involve dedifferentiation and possibly replication of myofiber nuclei (Walker, 1963). Only when later experiments showed that satellite cells, but not myonuclei, proliferated in situ after injury (Reznik, 1969) and that cells associated with cultured muscle fibers proliferated then fused to produce new fibers (Bischoff, 1975; Konigsberg, Lipton, & Konigsberg, 1975) did the concept of satellite cells as the source of new muscle (the term "stem cell" had not yet been coined) gain acceptance. A complication in both identifying satellite cells and evaluating their myogenic progenitor status was their apparent lack of myogenic lineage-specifying transcription factors during quiescence: following the identification of MyoD as a "master" transcriptional regulator for the muscle lineage (Tapscott et al., 1988) and the subsequent identification of myf-5 (Braun, Buschhausen-Denker, Bober, Tannich, & Arnold, 1989), myogenin (Wright, Sassoon, & Lin, 1989), and MRF4/herculin/myf-6 (Braun, Bober, Winter, Rosenthal, & Arnold, 1990; Miner & Wold, 1990; Rhodes & Konieczny, 1989), it was generally thought that expression of one or more of these proteins (collectively termed myogenic-regulatory factors or MRFs) was necessary for myogenic identity, yet satellite cells did not appear to possess mRNAs for any MRFs until after activation (Cornelison & Wold, 1997; Grounds, Garrett, Lai, Wright, & Beilharz, 1992). While later work indicated that myf-5 protein is indeed present and functional in quiescent satellite cells (Gayraud-Morel et al., 2012), with cells expressing lower levels of myf-5 exhibiting more stem-like behavior, the difficulty of detecting MRF expression made them unsuitable as markers for satellite cells.

Later, as the concept of the satellite cell as an adult stem cell developed, expression of markers common to other adult stem cells (particularly hematopoietic stem cells) was pursued both to facilitate satellite cell identification and purification and to evaluate their stem cell character. As a result of these studies, multiple cell surface antigens have been identified that mark either the majority of satellite cells (CD34, integrins α7 and β1, nestin, and CXCR4) (Beauchamp et al., 2000; Blanco-Bose, Yao, Kramer, & Blau, 2001; Shefer, Wleklinski-Lee, & Yablonka-Reuveni, 2004; Sherwood et al., 2004) or specific subsets with potentially distinct capacities and fates (Jagged-1, Frizzled-7, Vangl-2, and Sca-1) (Gnocchi, White, Ono, Ellis, & Zammit, 2009; Le Grand, Jones, Seale, Scime, & Rudnicki, 2009;

Mitchell et al., 2005). A great deal of current research focuses on using these and other markers to discriminate between subpopulations of activated satellite cells that will either maintain stem cell identity (i.e., continue to meet the definition of a satellite cell) or commit to the myogenic program and eventual terminal differentiation (see Section 4).

In particular, the question of whether the classical criteria for an adult stem cell can or should be applied to satellite cells is remains open. By the classical definition developed for hematopoietic stem cells, a true adult stem cell must be able to completely replace both the total differentiated cell population of an organism and the total stem cell population. Hematopoietic stem cells are also defined as multipotent, as they are competent to produce all of the distinct cell types in the blood lineage, but since muscle consists of a single cell type that proceeds through a differentiation program, this aspect of the definition may not be relevant to satellite cells. Regarding self-renewal, stem cells are considered to have the potential to divide symmetrically to produce two new stem cells, symmetrically to produce two progenitor cells, or asymmetrically to produce one stem cell and one daughter cell that, as a progenitor or transit-amplifying cell, will contribute to new differentiated cells as needed in the tissue.

Under nonpathological conditions, at least some satellite cells present in vivo appear to be competent to proliferate and provide daughter cells that will differentiate into new muscle for the life of the organism, although the efficiency of the process may be compromised by physiological changes in the environment associated with aging (see Section 6). Satellite cells can increase their numbers very rapidly, such that the satellite cells associated with a single myofiber will, within a few days in culture, multiply until there are more than enough myoblasts to completely replace the myofiber nuclei (Zammit et al., 2002). Satellite cell proliferation is notable in chronic injury conditions as well: in one experiment, rat muscles were severely injured by injection of a toxic drug every week for 6 months (Sadeh, Czyewski, & Stern, 1985) for an estimated minimum of 80 doubling events per satellite cell. While the repeatedly regenerated muscle was not morphologically normal, still there was not a substantial decrease in myonuclei.

The condition of complete replacement of both the stem and differentiated cell pools by a single stem cell, as exemplified by the HSC, is more difficult with satellite cells due to their more constrained location (as opposed to being either circulating or at least exposed constantly to the circulation). However, several studies have been published in which, if not one, then a very small number of engrafted satellite cells have been shown to contribute efficiently to replacement of both the stem and progenitor/differentiated cell pools (Hall, Banks, Chamberlain, & Olwin, 2010; Sacco, Doyonnas, Kraft, Vitorovic, & Blau, 2008; Tanaka et al., 2009). The question of satellite cell long-term self-renewal, particularly in the context of stem cell–niche interactions, is as yet not well answered, although intense effort in the community is focused on this question.

Thus, the current definition of a satellite cell is a cell-expressing Pax7 with the ability to generate progeny that will differentiate to replace damaged muscle tissue, while also maintaining a proliferative, MRF^{low} reserve population that is able to reenter quiescence after the damage is resolved. The status of satellite cells as the sole, obligate stem cell responsible for skeletal muscle regeneration in the adult has been resolved in the past few years, as simultaneous but independent genetic ablation studies from three different groups showed that if the Pax7-positive cell population is genetically ablated in fully adult mice, the competence of the muscle to regenerate following injury is lost (Lepper, Partridge, & Fan, 2011; Murphy, Lawson, Mathew, Hutcheson, & Kardon, 2011; Sambasivan et al., 2011). However, the same experimental paradigm failed to show a requirement for satellite cells in muscle hypertrophy (McCarthy et al., 2011), highlighting the degree of uncertainty that remains regarding satellite cells' precise role and requirement in adult muscle (see Section 5).

3. ORIGINS AND DEVELOPMENT

Skeletal muscle of the trunk and limbs in vertebrates develops from transient mesodermal structures called somites, which form along either side of the developing neural tube (Buckingham et al., 2003; Pourquie, 2001). The somites are originally unspecified, but in response to local signaling cues they become subdivided into the sclerotome, which will give rise to the axial skeleton (ribs and vertebrae); the myotome, which will give rise to skeletal muscle; the dermatome, which will give rise to the mesodermal layer of the skin; and the syndetome, which will form the tendons connecting muscle and bone (Brand-Saberi, Wilting, Ebensperger, & Christ, 1996; Brent, Schweitzer, & Tabin, 2003; Dubrulle & Pourquie, 2004; Pourquie, 2001). Other muscle lineages, such as the muscles of the head and neck, are formed from the unsegmented paraxial mesoderm rostral to the first somite (Noden, 1983; Pownall, Gustafsson, & Emerson, 2002). It is now well accepted that satellite cells of the trunk and limbs are, like the myofibers they associate with, derived from the myotome of the developing somite during embryogenesis (Relaix, Rocancourt, Mansouri, & Buckingham, 2005; Schienda et al., 2006), while satellite cells associated with head muscles derived from the cranial paraxial mesoderm also originate in the paraxial mesoderm (Harel et al., 2009).

Once the myotome is specified, it is then subdivided into two domains (epaxial and hypaxial) that will experience different local signaling environments, leading to different myogenic gene expression programs and different fates. In response to localized signals from the dorsal neural tube and the epidermis, the epaxial myotome will upregulate myf-5 and will begin to differentiate within the somite to lay down the muscles of the deep back. In contrast, the cells of the hypaxial myotome will initially upregulate MyoD and undergo an epithelial-to-mesenchymal transition, migrating as single cells around the trunk and into the forming limb buds, where they will coalesce into the precursors of adult muscles (Tajbakhsh & Cossu, 1997). These two myogenic lineages are initially equivalent in potential but will become restricted in competence rapidly after specification (Ordahl & Le-Douarin, 1992; Williams & Ordahl, 1997).

Satellite cells are defined by their ongoing expression of Pax7 (Seale et al., 2000), which is necessary for maintenance of the satellite cell population in postnatal life (Kassar-Duchossoy et al., 2005; Oustanina, Hause, & Braun, 2004). However, the timing and the genetic and cellular mechanisms by which the satellite cell lineage is set aside from embryonic and fetal myoblasts, which also express Pax7, and/or insulated from prodifferentiation signals during fusion and growth of the musculature remain elusive.

4. HETEROGENEITY

One of the most hotly pursued questions in the field at this time is the degree, nature, and functional implications of heterogeneity in the satellite cell population, as well as the existence and potential of nonsatellite stem cell populations that may contribute to muscle regeneration in vivo. Fundamental to the concept of satellite cells as adult stem cells is the requirement that most or all quiescent satellite cells, and at least a fraction of activated satellite cells, maintain stem cell identity and function over the life of the animal. These cells would presumably be lineage-negative, capable of both self-renewal and generation of differentiation-competent myoblasts, exempt from the effects of replicative senescence in vivo and able to organize and interact with an extracellular niche. As discussed earlier, ideally they would share the property of hematopoietic stem cells of being able to both fully repopulate the stem cell compartment and provide enough differentiation-competent progeny to fully replace the original tissue. Such stem cells in other tissues are quite rare, usually comprising less than 0.1% of the total number of mononuclear cells.

Multiple cell populations have been identified in muscle that are *not* satellite cells as they have classically been defined, but may be able to participate in muscle regeneration (Fig. 1). In particular, the identification in the early 2000s of cells sharing marker expression and stem cell engraftment capacity with "side population" or SP hematopoietic stem cells (Goodell, Brose, Paradis, Conner, & Mulligan, 1996; Liadaki et al., 2005) that could differentiate into skeletal muscle (Montanaro, Liadaki, Volinski, Flint, & Kunkel, 2003) as well as other mesodermal derivatives (Challen & Little, 2006) raised the possibility that satellite cells are themselves derived from a circulating progenitor, potentially one with origins in the bone marrow. While this hypothesis had been already been tested and disproved before the advent of convenient molecular markers for satellite cells (Grounds, 1983), the field's technical abilities been significantly expanded and refined over the course of 20 years, and initial descriptions of this mesodermal progenitor cell were promising both scientifically and therapeutically. However, multiple groups later found that pure populations of SP cells are not myogenic in vitro, although they can participate in the formation of myotubes when

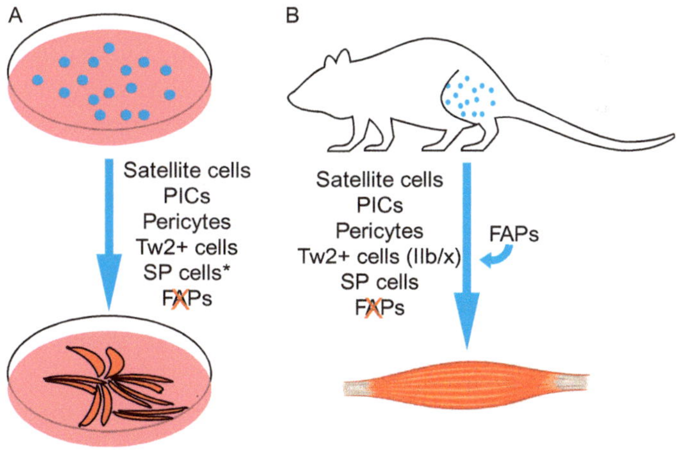

Fig. 1 In addition to satellite cells, several other cell types have been identified in the adult that are capable of myogenic differentiation in vitro (A; *asterisk*: SP cells require coculture with Pax7-expressing myoblasts) and/or in vivo (B). While FAPs do not directly incorporate into regenerated muscle, they are required for efficient satellite cell-mediated regeneration in vivo.

cocultured with Pax7-positive, *bona fide* satellite cells (Asakura, Seale, Girgis-Gabardo, & Rudnicki, 2002). An exception would be a subpopulation of conventional satellite cells (based on their expression of the cell-surface marker syndecan-4; Cornelison, Filla, Stanley, Rapraeger, & Olwin, 2001) that also expresses the SP marker ABCG2 (Tanaka et al., 2009). These cells actually have enhanced satellite stem cell characteristics, raising the possibility that these satellite-SP cells were present in early SP preparations from muscle and were responsible for the myogenic activity described.

Several populations of nonmuscle mesenchymal cells, such as PW1-positive, Pax7-negative interstitial cells (PICs) (Mitchell et al., 2010), and fibro–adipogenic progenitor cells (FAPs) (Joe et al., 2010) or muscle connective tissue fibroblasts (Mathew et al., 2011), are also present in skeletal muscle and appear to participate in muscle regeneration. PICs are defined by their expression of the p53-responsive transcription factor PW1, and in muscle they include the population of quiescent satellite cells ($Pax7^{+}$) and their descendants ($Pax7^{+}/MyoD^{+}$) as well as $Pax7^{-}$ interstitial cells that have both myogenic and adipogenic capacity (Pannerec, Formicola, Besson, Marazzi, & Sassoon, 2013). This is consistent with other data from the same group demonstrating that PW1 expression is a characteristic of all or most cycling and quiescent progenitor/stem cell populations (Besson et al., 2011). While not directly contributing to new muscle during regeneration in vivo, the contribution of FAPs to muscle regeneration and homeostasis is hypothesized to be as accessory cells that secrete trophic factors necessary for efficient and appropriate satellite cell activity (Paylor, Natarajan, Zhang, & Rossi, 2011). FAPs are also proposed to be responsive to satellite cell-derived soluble factors, such that if muscle regeneration is initiated but not successful FAPs will be recruited to "fill in" the gaps in the tissue with adipose and connective tissue (Natarajan, Lemos, & Rossi, 2010), which is characteristic of DMD muscle samples. Such cross-signaling is also apparent in the case of muscle connective tissue fibroblasts: genetic ablation of connective tissue fibroblasts, prior to acute injury leads to premature differentiation of satellite cell-derived myoblasts, resulting in smaller regenerated myofibers (Murphy et al., 2011). Because FAPs and muscle connective tissue fibroblasts were identified by different labs using different methodologies and functional definitions, they are described as unique cell types in the literature; however, overlap in marker expression and apparent function suggests that they may represent the same cellular population.

A nonmuscle population that is of particular interest due to its potential applications in cell-based therapies were originally termed mesangioblasts

(Ferrari et al., 1998) (see Section 8) and have more recently been identified as pericytes (Dellavalle et al., 2007). These blood vessel-associated cells are frequently found in close proximity to satellite cells (Christov et al., 2007) and can convert to a myogenic phenotype in vitro as well as in vivo in the context of acute or chronic injury or cell transplantation (Dellavalle et al., 2011). Significantly, these cells were also shown to contribute to the quiescent satellite cell pool in vivo, suggesting that they maintain stem cell potential in addition to myogenic differentiation.

A novel population of interstitial cells with the capacity to contribute to differentiated muscle in vivo has recently been described by the Olson group (Liu et al., 2017). These nonmuscle cells are marked by expression of the transcription factor *twist2* (Tw2) and will not only participate in myogenesis under in vitro and in vivo conditions but also appear to specifically contribute to fast (Type IIx/b) myofibers in vivo. Targeted ablation of the $Tw2^+$ population also leads to atrophy of Type IIx/b fibers, suggesting that these cells may specifically participate in the maintenance of fast myofibers during homeostasis.

Even among *bona fide* satellite cells, the intersection of intrinsic factors such as lineage and extrinsic influences such as niche signals generates significant potential for variation within the population (Fig. 2). Within the satellite cell population, significant heterogeneity in embryonic origin, lineage marker expression history, proliferation history, and gene expression have been noted in the literature (Biressi & Rando, 2010). For example, maintenance of Pax3 expression beyond development has been noted in satellite

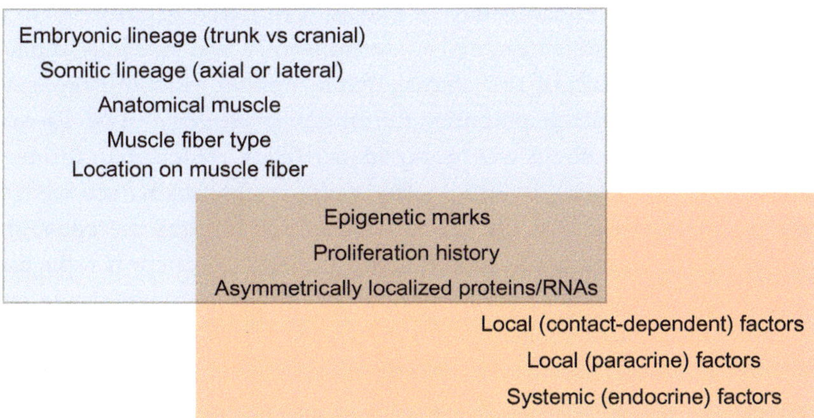

Fig. 2 The combined effects of inherited (*blue*) and environmental (*red*) determinants lead to intrinsic heterogeneity within and between satellite cell populations.

cells from the diaphragm, one hindlimb muscle, and ~50% of forelimb muscles (Relaix et al., 2006); interestingly, pericyte-derived myoblasts also tend to express Pax3 (Dellavalle et al., 2011). Differences in myogenic potential have also been noted between satellite cells associated with different muscles, most notably satellite cells of extraocular muscle vs trunk and limb muscles (Formicola, Marazzi, & Sassoon, 2014; Kallestad et al., 2011; Stuelsatz et al., 2015). Recent work also suggests that satellite cells associated with different lineages may be uniquely required in muscles from that lineage (Randolph et al., 2015).

Heterogeneity of potential and activity is also present even within satellite cells of the same muscle or resident on the same fiber; however, correlating different subpopulations with specific markers and defining the role of each have been problematic. The most frequently used subpopulation markers are again Pax7 and the myogenic-regulatory factors myf-5 and MyoD: MRF-negative cells in vitro tend to be both slower-cycling and more clonogenic than cells that have entered the myogenic program and possibly committed to differentiation by upregulating MRFs and concurrently downregulating Pax7 (Collins et al., 2005; Olguin & Pisconti, 2011). Using cell-surface markers and FACS followed by in vitro or in vivo analysis, several "recipes" for satellite cells with high engraftment and regeneration potential have been identified (Biressi & Rando, 2010; Tanaka et al., 2009). However, the extent to which the populations so isolated overlap, or the reason why expression of specific cell-surface molecules confers enhanced myogenic/stem cell potential, are not yet clearly understood.

An intriguing potential source of satellite cell heterogeneity lies in the requirement for an adult stem cell to continuously and faithfully replicate its genome for the life of the animal, while neither reaching replicative senescence or generating potential tumorigenic progeny. The so-called Immortal Strand hypothesis was proposed in 1975 as a solution to this problem (Cairns, 1975): if dividing stem cells preferentially retained the "template" DNA strand while the new daughter cell got the copy, then the potential for mutational errors that would lead to cancer is reduced or eliminated. While it is not a property of all satellite cells, a percentage ranging from 7% to 50% of satellite cells do appear to segregate their DNA nonrandomly (Conboy, Karasov, & Rando, 2007; Shinin, Gayraud-Morel, & Tajbakhsh, 2009). This property was later elegantly shown to correlate with asymmetric divisions generating a satellite stem cell and a committed progenitor cell, and to involve all of the chromatids during such divisions

(Rocheteau, Gayraud-Morel, Siegl-Cachedenier, Blasco, & Tajbakhsh, 2012) as would be expected of a mechanism to ensure genomic stability.

Because heterogeneity in the satellite cell population, and particularly asymmetry generated during cell division, is of such high interest, it should be noted that static preparations of satellite cells or muscle fibers should not be overinterpreted with respect to apparently asymmetric distribution of mRNA or protein. Timelapse video analysis of both monoculture and myofiber cultures shows that satellite cells engage in multiple close associations with other satellite cells that may or may not be related by recent lineage, and that in some cases these physical associations are extremely long-lasting (Siegel, Atchison, Fisher, Davis, & Cornelison, 2009; Siegel, Kuhlmann, & Cornelison, 2011). Thus, observation of two adjacent satellite cells on a myofiber with different staining patterns indicates heterogeneity but not necessarily asymmetry.

Functional heterogeneity of satellite cell populations can also occur due to factors other than cell lineage, such as the surrounding extracellular environment. While chronic conditions such as aging or muscle disease lead to multifactorial changes in the satellite cell extracellular environment and to modifications in satellite cell potential (see Sections 6 and 7), it has recently been shown that acute injury can also promote persistent changes in satellite cell activity at the population level. The Rando group has coined the term "G_{alert}" to describe the state of quiescent satellite cells that are "primed" to activate more rapidly due to an injury event distal to them in time and/or space (Rodgers et al., 2014). The population used to show this distinct state was satellite cells isolated from the uninjured contralateral legs of mice that had undergone a standard acute muscle injury consisting of injection of a small volume of 1.2% barium chloride into one TA (tibialis anterior) muscle. They found that the contralateral cells, while remaining quiescent until explanted into culture, exhibit changes in the kinetics of cell cycle entry (earlier), propensity to cycle (greater), cell size (larger), transcriptional activity and mitochondrial metabolism (both higher) when compared to quiescent satellite cells from an uninjured mouse, all of which contribute to enhanced regenerative potential. Not only do satellite cells enter the G_{alert} state in response to muscle injury, the transition can also be triggered by injuries to other tissues such as skin. Induction of the G_{alert} state was shown to require signaling through the HGF receptor c-met, and more recent work has identified HGF activator as a systemic, circulating signal that promotes the alert state in multiple stem cell populations (Rodgers, Schroeder, Ma, & Rando, 2017).

These data, in addition to suggesting that satellite cells can develop cellular "memory" based on their exposure to transient changes in the extracellular signaling environment, are also consistent with the experience of many in the field that satellite cells from the contralateral limb in an acute injury are not ideal as a control because they do not behave the same as cells from an uninjured animal. This transient and systemic change in satellite cell potential due to a distant injury may be the only source of acquired, acute heterogeneity between satellite cell populations or it may not; in either case one should bear in mind that even if the age, sex, and genotype of experimental animals are identical, there may yet be functional differences in the satellite cells due to different treatments, experimental or otherwise.

5. EXERCISE AND ADAPTATION

Skeletal muscle fibers are syncytial cells, with many myonuclei serving a large volume of cytoplasm. Myonuclei are arranged across the myofiber, possibly to optimize the amount of cytoplasm near each myonucleus (Bruusgaard, Liestol, Ekmark, Kollstad, & Gundersen, 2003). It is well established that satellite cells (cells of the Pax7 + lineage) are required during early postnatal growth to supply new myonuclei to hypertrophying myofibers (Moss & Leblond, 1970), and that blocking satellite cell proliferation or fusion during this timeframe will result in decreased myofiber size (Schiaffino, Dyar, Ciciliot, Blaauw, & Sandri, 2013). Therefore, when myofibers hypertrophy due to exercise, presumably at some point additional nuclei would be needed to serve the expanded cytoplasm, and these nuclei would derive from activated satellite cells (Pallafacchina, Blaauw, & Schiaffino, 2012). In addition, exercise is thought to produce slight damage in myofibers, which would theoretically be sufficient to activate local satellite cells (Darr & Schultz, 1987).

The contribution and requirement of satellite cells for muscle hypertrophy following exercise have been difficult to assess, however, due to technical difficulties tracking satellite cell incorporation and inducing hypertrophy in the absence of myofiber damage. Experiments using the same satellite cell ablation paradigm used to establish a requirement for satellite cells in muscle regeneration showed no similar requirement for hypertrophy (Jackson et al., 2012; McCarthy et al., 2011); however, this result is balanced by multiple studies that see significant expansion of the satellite cell pool with exercise (Cermak et al., 2013; Shefer, Rauner, Stuelsatz, Benayahu, & Yablonka-Reuveni, 2013; Snijders et al., 2012) as well as more recent work which suggests that satellite cell recruitment is essential for increases in myofiber cross-sectional area (Egner, Bruusgaard, & Gundersen, 2016). More recently, in vivo imaging has also suggested a requirement for satellite cells in hypertrophy (Egner et al., 2016). Signaling pathways that are consistent with active, satellite cell-mediated myogenesis are also induced during exercise (Leiter, Peeler, & Anderson, 2011; Paulsen, Mikkelsen, Raastad, & Peake, 2012). A distinction between these opposing experimental results may therefore be the question of whether satellite cells are absolutely required for exercise-induced hypertrophy, vs being routinely involved when they are available ideally, this question will be resolved by future studies.

6. AGING

Loss of muscle mass and strength (sarcopenia) is one of the primary adverse physiological consequences of aging and accounts for ~1.5% of total healthcare expenditures in the United States (Janssen, Shepard, Katzmarzyk, & Roubenoff, 2004). Many of the molecular mechanisms implicated in muscle loss with aging are the same ones involved in muscle loss due to disuse (atrophy); however, the capacity of aged muscle to *restore* lost muscle mass and function is even more dramatically decreased. While it had been proposed for many years that this could be due to an age-dependent decline in satellite cell number, the relatively small and highly variable decreases in the size of the quiescent satellite cell pool across multiple muscle types and species would not be enough to account for the loss of regenerative capacity (Brack & Rando, 2007). This would seem to imply an age-associated defect in satellite cell *function*, yet identifying a cell-autonomous defect in aged satellite cells has been elusive. Instead, the majority of factors that have been demonstrated to affect satellite cell function in young vs aged animals have been properties not of the satellite cells themselves, but of their extracellular environment.

Early experiments by Gutmann and Carlson in the 1970s made the surprising observation that when muscles from old rats were transplanted into young animals, or vice versa, the speed and efficiency of regeneration were determined by the host environment rather than the donor muscle (Gutmann & Carlson, 1976). This agrees with several later in vivo results using slightly different methods and measurements (Carlson & Faulkner, 1989; Grounds, 1987). A similar dominant effect of the host environment was seen when isolated murine satellite cells were stimulated in vitro with crushed muscle extract derived from either young or old mice, or with conditioned medium from myotubes differentiated in vitro from cells of young or old mice. In each case, paracrine stimulation by young muscle extract or conditioned medium efficiently stimulated proliferation in both young and old myoblasts, while old extract or conditioned medium failed to do so for either young or old myoblasts (Mezzogiorno, Coletta, Zani, Cossu, & Molinaro, 1993). A more dramatic demonstration of this effect was seen when Rando's group created parabiosed pairs of mice (two mice sharing a circulatory system), then injured one mouse in each pair. While pairs of old mice and pairs of young mice regenerated as would be expected based on their age, in young:old pairs where the old mouse was injured, satellite cells proliferated and muscle was regenerated similar to what was seen if the young mouse of the pair was injured, which was still less than what would be seen in a young:young pair (Conboy et al., 2005). These results suggested that a titratable factor is present in the serum of young animals that conveys a highly regenerative phenotype on the satellite cells exposed to it. The molecular basis for this effect was proposed to be decreased levels of circulating Notch ligand in aged serum, possibly contributing to repressive chromatin-level changes that inhibit satellite cell activation (Conboy, Conboy, Smythe, & Rando, 2003; Conboy et al., 2005). More recent work from this group has confirmed that quiescent satellite cells from old mice do indeed carry histone marks characteristic of Polycomb group-mediated repressive methylation (Liu et al., 2013).

Other extracellular signaling pathways have been shown to alter aged satellite cell response to injury, and the majority of them are derived from non-muscle sources such as neurons and their associated cells, inflammatory cells, connective tissue, or the extracellular matrix (Grounds, 1998). Age-associated changes in signaling between the environment and satellite cells can generally be divided into a decrease in promyogenic signals such as IGF-1 (Goldspink & Harridge, 2004) or an increase in antimyogenic/profibrotic signals such as Wnt3A (Brack et al., 2007) or TGF-β (Carlson et al., 2009). Inflammation markers that are more prevalent during aging also appear to decrease the satellite cell response: elevated IL-6 levels in aged serum are described as "blunting" the satellite cell response to injury (McKay et al., 2013), and proinflammatory cytokines such as IL-1α and TNF- α block myogenic differentiation (Degens, 2007; Trendelenburg, Meyer, Jacobi, Feige, & Glass, 2012).

Most recently, FGF2, which has been a candidate for dysregulated signaling in aging muscle for some time (Grounds, 1998), has emerged as a highly promising axis for potential therapeutic manipulation. The Brack group has shown that expression of Sprouty1, a negative regulator of FGF signaling, is required for activated satellite cells to reenter a quiescent state (Shea et al., 2010). They also found that in aged muscle, increased FGF2 released by host myofibers appears to either promote inappropriate satellite cell activation in the absence of damage or prevent a return to quiescence following damage, both of which lead to a decrease in the satellite stem cell pool (Chakkalakal, Jones, Basson, & Brack, 2012). Concurrently, the Olwin group showed that injection of a single myofiber (with its associated satellite cells) would lead to not only participation of those satellite cells in acute muscle repair but a decrease in muscle atrophy with age and a near-lifelong enhancement of muscle mass, stem cell number, and force generation (Hall et al., 2010). These surprising results were later shown to be a response to FGF signaling (Bernet et al., 2014).

Conflicting reports have been published recently regarding the requirement for satellite cells during adult homeostasis. Genetic labeling and ablation experiments done in the Kardon lab (Keefe et al., 2015) show that satellite cells contribute to mature myofibers during adulthood to a greater or lesser extent depending on the muscle (as measured at 20 months of age), and that in the case of at least one muscle (the EDL) this contribution is necessary to maintain myofiber cross-sectional area. However, similar experiments in the Peterson lab (Fry et al., 2015) found no such requirement for satellite cells during adult homeostasis in the absence of injury (as measured at 24 months of age). More recently, in vivo imaging has also suggested a requirement for satellite cells in hypertrophy (Egner et al., 2016). It is likely that these disparate results are due to differences in the experimental design such as efficiency of the recombination driver line, and future work in aging models will hopefully reconcile the two datasets and generate a more complete model of satellite cell activity during aging.

7. DISEASE

The most intense clinical focus on satellite cells is in the context of Duchenne's muscular dystrophy, a neuromuscular disease caused by mutations in the *dystrophin* gene that lead to poor integrity of the myofiber plasma membrane and basal lamina (Blau, Webster, & Pavlath, 1983; Grounds, 2008; Hoffman, Brown, & Kunkel, 1987; Kunkel & Hoffman, 1989). This in turn leads to chronic damage and repair of the muscle by satellite cells, which eventually becomes inefficient and subsequently fails, leading to death (generally from respiratory and/or cardiac dysfunction). A confounding aspect of the condition has been that, although there are numerous satellite cell phenotypes described in animal models and patients consistent with chronic satellite cell activation and muscle remodeling (Decary et al., 2000; Kottlors & Kirschner, 2010; Morgan & Zammit, 2010; Sacco et al., 2010; Serrano et al., 2011), satellite cells themselves have not been thought to express dystrophin. Thus, the extent to which this deficit is an inherent or accumulated defect in the satellite cells themselves, vs a response to a pathological environment, has been difficult to ascertain.

One possible cell-intrinsic effect is suggested by studies showing that in young human DMD patients, satellite cell telomeres have been shortened nearly 15-fold more than in normal muscles (Decary et al., 2000). This would be consistent with a dramatically increased demand on the satellite cells, as they try to keep up with the ongoing muscle damage and repair that is characteristic of the disease. Similar telomere shortening is seen in some extreme cases of endurance athletes, reinforcing the hypothesis that unusually high requirements for muscle regeneration place a replicative strain on satellite cells (Collins et al., 2003). Since telomeres are required for DNA replication during cell division, this premature and accelerated shortening (187 bp/year vs 13 bp/year) could result in early replicative senescence of DMD satellite cells.

Recent work from the Rudnicki lab to address the question of whether there is a satellite cell phenotype directly related to the loss of dystrophin has produced the surprising result that, unlike most myoblasts, satellite cells do express dystrophin protein and localize it asymmetrically (Dumont et al., 2015). Further, in the absence of dystrophin other proteins whose asymmetric localization drives asymmetric cell divisions are not sequestered appropriately, and cells attempting to undergo asymmetric division instead undergo mitotic crisis. This exciting new result may solve the longstanding puzzle of satellite cell-intrinsic defects in individuals lacking dystrophin, but it also raises the concern that therapies must now be designed to take both the differentiated tissue and the stem cell into account.

Acute inflammation generally accompanies muscle damage and repair, and signals derived from monocytes and macrophages are critical for successful satellite cell-mediated muscle regeneration (reviewed in Saclier, Cuvellier, Magnan, Mounier, & Chazaud, 2013; Tidball, 2011), but these signals are altered or dysregulated in the context of chronic muscle regeneration. Because both chronic inflammation and accumulating fibrosis are hallmarks of dystrophic muscle, the progressively more pathological environment DMD satellite cells are exposed to would also be expected to alter their physiology (Acharyya et al., 2010; Onofre-Oliveira, Santos, Martins, Ayub-Guerrieri, & Vainzof, 2012; Schuierer, Mann, Bildsoe, Huxley, & Hughes, 2005). As has been observed in aging, the extracellular environment appears to have a dominant effect over cell-intrinsic defects (if any) due to the lack of functional dystrophin protein, at least in the mouse (Schuierer et al., 2005) and canine (Berg, Beffa, Cook, & Cornelison, 2011) disease models. Accordingly, many of the treatments proposed to ameliorate the severity of the disease in humans are designed to decrease inflammation, fibrosis, or both (Eghtesad, Jhunjhunwala, Little, & Clemens, 2011; Fairclough, Perkins, & Davies, 2012; Grounds & Torrisi, 2004; Li, Mittal, Makonchuk, Bhatnagar, & Kumar, 2009; Ruegg, 2013; Vidal et al., 2012).

8. THERAPY

In theory, satellite cells would be an ideal vector for cell-based therapies in both neuromuscular diseases such as Duchenne's muscular dystrophy and, potentially, other genetic conditions caused by lack of a functional protein. They can be isolated and expanded then (if they are autologous, or derived from the patient) made to express the therapeutic protein via viral transduction or other methods. When reinjected into the patient, the satellite cells would fuse with existing muscle fibers; because the myofibers are syncytial, it would be unnecessary to correct *all* myonuclei, merely a sufficient amount to restore adequate dystrophin expression across the myofiber (which may be as little as 30% of the normal levels) (Neri et al., 2007). Proof-of-principle was provided in 1978 when it was shown that donor myoblasts could be injected into a host, where they would fuse with existing myofibers (Partridge, Grounds, & Sloper, 1978). Later work demonstrated that if dystrophin-expressing wild-type myoblasts were injected into a dystrophic host, some host fibers would later stain positive for dystrophin (Partridge, Morgan, Coulton, Hoffman, & Kunkel, 1989).

Unfortunately, in practice satellite cell-based therapies have proven to be much more problematic than they are in principle: many clinical trials in the 1990s showed neither substantially increased dystrophin expression nor functional improvement in patients (Briggs & Morgan, 2013). The major sources of failure have been cell death due to host immune response, cryptic changes in extracted cells during culture that make them less able to survive and engraft as stem cells, and lack of dispersal of injected satellite cells (Peault et al., 2007). To overcome these obstacles, studies have been designed to improve host immunosuppression; isolate and purify smaller, more efficient cell populations; and use large numbers of spaced injection sites to spread the donor cells (Briggs & Morgan, 2013). Some success has been seen with these modifications: a recent phase I clinical trial for DMD including continuous immunosuppression and several hundred injections into a single muscle (spaced 1–2 mm apart) resulted in about 30% of host myofibers staining positive for dystrophin (Skuk et al., 2006). However, the positive results were in only one patient out of nine, in only the small area of muscle that was injected, and required long-term immunosuppression even after the procedure. Currently, the most successful cell-based therapies in animal models involve nondonor-derived, nonsatellite mesenchymal stem cells termed mesangioblasts (Galvez et al., 2006; Sampaolesi et al., 2006); a human clinical study by the same group is currently underway (Tedesco, Dellavalle, Diaz-Manera, Messina, & Cossu, 2010); however, preliminary results highlight unanticipated difficulties that may impede future applications of this approach.

As an alternative to cell-based therapies, other groups have attempted to introduce functional dystrophin into patient's satellite cells, in the hope that as they fuse with dystrophic muscle, their "good" dystrophin protein will be expressed throughout the myofiber. A major difficulty in this approach is the extremely large size of the *dystrophin* gene itself: at approximately 2.2 Mb (0.07% of the entire human genome) it is one of the largest human genes (Hoffman et al., 1987). The fully processed *dystrophin* mRNA is approximately 14 kb, which makes it much too large to package into any existing transfer or expression vector. Thus, candidate therapies aimed at inserting an exogenous copy of the coding sequence into existing muscle tissue and resident satellite cells are done using various truncated minidystrophins (Li et al., 2005; Odom, Gregorevic, Allen, & Chamberlain, 2011; Ragot et al., 1993; Wang, Li, & Xiao, 2000) or microdystrophins (Kimura, Li, Gregorevic, Fall, & Chamberlain, 2010; Shin et al., 2013; Yoshimura et al., 2004). These have shown some promise in animal models but have not yet moved into clinical trials. An additional approach that has shown promise is antisense oligonucleotide-mediated exon skipping, which would also generate incomplete but functional dystrophin proteins by preventing inclusion of exons harboring mutations that generate nonfunctional dystrophin (Dunckley, Manoharan, Villiet, Eperon, & Dickson, 1998; Lu et al., 2003; van Deutekom et al., 2001), although to date it has only been targeted at differentiated myofibers rather than satellite cells. More recently, the potential for genome editing with TALENs or Crispr–Cas9 to directly correct mutations in *dystrophin* in germ cells, induced pluripotent stem cells, or potentially in satellite cells themselves in vivo (Li et al., 2015; Long et al., 2014; Ousterout et al., 2015) has opened a promising new avenue to potential therapies.

9. OF MICE AND MEN

A significant impetus for trying to understand satellite cell function and dysfunction at the molecular and cellular level is obviously the goal of improving human health, yet the vast majority of studies described here and elsewhere are, necessarily, done on animal models such as mouse. As is usual in studies done in model systems, the prevailing assumption is that the genes, pathways, and cellular processes uncovered in model organisms, particularly mammals, will parallel those in humans. However, there are multiple caveats to this approach (Boldrin, Muntoni, & Morgan, 2010).

One area in which mice are a particularly poor model for human muscle function is the case of dystrophin deficiency (DMD vs mdx) (Banks & Chamberlain, 2008; Vainzof et al., 2008). While mice and humans are both mammals, mice weigh 20–30 g and have a lifespan of ~2 years at most, so neither the physical nor the temporal forces that would dictate key aspects of the human satellite cell response are present. Mice, unlike humans, are also commonly analyzed as inbred strains, which can possess different intrinsic characteristics that would be expected to modify experimental results (Grounds & McGeachie, 1989; Irintchev & Wernig, 1987). Mice also have higher expression of the utrophin protein, which may compensate for a defective dystrophin (Pons, Robert, Marini, & Leger, 1994), and their satellite cells (like all mouse cells) have telomeres that are ~50-fold longer than a human's (Prowse & Greider, 1995). Evidence for each of these as a disease-modifying factor in mouse comes from mutant strains carrying multiple mutations: mdx/utrophin knockout mice (Grady et al., 1997) or mdx/telomerase knockout mice (Sacco et al., 2010) have a much more severe phenotype, approaching that of the human disease state. Particularly in the case of the telomerase knockout, these results support a role for replicative senescence of satellite cells as a contributing factor to disease pathology.

10. DIRECTIONS

This brief overview is intended to highlight the areas in which open questions in satellite cell biology are currently known. While many frontiers remain, advances in the field have accelerated continuously over the past 50, 20, and 5 years—as this trend continues, the questions of origin, potential, and heterogeneity; role in exercise, aging, and disease; and applications to therapy will lead to more answers in the coming years, and present additional questions.

Acknowledgment

D.D.W.C. is funded by NIH Grants AR056814 and AR067450.

REFERENCES

Acharyya, S., Sharma, S. M., Cheng, A. S., Ladner, K. J., He, W., Kline, W., et al. (2010). TNF inhibits notch-1 in skeletal muscle cells by Ezh2 and DNA methylation mediated repression: Implications in duchenne muscular dystrophy. *PLoS One*, *5*, e12479.

Allbrook, D. (1962). An electron microscopic study of regenerating skeletal muscle. *Journal of Anatomy*, *96*, 137–152.

Asakura, A., Seale, P., Girgis-Gabardo, A., & Rudnicki, M. A. (2002). Myogenic specification of side population cells in skeletal muscle. *The Journal of Cell Biology*, *159*, 123–134.

Banks, G. B., & Chamberlain, J. S. (2008). The value of mammalian models for duchenne muscular dystrophy in developing therapeutic strategies. *Current Topics in Developmental Biology*, *84*, 431–453.

Beauchamp, J. R., Heslop, L., Yu, D. S., Tajbakhsh, S., Kelly, R. G., Wernig, A., et al. (2000). Expression of CD34 and Myf5 defines the majority of quiescent adult skeletal muscle satellite cells. *The Journal of Cell Biology*, *151*, 1221–1234.

Berg, Z., Beffa, L. R., Cook, D. P., & Cornelison, DDW (2011). Muscle satellite cells from GRMD dystrophic dogs are not phenotypically distinguishable from wild type satellite cells in ex vivo culture. *Neuromuscular Disorders*, *21*, 282–290.

Bernet, J. D., Doles, J. D., Hall, J. K., Kelly Tanaka, K., Carter, T. A., & Olwin, B. B. (2014). p38 MAPK signaling underlies a cell-autonomous loss of stem cell self-renewal in skeletal muscle of aged mice. *Nature Medicine*, *20*, 265–271.

Besson, V., Smeriglio, P., Wegener, A., Relaix, F., Nait Oumesmar, B., Sassoon, D. A., et al. (2011). PW1 gene/paternally expressed gene 3 (PW1/Peg3) identifies multiple adult stem and progenitor cell populations. *Proceedings of the National Academy of Sciences of the United States of America*, *108*, 11470–11475.

Biressi, S., & Rando, T. A. (2010). Heterogeneity in the muscle satellite cell population. *Seminars in Cell & Developmental Biology*, *21*, 845–854.

Bischoff, R. (1975). Regeneration of single skeletal muscle fibers in vitro. *The Anatomical Record*, *182*, 215–235.

Blanco-Bose, W. E., Yao, C. C., Kramer, R. H., & Blau, H. M. (2001). Purification of mouse primary myoblasts based on alpha 7 integrin expression. *Experimental Cell Research*, *265*, 212–220.

Blau, H. M., Webster, C., & Pavlath, G. K. (1983). Defective myoblasts identified in Duchenne muscular dystrophy. *Proceedings of the National Academy of Sciences of the United States of America*, *80*, 4856–4860.

Boldrin, L., Muntoni, F., & Morgan, J. E. (2010). Are human and mouse satellite cells really the same? *The Journal of Histochemistry and Cytochemistry*, *58*, 941–955.

Brack, A. S., Conboy, M. J., Roy, S., Lee, M., Kuo, C. J., Keller, C., et al. (2007). Increased Wnt signaling during aging alters muscle stem cell fate and increases fibrosis. *Science*, *317*, 807–810.

Brack, A. S., & Rando, T. A. (2007). Intrinsic changes and extrinsic influences of myogenic stem cell function during aging. *Stem Cell Reviews*, *3*, 226–237.
Brand-Saberi, B., Wilting, J., Ebensperger, C., & Christ, B. (1996). The formation of somite compartments in the avian embryo. *International Journal of Developmental Biology*, *40*, 411–420.
Braun, T., Bober, E., Winter, B., Rosenthal, N., & Arnold, H. H. (1990). Myf-6, a new member of the human gene family of myogenic determination factors: Evidence for a gene cluster on chromosome 12. *The EMBO Journal*, *9*, 821–831.
Braun, T., Buschhausen-Denker, G., Bober, E., Tannich, E., & Arnold, H. H. (1989). A novel human muscle factor related to but distinct from MyoD1 induces myogenic conversion in 10T1/2 fibroblasts. *The EMBO Journal*, *8*, 701–709.
Brent, A. E., Schweitzer, R., & Tabin, C. J. (2003). A somitic compartment of tendon progenitors. *Cell*, *113*, 235–248.
Briggs, D., & Morgan, J. E. (2013). Recent progress in satellite cell/myoblast engraftment—Relevance for therapy. *The FEBS Journal*, *280*, 4281–4293.
Bruusgaard, J. C., Liestol, K., Ekmark, M., Kollstad, K., & Gundersen, K. (2003). Number and spatial distribution of nuclei in the muscle fibres of normal mice studied in vivo. *The Journal of Physiology*, *551*, 467–478.
Buckingham, M., Bajard, L., Chang, T., Daubas, P., Hadchouel, J., Meilhac, S., et al. (2003). The formation of skeletal muscle: From somite to limb. *Journal of Anatomy*, *202*, 59–68.
Cairns, J. (1975). Mutation selection and the natural history of cancer. *Nature*, *255*, 197–200.
Carlson, M. E., Conboy, M. J., Hsu, M., Barchas, L., Jeong, J., Agrawal, A., et al. (2009). Relative roles of TGF-beta1 and Wnt in the systemic regulation and aging of satellite cell responses. *Aging Cell*, *8*, 676–689.
Carlson, B. M., & Faulkner, J. A. (1989). Muscle transplantation between young and old rats: Age of host determines recovery. *The American Journal of Physiology*, *256*, C1262–6.
Cermak, N. M., Snijders, T., McKay, B. R., Parise, G., Verdijk, L. B., Tarnopolsky, M. A., et al. (2013). Eccentric exercise increases satellite cell content in type II muscle fibers. *Medicine and Science in Sports and Exercise*, *45*, 230–237.
Chakkalakal, J. V., Jones, K. M., Basson, M. A., & Brack, A. S. (2012). The aged niche disrupts muscle stem cell quiescence. *Nature*, *490*, 355–360.
Challen, G. A., & Little, M. H. (2006). A side order of stem cells: The SP phenotype. *Stem Cells*, *24*, 3–12.
Christov, C., Chretien, F., Abou-Khalil, R., Bassez, G., Vallet, G., Authier, F. J., et al. (2007). Muscle satellite cells and endothelial cells: Close neighbors and privileged partners. *Molecular Biology of the Cell*, *18*, 1397–1409.
Collins, C. A., Olsen, I., Zammit, P. S., Heslop, L., Petrie, A., Partridge, T. A., et al. (2005). Stem cell function, self-renewal, and behavioral heterogeneity of cells from the adult muscle satellite cell niche. *Cell*, *122*, 289–301.
Collins, M., Renault, V., Grobler, L. A., St. Clair Gibson, A., Lambert, M. I., Wayne Derman, E., et al. (2003). Athletes with exercise-associated fatigue have abnormally short muscle DNA telomeres. *Medicine and Science in Sports and Exercise*, *35*, 1524–1528.
Conboy, I. M., Conboy, M. J., Smythe, G. M., & Rando, T. A. (2003). Notch-mediated restoration of regenerative potential to aged muscle. *Science*, *302*, 1575–1577.
Conboy, I. M., Conboy, M. J., Wagers, A. J., Girma, E. R., Weissman, I. L., & Rando, T. A. (2005). Rejuvenation of aged progenitor cells by exposure to a young systemic environment. *Nature*, *433*, 760–764.
Conboy, M. J., Karasov, A. O., & Rando, T. A. (2007). High incidence of non-random template strand segregation and asymmetric fate determination in dividing stem cells and their progeny. *PLoS Biology*, *5*, e102.

Cornelison, DDW, Filla, M. S., Stanley, H. M., Rapraeger, A. C., & Olwin, B. B. (2001). Syndecan-3 and syndecan-4 specifically mark skeletal muscle satellite cells and are implicated in satellite cell maintenance and muscle regeneration. *Developmental Biology, 239*, 79–94.

Cornelison, DDW, & Wold, B. J. (1997). Single-cell analysis of regulatory gene expression in quiescent and activated mouse skeletal muscle satellite cells. *Developmental Biology, 191*, 270–283.

Darr, K. C., & Schultz, E. (1987). Exercise-induced satellite cell activation in growing and mature skeletal muscle. *Journal of Applied Physiology, 63*, 1816–1821.

Decary, S., Hamida, C. B., Mouly, V., Barbet, J. P., Hentati, F., & Butler-Browne, G. S. (2000). Shorter telomeres in dystrophic muscle consistent with extensive regeneration in young children. *Neuromuscular Disorders, 10*, 113–120.

Degens, H. (2007). Age-related skeletal muscle dysfunction: Causes and mechanisms. *Journal of Musculoskeletal & Neuronal Interactions*, 7, 246–252.

Dellavalle, A., Maroli, G., Covarello, D., Azzoni, E., Innocenzi, A., Perani, L., et al. (2011). Pericytes resident in postnatal skeletal muscle differentiate into muscle fibres and generate satellite cells. *Nature Communications, 2*, 499.

Dellavalle, A., Sampaolesi, M., Tonlorenzi, R., Tagliafico, E., Sacchetti, B., Perani, L., et al. (2007). Pericytes of human skeletal muscle are myogenic precursors distinct from satellite cells. *Nature Cell Biology, 9*, 255–267.

Dubrulle, J., & Pourquie, O. (2004). Coupling segmentation to axis formation. *Development, 131*, 5783–5793.

Dumont, N. A., Wang, Y. X., von Maltzahn, J., Pasut, A., Bentzinger, C. F., Brun, C. E., et al. (2015). Dystrophin expression in muscle stem cells regulates their polarity and asymmetric division. *Nature Medicine, 21*, 1455–1463.

Dunckley, M. G., Manoharan, M., Villiet, P., Eperon, I. C., & Dickson, G. (1998). Modification of splicing in the dystrophin gene in cultured Mdx muscle cells by antisense oligoribonucleotides. *Human Molecular Genetics*, 7, 1083–1090.

Eghtesad, S., Jhunjhunwala, S., Little, S. R., & Clemens, P. R. (2011). Rapamycin ameliorates dystrophic phenotype in mdx mouse skeletal muscle. *Molecular Medicine, 17*, 917–924.

Egner, I. M., Bruusgaard, J. C., & Gundersen, K. (2016). Satellite cell depletion prevents fiber hypertrophy in skeletal muscle. *Development, 143*, 2898–2906.

Fairclough, R. J., Perkins, K. J., & Davies, K. E. (2012). Pharmacologically targeting the primary defect and downstream pathology in Duchenne muscular dystrophy. *Current Gene Therapy, 12*, 206–244.

Ferrari, G., Cusella-De Angelis, G., Coletta, M., Paolucci, E., Stornaiuolo, A., Cossu, G., et al. (1998). Muscle regeneration by bone marrow-derived myogenic progenitors. *Science, 279*, 1528–1530.

Formicola, L., Marazzi, G., & Sassoon, D. A. (2014). The extraocular muscle stem cell niche is resistant to ageing and disease. *Frontiers in Aging Neuroscience, 6*, 328.

Fry, C. S., Lee, J. D., Mula, J., Kirby, T. J., Jackson, J. R., Liu, F., et al. (2015). Inducible depletion of satellite cells in adult, sedentary mice impairs muscle regenerative capacity without affecting sarcopenia. *Nature Medicine, 21*, 76–80.

Galvez, B. G., Sampaolesi, M., Brunelli, S., Covarello, D., Gavina, M., Rossi, B., et al. (2006). Complete repair of dystrophic skeletal muscle by mesoangioblasts with enhanced migration ability. *The Journal of Cell Biology, 174*, 231–243.

Gayraud-Morel, B., Chretien, F., Jory, A., Sambasivan, R., Negroni, E., Flamant, P., et al. (2012). Myf5 haploinsufficiency reveals distinct cell fate potentials for adult skeletal muscle stem cells. *Journal of Cell Science, 125*, 1738–1749.

Gnocchi, V. F., White, R. B., Ono, Y., Ellis, J. A., & Zammit, P. S. (2009). Further characterisation of the molecular signature of quiescent and activated mouse muscle satellite cells. *PLoS One, 4*, e5205.

Goldspink, G., & Harridge, S. D. (2004). Growth factors and muscle ageing. *Experimental Gerontology*, *39*, 1433–1438.
Goodell, M. A., Brose, K., Paradis, G., Conner, A. S., & Mulligan, R. C. (1996). Isolation and functional properties of murine hematopoietic stem cells that are replicating in vivo. *The Journal of Experimental Medicine*, *183*, 1797–1806.
Grady, R. M., Teng, H., Nichol, M. C., Cunningham, J. C., Wilkinson, R. S., & Sanes, J. R. (1997). Skeletal and cardiac myopathies in mice lacking utrophin and dystrophin: A model for Duchenne muscular dystrophy. *Cell*, *90*, 729–738.
Grounds, M. D. (1983). Skeletal muscle precursors do not arise from bone marrow cells. *Cell and Tissue Research*, *234*, 713–722.
Grounds, M. D. (1987). Phagocytosis of necrotic muscle in muscle isografts is influenced by the strain, age, and sex of host mice. *The Journal of Pathology*, *153*, 71–82.
Grounds, M. D. (1998). Age-associated changes in the response of skeletal muscle cells to exercise and regeneration. *Annals of the New York Academy of Sciences*, *854*, 78–91.
Grounds, M. D. (2008). Two-tiered hypotheses for Duchenne muscular dystrophy. *Cellular and Molecular Life Sciences*, *65*, 1621–1625.
Grounds, M. D., Garrett, K. L., Lai, M. C., Wright, W. E., & Beilharz, M. W. (1992). Identification of skeletal muscle precursor cells in vivo by use of MyoD1 and myogenin probes. *Cell and Tissue Research*, *267*, 99–104.
Grounds, M. D., & McGeachie, J. K. (1989). A comparison of muscle precursor replication in crush-injured skeletal muscle of Swiss and BALBc mice. *Cell and Tissue Research*, *255*, 385–391.
Grounds, M. D., & Torrisi, J. (2004). Anti-TNFalpha (Remicade) therapy protects dystrophic skeletal muscle from necrosis. *The FASEB Journal*, *18*, 676–682.
Gutmann, E., & Carlson, B. M. (1976). Regeneration and transplantation of muscles in old rats and between young and old rats. *Life Sciences*, *18*, 109–114.
Hall, J. K., Banks, G. B., Chamberlain, J. S., & Olwin, B. B. (2010). Prevention of muscle aging by myofiber-associated satellite cell transplantation. *Science Translational Medicine*, *2*, 57ra83.
Harel, I., Nathan, E., Tirosh-Finkel, L., Zigdon, H., Guimaraes-Camboa, N., Evans, S. M., et al. (2009). Distinct origins and genetic programs of head muscle satellite cells. *Developmental Cell*, *16*, 822–832.
Hoffman, E. P., Brown, R. H. J., & Kunkel, L. M. (1987). Dystrophin: The protein product of the duchenne muscular dystrophy locus. *Cell*, *51*, 919–928.
Irintchev, A., & Wernig, A. (1987). Muscle damage and repair in voluntarily running mice: Strain and muscle differences. *Cell and Tissue Research*, *249*, 509–521.
Jackson, J. R., Mula, J., Kirby, T. J., Fry, C. S., Lee, J. D., Ubele, M. F., et al. (2012). Satellite cell depletion does not inhibit adult skeletal muscle regrowth following unloading-induced atrophy. *American Journal of Physiology Cell Physiology*, *303*, C854–61.
Janssen, I., Shepard, D. S., Katzmarzyk, P. T., & Roubenoff, R. (2004). The healthcare costs of sarcopenia in the United States. *Journal of the American Geriatrics Society*, *52*, 80–85.
Joe, A. W., Yi, L., Natarajan, A., Le Grand, F., So, L., Wang, J., et al. (2010). Muscle injury activates resident fibro/adipogenic progenitors that facilitate myogenesis. *Nature Cell Biology*, *12*, 153–163.
Kallestad, K. M., Hebert, S. L., McDonald, A. A., Daniel, M. L., Cu, S. R., & McLoon, L. K. (2011). Sparing of extraocular muscle in aging and muscular dystrophies: A myogenic precursor cell hypothesis. *Experimental Cell Research*, *317*, 873–885.
Kassar-Duchossoy, L., Giacone, E., Gayraud-Morel, B., Jory, A., Gomes, D., & Tajbakhsh, S. (2005). Pax3/Pax7 mark a novel population of primitive myogenic cells during development. *Genes & Development*, *19*, 1426–1431.

Katz, F. R. S. (1961). The termination of the afferent nerve fiber in the muscle spindle of the frog. *Philosophical Transactions of the Royal Society of London Series B, Biological Sciences, 243*, 221–225.

Keefe, A. C., Lawson, J. A., Flygare, S. D., Fox, Z. D., Colasanto, M. P., Mathew, S. J., et al. (2015). Muscle stem cells contribute to myofibres in sedentary adult mice. *Nature Communications, 6*, 7087.

Kimura, E., Li, S., Gregorevic, P., Fall, B. M., & Chamberlain, J. S. (2010). Dystrophin delivery to muscles of mdx mice using lentiviral vectors leads to myogenic progenitor targeting and stable gene expression. *Molecular Therapy, 18*, 206–213.

Konigsberg, U. R., Lipton, R. H., & Konigsberg, I. R. (1975). The regenerative response of single mature muscle fibers isolated in vitro. *Developmental Biology, 45*, 260–275.

Kottlors, M., & Kirschner, J. (2010). Elevated satellite cell number in Duchenne muscular dystrophy. *Cell and Tissue Research, 340*, 541–548.

Kunkel, L. M., & Hoffman, E. P. (1989). Duchenne/Becker muscular dystrophy: A short overview of the gene, the protein, and current diagnostics. *British Medical Bulletin, 45*, 630–643.

Le Grand, F., Jones, A. E., Seale, V., Scime, A., & Rudnicki, M. A. (2009). Wnt7a activates the planar cell polarity pathway to drive the symmetric expansion of satellite stem cells. *Cell Stem Cell, 4*, 535–547.

Leiter, J. R., Peeler, J., & Anderson, J. E. (2011). Exercise-induced muscle growth is muscle-specific and age-dependent. *Muscle & Nerve, 43*, 828–838.

Lepper, C., Partridge, T. A., & Fan, C. M. (2011). An absolute requirement for Pax7-positive satellite cells in acute injury-induced skeletal muscle regeneration. *Development, 138*, 3639–3646.

Li, H. L., Fujimoto, N., Sasakawa, N., Shirai, S., Ohkame, T., Sakuma, T., et al. (2015). Precise correction of the dystrophin gene in duchenne muscular dystrophy patient induced pluripotent stem cells by TALEN and CRISPR-Cas9. *Stem Cell Reports, 4*, 143–154.

Li, S., Kimura, E., Fall, B. M., Reyes, M., Angello, J. C., Welikson, R., et al. (2005). Stable transduction of myogenic cells with lentiviral vectors expressing a minidystrophin. *Gene Therapy, 12*, 1099–1108.

Li, H., Mittal, A., Makonchuk, D. Y., Bhatnagar, S., & Kumar, A. (2009). Matrix metalloproteinase-9 inhibition ameliorates pathogenesis and improves skeletal muscle regeneration in muscular dystrophy. *Human Molecular Genetics, 18*, 2584–2598.

Liadaki, K., Kho, A. T., Sanoudou, D., Schienda, J., Flint, A., Beggs, A. H., et al. (2005). Side population cells isolated from different tissues share transcriptome signatures and express tissue-specific markers. *Experimental Cell Research, 303*, 360–374.

Liu, L., Cheung, T. H., Charville, G. W., Hurgo, B. M., Leavitt, T., Shih, J., et al. (2013). Chromatin modifications as determinants of muscle stem cell quiescence and chronological aging. *Cell Reports, 4*, 189–204.

Liu, N., Garry, G. A., Li, S., Bezprozvannaya, S., Sanchez-Ortiz, E., Chen, B., et al. (2017). A Twist2-dependent progenitor cell contributes to adult skeletal muscle. *Nature Cell Biology, 19*, 202–213.

Long, C., McAnally, J. R., Shelton, J. M., Mireault, A. A., Bassel-Duby, R., & Olson, E. N. (2014). Prevention of muscular dystrophy in mice by CRISPR/Cas9-mediated editing of germline DNA. *Science, 345*, 1184–1188.

Lu, Q. L., Mann, C. J., Lou, F., Bou-Gharios, G., Morris, G. E., Xue, S. A., et al. (2003). Functional amounts of dystrophin produced by skipping the mutated exon in the mdx dystrophic mouse. *Nature Medicine, 9*, 1009–1014.

Mathew, S. J., Hansen, J. M., Merrell, A. J., Murphy, M. M., Lawson, J. A., Hutcheson, D. A., et al. (2011). Connective tissue fibroblasts and Tcf4 regulate myogenesis. *Development, 138*, 371–384.

Mauro, A. (1961). Satellite cells of skeletal muscle fibers. *The Journal of Biophysical and Biochemical Cytology, 9*, 493–495.
McCarthy, J. J., Mula, J., Miyazaki, M., Erfani, R., Garrison, K., Farooqui, A. B., et al. (2011). Effective fiber hypertrophy in satellite cell-depleted skeletal muscle. *Development, 138*, 3657–3666.
McKay, B. R., Ogborn, D. I., Baker, J. M., Toth, K. G., Tarnopolsky, M. A., & Parise, G. (2013). Elevated SOCS3 and altered IL-6 signaling is associated with age-related human muscle stem cell dysfunction. *American Journal of Physiology. Cell Physiology, 304*, C717–28.
Mezzogiorno, A., Coletta, M., Zani, B. M., Cossu, G., & Molinaro, M. (1993). Paracrine stimulation of senescent satellite cell proliferation by factors released by muscle or myotubes from young mice. *Mechanisms of Ageing and Development, 70*, 35–44.
Miner, J. H., & Wold, B. (1990). Herculin, a fourth member of the MyoD family of myogenic regulatory genes. *Proceedings of the National Academy of Sciences of the United States of America, 87*, 1089–1093.
Mitchell, P. O., Mills, T., O'Connor, R. S., Kline, E. R., Graubert, T., Dzierzak, E., et al. (2005). Sca-1 negatively regulates proliferation and differentiation of muscle cells. *Developmental Biology, 283*, 240–252.
Mitchell, K. J., Pannerec, A., Cadot, B., Parlakian, A., Besson, V., Gomes, E. R., et al. (2010). Identification and characterization of a non-satellite cell muscle resident progenitor during postnatal development. *Nature Cell Biology, 12*, 257–266.
Montanaro, F., Liadaki, K., Volinski, J., Flint, A., & Kunkel, L. M. (2003). Skeletal muscle engraftment potential of adult mouse skin side population cells. *Proceedings of the National Academy of Sciences of the United States of America, 100*, 9336–9341.
Morgan, J. E., & Zammit, P. S. (2010). Direct effects of the pathogenic mutation on satellite cell function in muscular dystrophy. *Experimental Cell Research, 316*, 3100–3108.
Moss, F. P., & Leblond, C. P. (1970). Satellite cells as the source of nuclei in muscles of growing rats. *The Anatomical Record, 170*, 421–436.
Murphy, M. M., Lawson, J. A., Mathew, S. J., Hutcheson, D. A., & Kardon, G. (2011). Satellite cells, connective tissue fibroblasts and their interactions are crucial for muscle regeneration. *Development, 138*, 3625–3637.
Natarajan, A., Lemos, D. R., & Rossi, F. M. (2010). Fibro/adipogenic progenitors: A double-edged sword in skeletal muscle regeneration. *Cell Cycle, 9*, 2045–2046.
Neri, M., Torelli, S., Brown, S., Ugo, I., Sabatelli, P., Merlini, L., et al. (2007). Dystrophin levels as low as 30% are sufficient to avoid muscular dystrophy in the human. *Neuromuscular Disorders, 17*, 913–918.
Noden, D. M. (1983). The embryonic origins of avian cephalic and cervical muscles and associated connective tissues. *The American Journal of Anatomy, 168*, 257–276.
Odom, G. L., Gregorevic, P., Allen, J. M., & Chamberlain, J. S. (2011). Gene therapy of mdx mice with large truncated dystrophins generated by recombination using rAAV6. *Molecular Therapy, 19*, 36–45.
Olguin, H. C., & Pisconti, A. (2011). Marking the tempo for myogenesis: Pax7 and the regulation of muscle stem cell fate decisions. *Journal of Cellular and Molecular Medicine*, 1013–1025.
Onofre-Oliveira, P. C., Santos, A. L., Martins, P. M., Ayub-Guerrieri, D., & Vainzof, M. (2012). Differential expression of genes involved in the degeneration and regeneration pathways in mouse models for muscular dystrophies. *Neuromolecular Medicine, 14*, 74–83.
Ordahl, C. P., & Le-Douarin, N. M. (1992). Two myogenic lineages within the developing somite. *Development, 114*, 339–353.
Oustanina, S., Hause, G., & Braun, T. (2004). Pax7 directs postnatal renewal and propagation of myogenic satellite cells but not their specification. *The EMBO Journal, 23*, 3430–3439.
Ousterout, D. G., Kabadi, A. M., Thakore, P. I., Majoros, W. H., Reddy, T. E., & Gersbach, C. A. (2015). Multiplex CRISPR/Cas9-based genome editing for correction

of dystrophin mutations that cause Duchenne muscular dystrophy. *Nature Communications, 6*, 6244.
Pallafacchina, G., Blaauw, B., & Schiaffino, S. (2012). Role of satellite cells in muscle growth and maintenance of muscle mass. *Nutrition, Metabolism, and Cardiovascular Diseases, 23*(Suppl. 1), S12–S18.
Pannerec, A., Formicola, L., Besson, V., Marazzi, G., & Sassoon, D. A. (2013). Defining skeletal muscle resident progenitors and their cell fate potentials. *Development, 140*, 2879–2891.
Partridge, T. A., Grounds, M., & Sloper, J. C. (1978). Evidence of fusion between host and donor myoblasts in skeletal muscle grafts. *Nature, 273*, 306–308.
Partridge, T. A., Morgan, J. E., Coulton, G. R., Hoffman, E. P., & Kunkel, L. M. (1989). Conversion of mdx myofibres from dystrophin-negative to -positive by injection of normal myoblasts. *Nature, 337*, 176–179.
Paulsen, G., Mikkelsen, U. R., Raastad, T., & Peake, J. M. (2012). Leucocytes, cytokines and satellite cells: What role do they play in muscle damage and regeneration following eccentric exercise? *Exercise Immunology Review, 18*, 42–97.
Paylor, B., Natarajan, A., Zhang, R. H., & Rossi, F. (2011). Nonmyogenic cells in skeletal muscle regeneration. *Current Topics in Developmental Biology, 96*, 139–165.
Peault, B., Rudnicki, M., Torrente, Y., Cossu, G., Tremblay, J. P., Partridge, T., et al. (2007). Stem and progenitor cells in skeletal muscle development, maintenance, and therapy. *Molecular Therapy, 15*, 867–877.
Pons, F., Robert, A., Marini, J. F., & Leger, J. J. (1994). Does utrophin expression in muscles of mdx mice during postnatal development functionally compensate for dystrophin deficiency. *Journal of the Neurological Sciences, 122*, 162–170.
Pourquie, O. (2001). Vertebrate somitogenesis. *Annual Review of Cell and Developmental Biology, 17*, 311–350.
Pownall, M. E., Gustafsson, M. K., & Emerson, C. P. J. (2002). Myogenic regulatory factors and the specification of muscle progenitors in vertebrate embryos. *Annual Review of Cell and Developmental Biology, 18*, 747–783.
Price, H. M., Howes, E. L. J., & Blumberg, J. M. (1964). Ultrastructural modifications in skeletal muscle fibers injured by cold. II. Cells on the sarcolemmal tube: Observations on "discontinuous" regeneration and myofibril formation. *Laboratory Investigation, 13*, 1279–1302.
Prowse, K. R., & Greider, C. W. (1995). Developmental and tissue-specific regulation of mouse telomerase and telomere length. *Proceedings of the National Academy of Sciences of the United States of America, 92*, 4818–4822.
Ragot, T., Vincent, N., Chafey, P., Vigne, E., Gilgenkrantz, H., Couton, D., et al. (1993). Efficient adenovirus-mediated transfer of a human minidystrophin gene to skeletal muscle of mdx mice. *Nature, 361*, 647–650.
Randolph, M. E., Phillips, B. L., Choo, H. J., Vest, K. E., Vera, Y., & Pavlath, G. K. (2015). Pharyngeal satellite cells undergo myogenesis under basal conditions and are required for pharyngeal muscle maintenance. *Stem Cells, 33*, 3581–3595.
Relaix, F., Montarras, D., Zaffran, S., Gayraud-Morel, B., Rocancourt, D., Tajbakhsh, S., et al. (2006). Pax3 and Pax7 have distinct and overlapping functions in adult muscle progenitor cells. *The Journal of Cell Biology, 172*, 91–102.
Relaix, F., Rocancourt, D., Mansouri, A., & Buckingham, M. (2005). A Pax3/Pax7-dependent population of skeletal muscle progenitor cells. *Nature, 435*, 948–953.
Reznik, M. (1969). Thymidine-3H uptake by satellite cells of regenerating skeletal muscle. *The Journal of Cell Biology, 40*, 568–571.
Rhodes, S. J., & Konieczny, S. F. (1989). Identification of MRF4: A new member of the muscle regulatory factor gene family. *Genes & Development, 3*, 2050–2061.

Rocheteau, P., Gayraud-Morel, B., Siegl-Cachedenier, I., Blasco, M. A., & Tajbakhsh, S. (2012). A subpopulation of adult skeletal muscle stem cells retains all template DNA strands after cell division. *Cell*, *148*, 112–125.

Rodgers, J. T., King, K. Y., Brett, J. O., Cromie, M. J., Charville, G. W., Maguire, K. K., et al. (2014). mTORC1 controls the adaptive transition of quiescent stem cells from G0 to G(Alert). *Nature*, *510*, 393–396.

Rodgers, J. T., Schroeder, M. D., Ma, C., & Rando, T. A. (2017). HGFA is an injury-regulated systemic factor that induces the transition of stem cells into GAlert. *Cell Reports*, *19*, 479–486.

Ruegg, U. T. (2013). Pharmacological prospects in the treatment of Duchenne muscular dystrophy. *Current Opinion in Neurology*, *26*, 577–584.

Sacco, A., Doyonnas, R., Kraft, P., Vitorovic, S., & Blau, H. M. (2008). Self-renewal and expansion of single transplanted muscle stem cells. *Nature*, *456*, 502–506.

Sacco, A., Mourkioti, F., Tran, R., Choi, J., Llewellyn, M., Kraft, P., et al. (2010). Short telomeres and stem cell exhaustion model Duchenne muscular dystrophy in mdx/mTR mice. *Cell*, *143*, 1059–1071.

Saclier, M., Cuvellier, S., Magnan, M., Mounier, R., & Chazaud, B. (2013). Monocyte/macrophage interactions with myogenic precursor cells during skeletal muscle regeneration. *The FEBS Journal*, *280*, 4118–4130.

Sadeh, M., Czyewski, K., & Stern, L. Z. (1985). Chronic myopathy induced by repeated bupivacaine injections. *Journal of the Neurological Sciences*, *67*, 229–238.

Sambasivan, R., Yao, R., Kissenpfennig, A., Van Wittenberghe, L., Paldi, A., Gayraud-Morel, B., et al. (2011). Pax7-expressing satellite cells are indispensable for adult skeletal muscle regeneration. *Development*, *138*, 3647–3656.

Sampaolesi, M., Blot, S., D'Antona, G., Granger, N., Tonlorenzi, R., Innocenzi, A., et al. (2006). Mesoangioblast stem cells ameliorate muscle function in dystrophic dogs. *Nature*, *444*, 574–579.

Schiaffino, S., Dyar, K. A., Ciciliot, S., Blaauw, B., & Sandri, M. (2013). Mechanisms regulating skeletal muscle growth and atrophy. *The FEBS Journal*, *280*, 4294–4314.

Schienda, J., Engleka, K. A., Jun, S., Hansen, M. S., Epstein, J. A., Tabin, C. J., et al. (2006). Somitic origin of limb muscle satellite and side population cells. *Proceedings of the National Academy of Sciences of the United States of America*, *103*, 945–950.

Schuierer, M. M., Mann, C. J., Bildsoe, H., Huxley, C., & Hughes, S. M. (2005). Analyses of the differentiation potential of satellite cells from myoD-/-, mdx, and PMP22 C22 mice. *BMC Musculoskeletal Disorders*, *6*, 15.

Seale, P., Sabourin, L. A., Girgis-Gabardo, A., Mansouri, A., Gruss, P., & Rudnicki, M. A. (2000). Pax7 is required for the specification of myogenic satellite cells. *Cell*, *102*, 777–786.

Serrano, A. L., Mann, C. J., Vidal, B., Ardite, E., Perdiguero, E., & Munoz-Canoves, P. (2011). Cellular and molecular mechanisms regulating fibrosis in skeletal muscle repair and disease. *Current Topics in Developmental Biology*, *96*, 167–201.

Shea, K. L., Xiang, W., LaPorta, V. S., Licht, J. D., Keller, C., Basson, M. A., et al. (2010). Sprouty1 regulates reversible quiescence of a self-renewing adult muscle stem cell pool during regeneration. *Cell Stem Cell*, *6*, 117–129.

Shefer, G., Rauner, G., Stuelsatz, P., Benayahu, D., & Yablonka-Reuveni, Z. (2013). Moderate-intensity treadmill running promotes expansion of the satellite cell pool in young and old mice. *The FEBS Journal*, 4063–4073.

Shefer, G., Wleklinski-Lee, M., & Yablonka-Reuveni, Z. (2004). Skeletal muscle satellite cells can spontaneously enter an alternative mesenchymal pathway. *Journal of Cell Science*, *117*, 5393–5404.

Sherwood, R. I., Christensen, J. L., Conboy, I. M., Conboy, M. J., Rando, T. A., Weissman, I. L., et al. (2004). Isolation of adult mouse myogenic progenitors: Functional heterogeneity of cells within and engrafting skeletal muscle. *Cell, 119*, 543–554.
Shin, J. H., Pan, X., Hakim, C. H., Yang, H. T., Yue, Y., Zhang, K., et al. (2013). Micro-dystrophin ameliorates muscular dystrophy in the canine model of duchenne muscular dystrophy. *Molecular Therapy, 21*, 750–757.
Shinin, V., Gayraud-Morel, B., & Tajbakhsh, S. (2009). Template DNA-strand co-segregation and asymmetric cell division in skeletal muscle stem cells. *Methods in Molecular Biology, 482*, 295–317.
Siegel, A. L., Atchison, K., Fisher, K. E., Davis, G. E., & Cornelison, DDW (2009). 3D timelapse analysis of muscle satellite cell motility. *Stem Cells, 27*, 2527–2538.
Siegel, A. L., Kuhlmann, P. K., & Cornelison, DDW (2011). Muscle satellite cell proliferation and association: New insights from myofiber time-lapse imaging. *Skeletal Muscle, 1*, 7.
Skuk, D., Goulet, M., Roy, B., Chapdelaine, P., Bouchard, J. P., Roy, R., et al. (2006). Dystrophin expression in muscles of duchenne muscular dystrophy patients after high-density injections of normal myogenic cells. *Journal of Neuropathology and Experimental Neurology, 65*, 371–386.
Snijders, T., Verdijk, L. B., Beelen, M., McKay, B. R., Parise, G., Kadi, F., et al. (2012). A single bout of exercise activates skeletal muscle satellite cells during subsequent overnight recovery. *Experimental Physiology, 97*, 762–773.
Stuelsatz, P., Shearer, A., Li, Y., Muir, L. A., Ieronimakis, N., Shen, Q. W., et al. (2015). Extraocular muscle satellite cells are high performance myo-engines retaining efficient regenerative capacity in dystrophin deficiency. *Developmental Biology, 397*, 31–44.
Tajbakhsh, S., & Cossu, G. (1997). Establishing myogenic identity during somitogenesis. *Current Opinion in Genetics & Development, 7*, 634–641.
Tanaka, K. K., Hall, J. K., Troy, A. A., Cornelison, DDW, Majka, S. M., & Olwin, B. B. (2009). Syndecan-4-expressing muscle progenitor cells in the SP engraft as satellite cells during muscle regeneration. *Cell Stem Cell, 4*, 217–225.
Tapscott, S. J., Davis, R. L., Thayer, M. J., Cheng, P. F., Weintraub, H., & Lassar, A. B. (1988). MyoD1: A nuclear phosphoprotein requiring a Myc homology region to convert fibroblasts to myoblasts. *Science, 242*, 405–411.
Tedesco, F. S., Dellavalle, A., Diaz-Manera, J., Messina, G., & Cossu, G. (2010). Repairing skeletal muscle: Regenerative potential of skeletal muscle stem cells. *The Journal of Clinical Investigation, 120*, 11–19.
Tidball, J. G. (2011). Mechanisms of muscle injury, repair, and regeneration. *Comprehensive Physiology, 1*, 2029–2062.
Trendelenburg, A. U., Meyer, A., Jacobi, C., Feige, J. N., & Glass, D. J. (2012). TAK-1/p38/nNFkappaB signaling inhibits myoblast differentiation by increasing levels of Activin A. *Skeletal Muscle, 2*, 3.
Vainzof, M., Ayub-Guerrieri, D., Onofre, P. C., Martins, P. C., Lopes, V. F., Zilberztajn, D., et al. (2008). Animal models for genetic neuromuscular diseases. *Journal of Molecular Neuroscience, 34*, 241–248.
van Deutekom, J. C., Bremmer-Bout, M., Janson, A. A., Ginjaar, I. B., Baas, F., den Dunnen, J. T., et al. (2001). Antisense-induced exon skipping restores dystrophin expression in DMD patient derived muscle cells. *Human Molecular Genetics, 10*, 1547–1554.
Vidal, B., Ardite, E., Suelves, M., Ruiz-Bonilla, V., Janue, A., Flick, M. J., et al. (2012). Amelioration of Duchenne muscular dystrophy in mdx mice by elimination of matrix-associated fibrin-driven inflammation coupled to the alphaMbeta2 leukocyte integrin receptor. *Human Molecular Genetics, 21*, 1989–2004.

Walker, B. E. (1963). The origin of myoblasts and the problem of dedifferentiation. *Experimental Cell Research*, *30*, 80–92.

Wang, B., Li, J., & Xiao, X. (2000). Adeno-associated virus vector carrying human minidystrophin genes effectively ameliorates muscular dystrophy in mdx mouse model. *Proceedings of the National Academy of Sciences of the United States of America*, *97*, 13714–13719.

Williams, B. A., & Ordahl, C. P. (1997). Emergence of determined myotome precursor cells in the somite. *Development*, *124*, 4983–4997.

Wright, W. E., Sassoon, D. A., & Lin, V. K. (1989). Myogenin, a factor regulating myogenesis, has a domain homologous to MyoD. *Cell*, *56*, 607–617.

Yin, H., Price, F., & Rudnicki, M. A. (2013). Satellite cells and the muscle stem cell niche. *Physiological Reviews*, *93*, 23–67.

Yoshimura, M., Sakamoto, M., Ikemoto, M., Mochizuki, Y., Yuasa, K., Miyagoe-Suzuki,- Y., et al. (2004). AAV vector-mediated microdystrophin expression in a relatively small percentage of mdx myofibers improved the mdx phenotype. *Molecular Therapy*, *10*, 821–828.

Zammit, P. S., Golding, J. P., Nagata, Y., Hudon, V., Partridge, T. A., & Beauchamp, J. R. (2004). Muscle satellite cells adopt divergent fates: A mechanism for self-renewal? *The Journal of Cell Biology*, *166*, 347–357.

Zammit, P. S., Heslop, L., Hudon, V., Rosenblatt, J. D., Tajbakhsh, S., Buckingham, M. E., et al. (2002). Kinetics of myoblast proliferation show that resident satellite cells are competent to fully regenerate skeletal muscle fibers. *Experimental Cell Research*, *281*, 39–49.

CHAPTER EIGHT

Epigenetic Regulation of Adult Myogenesis

Daniel C.L. Robinson*,†, Francis J. Dilworth*,†,1
*Sprott Centre for Stem Cell Research, Regenerative Medicine Program, Ottawa Hospital Research Institute, Ottawa, ON, Canada
†University of Ottawa, Ottawa, ON, Canada
[1]Corresponding author: e-mail address: jdilworth@ohri.ca

Contents

Abstract

Skeletal muscle regeneration is an efficient stem cell-based repair system that ensures healthy musculature. For this repair system to function continuously throughout life, muscle stem cells must contribute to the process of myofiber repair as well as repopulation of the stem cell niche. The decision made by the muscle stem cells to

Current Topics in Developmental Biology, Volume 126
ISSN 0070-2153
https://doi.org/10.1016/bs.ctdb.2017.08.002
© 2018 Elsevier Inc.
All rights reserved.

commit to the muscle repair or to remain a stem cell depends upon patterns of gene expression, a process regulated at the epigenetic level. Indeed, it is well accepted that dynamic changes in epigenetic landscapes to control DNA accessibility and expression is a critical component during myogenesis for the effective repair of damaged muscle. Changes in the epigenetic landscape are governed by various posttranslational histone tail modifications, nucleosome repositioning, and DNA methylation events which collectively allow the control of changes in transcription networks during transitions of satellite cells from a dormant quiescent state toward terminal differentiation. This chapter focuses upon the specific epigenetic changes that occur during muscle stem cell-mediated regeneration to ensure myofiber repair and continuity of the stem cell compartment. Furthermore, we explore open questions in the field that are expected to be important areas of exploration as we move toward a more thorough understanding of the epigenetic mechanism regulating muscle regeneration.

1. INTRODUCTION

1.1 Master Regulators in Adult Skeletal Myogenesis for Muscle Repair

Efficient skeletal muscle regeneration depends upon the ability of muscle stem cells (satellite cells) to transit between distinct cellular fates in response to signals from injured muscle for the repair of damaged myofibers. In healthy skeletal muscle, satellite cells reside predominantly in a dormant or quiescent state within their niche situated between the muscle fiber and the basal lamina (Charge & Rudnicki, 2004; Mauro, 1961). In the niche, satellite cell quiescence is periodically interrupted for homeostatic maintenance of the muscle fibers (Pawlikowski, Pulliam, Betta, Kardon, & Olwin, 2015). Under such conditions, satellite cells likely undergo asymmetric cell divisions that allow the commitment of one daughter cell to the myogenic lineage that are used to reestablish the finite number of nuclei needed for health of the myofiber, and one daughter cell that maintains the satellite cell fate to repopulate the niche. In the case of traumatic muscle injury, satellite cells activation leads to alternating cell fate outcomes where cells either undergo asymmetric cell division as described earlier, or symmetric cell divisions giving rise to two daughter cells committed to a massive expansion of the proliferating myogenic progenitor cell pool (Holterman & Rudnicki, 2005). As the expanding muscle progenitor cell population attains numbers sufficient to regenerate new myofibers, cells become committed to terminal differentiation where mononucleated myocytes fuse with one another to form healthy new multinucleate myofibers.

Each of the myogenic cell fates (satellite cell, myogenic progenitor cell, myofibers) associated with muscle regeneration is characterized by a distinct gene expression program. As the genome of all cells within each cell fate is identical, the decision of which genes to express is regulated at the epigenetic level. More precisely, the cell fate transitions are regulated by changes to the genome packaging within the nucleus to influence the set of genes to be expressed. Directing the epigenetic program of each myogenic cell fate is a series of cell-type-specific transcription factors. In the uncommitted satellite cell, the paired-box gene, Pax7, plays an essential role in maintaining the stem cell identity of this population (Seale et al., 2000). Upon activation, a proportion of the satellite cells will become committed to the myogenic lineage through the expression of a family of muscle-specific basic helix–loop–helix transcription factors that have been termed muscle regulatory factors or MRFs. Among the MRF family of transcription factors, MyoD and Myf5 play essential roles in expansion of the muscle progenitor cell population, while MyoD and myogenin are required for terminal differentiation and activation of muscle-specific genes (Singh & Dilworth, 2013), as reviewed in Fig. 1. While the aforementioned muscle transcription factors play key roles in regulating cell fate transitions during regeneration, their ability to regulate gene expression does depend upon accessory transcription factors such as myocyte enhancer factor 2 (MEF2) (Black & Olson, 1998) and Sineoculis homeobox homolog (SIX) families (Liu, Chu, Chakroun, Islam, & Blais, 2010; Niro et al., 2010). Together with the muscle regulatory factors (MyoD, Myf5, Myogenin), the more ubiquitously expressed SIX1/SIX4 and MEF2C/MEF2D transcription factors act to recruit the epigenetic machinery to specific loci to regulate gene expression (Aziz, Liu, & Dilworth, 2010). In this chapter, we will focus on the epigenetic processes, along with the epigenetic machinery that function with the myogenic transcription factors to control muscle regeneration.

1.2 Epigenetic Regulation of Gene Expression

Epigenetics is broadly categorized as heritable, dynamic changes in gene expression that are not dependent on DNA sequence, but rather the ability of different genes to be expressed in a given cellular context (Heerboth et al., 2014). By this definition, the ability of a gene to be expressed is determined by packaging of the genome in the nucleus, which will make specific DNA sequences more or less accessible to the transcriptional machinery required for establishing cell-specific transcription networks. The overall accessibility

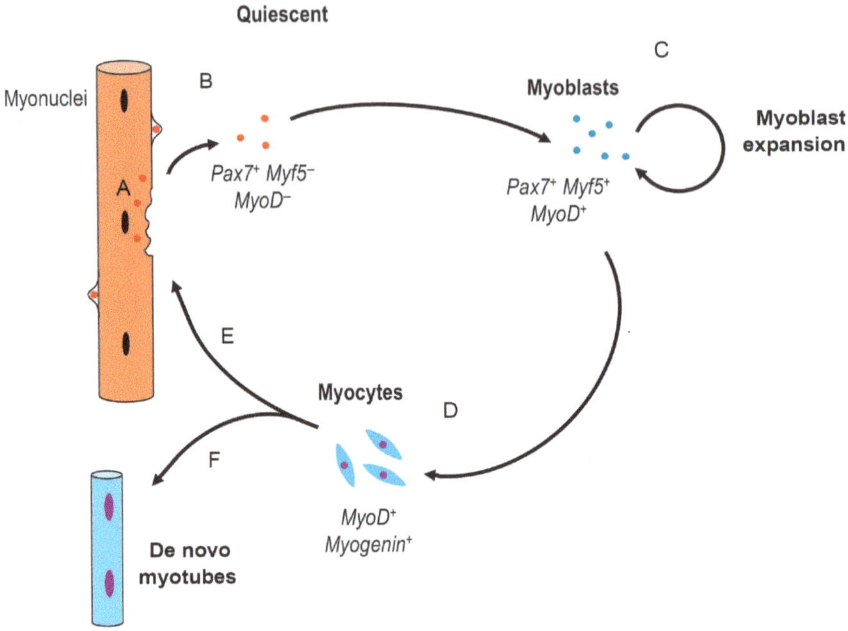

Fig. 1 After muscle damage, (A) quiescent satellite cells ($Pax7^{+}Myf5^{-}MyoD^{-}$) localize to the site of injury. (B) Here they become activated to enter the myogenic cycle. (C) Activated satellite cells become proliferating myoblasts to rapidly expand the pool of muscle progenitors ($Pax7^{+}Myf5^{+}MyoD^{+}$). Once the pool of myoblasts is sufficiently enlarged, (D) myoblasts become myocytes committed toward terminal differentiation ($MyoD^{+}$ $Myogenin^{+}$). (E) Myocytes may fuse with damaged myofibers to repair them, or (F) may fuse with one another for de novo myofiber formation.

of DNA for being bound by transcription factors to affect downstream gene expression is regulated by many different processes that globally comprise epigenetic control in cells (Zhang & Pradhan, 2014). As such, genes may become activated upon specific spatial, temporal, and other environmental signals such as presence of cytokines in its surrounding environment. Importantly, epigenetic processes are dynamic, allowing transitions between active and repressive states in both coding and noncoding genomic regions.

To better understand epigenetic regulations in cells, we must first understand the organization of the genome in the nucleus. The genome itself, and the genes it encodes, remains relatively static (except B- and T-cells which undergo V(D)J-recombination to develop our immune system) throughout the life span of an individual. Despite the physical size of the genome, it is effectively packaged into the nucleus by a direct association of DNA with histone octamers (histones H2A, H2B, H3, and H4) to form nucleosomes.

Wrapping of 147 basepairs of DNA around the histone octamer forms the first level of DNA compaction, which is referred to as the 10-nm fiber and has the allure of "beads on a string" (Zhu & Li, 2016). A higher level of compaction is achieved through binding of the histone linker H1 that is able to interact with adjacent nucleosomes to form the more condensed 30-nm fiber (Zhu & Li, 2016). Following formation of the 30-nm fiber, chromatin undergoes distinct levels of looping which culminates in the most compact state of DNA, the chromosome (Zhu & Li, 2016). It is known that chromatin looping allows effective chromatin compaction and regulation, but their specific higher order chromatin organization is not yet well understood (Kadauke & Blobel, 2009). Although some new techniques such as chromosome conformation capture have started to show the physical interaction of distant genomic loci with one another (Dostie & Dekker, 2007; Hackett & Surani, 2013), much more work is required to understand higher order chromatin compaction in establishing the chromosomal structure.

The level of chromatin compaction at specific sites in the genome will dictate whether a gene lies within the more accessible euchromatin or the tightly packed heterochromatin. Heterochromatin compartments are subdivided into two categories: (1) constitutive heterochromatin that is always densely packed, and typically found in gene poor regions of the genome; or (2) facultative heterochromatin that is condensed and transcriptionally silent but is readily converted to an open chromatin state to allow for gene expression (Trojer & Reinberg, 2007). In contrast, euchromatin is characterized as an "open" chromatin structure that is transcriptionally permissive.

Our current understanding of the epigenetic processes controlling gene expression centers on the ability of proteins and RNA species to modify the lowest level of chromatin organization, which is the interaction of DNA with histones to regulate accessibility and subsequent ability to be transcribed. Four main categories regulate gene accessibility, including methylation of DNA, posttranslational modifications of histones, nucleosome positioning, and noncoding RNAs (Zhang & Pradhan, 2014). DNA methylation patterns are heritable marks maintained during cellular division and are directly involved in regulating gene expression (Hackett & Surani, 2013). Recently, it has been shown that dynamic changes in DNA methylation patterns are implicated in regulating expression of cellular transcription networks (Ziller et al., 2013). Posttranslational modification of histone proteins is another method of epigenetic control, wherein direct modifications of the lysine/arginine-rich histone tails alter the relative charge of the chromatin to modify its ability to interact with the underlying DNA and/or proteins needed to regulate gene expression. These posttranslational modifications may cause chromatin to become more open or condensed which will have a direct effect on the propensity of DNA to be recognized by transcription factors (Zhang & Pradhan, 2014). Posttranslational histone modifications occur most often on the protruding tails of histones through direct methylation, acetylation, phosphorylation, ubiquitination, and/or sumoylation (Jenuwein & Allis, 2001). Nucleosome assembly and histone variants in nucleosomes also affect DNA accessibility, wherein specific histone chaperones may deposit histone variants to promote more permissive or condensed chromatin structures (Yang, Song, et al., 2011). Lastly, noncoding RNAs have emerged as molecules important for epigenetic regulation of regulating gene expression by promoting formation of heterochromatin (Majid et al., 2010), targeting specific enzymatic proteins for their degradation (Lafontaine & Tollervey, 1998), and X inactivation (Jeon & Lee, 2011). Collectively, these epigenetic pathways work together to allow intricate control on gene expression to regulate cell-specific transcriptional networks.

1.3 Epigenetic Control in Adult Myogenesis

Gene expression profiles in adult myogenesis have been well established, but the specific events regulating transitions in cellular states is less understood. It is becoming increasingly evident that specific epigenetic mechanisms regulate DNA accessibility to control dynamic changes in gene expression profiles during myogenesis to promote effective transitions between cellular states. We focus here on the dynamic epigenetic changes that occur in satellite cells as they undergo cellular fate changes in adult myogenesis for the effective repair of damaged skeletal muscle. These highlight the importance of understanding the epigenetic landscapes in cellular processes and remain a fascinating field with promising applications. A complete understanding of the epigenetic mechanisms that regulate myogenic cellular fate transitions could allow the development novel therapeutics to treat muscle-wasting diseases such as the muscular dystrophies and age-related sarcopenia. An ability to control endogenous epigenetic modifiers would allow us to induce specific changes in transcriptional activity at various genes implicated in myogenesis to promote regeneration in persons afflicted with muscle-wasting disorders.

2. POSTTRANSLATIONAL HISTONE TAIL MODIFICATIONS AS AN EPIGENETIC REGULATOR

Proper epigenetic regulation requires the ability of histone tails to undergo dynamic posttranslational modifications which confer specific chemical environments to nucleosome assemblies of histone octamers. This allows changes in the interaction between the core histones with its surrounding DNA to effectively regulate accessibility of specific genomic regions to DNA-binding proteins. A looser DNA–nucleosome assembly is more permissive for binding of transcription cofactors and the recruitment of RNA Polymerase II machinery for the subsequent active expression of the gene. On the contrary, a tighter histone interaction with its associated DNA renders the DNA less accessible to specific transcription factors, conferring a repressed state that prevents downstream gene expression. In addition, the posttranslational modification of histones can alter chromatin binding specificity of regulatory proteins by creating increased or decreased binding ability.

Specific posttranslational histone marks that may be gained include the well-studied methylation and acetylation, but also ubiquitination, sumoylation, and phosphorylation patterns may undergo dynamic changes to control the overall charge distribution on the histone tails, affecting their ability to interact with DNA. Acetylation events are targeted to the ε-acetyl group on lysine residues but can also occur on the α-amino group of the first amino acid in a protein. Similarly, lysines can be modified by ubiquitin or sumolyation. Methylation of histones usually occur on ε-amino groups of lysine residues which can be mono-, di-, and trimethylated, or the ε-amino groups of arginine which can be mono- or dimethylated. In the case of dimethylated arginine, the methyl groups can be positioned symmetrically or asymmetrically which have different outcomes on the ability of chromatin-binding proteins to intact with specific genes. Phosphorylation of the histones can occur on serine, threonine, or tyrosine residues. While it can sometimes be easy to lump a particular modification with a transcriptional outcome, there are always exceptions to the rule. For instance, histone acetylation is widely regarded as a mark that is associated with actively transcribed euchromatin, recent studies have shown that acetylation of histone H4 lysine 20 (H4K20ac) is a mark associated with transcriptional repression (Kaimori et al., 2016). Thus, the interpretation of specific histone modifications must be viewed in the context within the nucleosome.

At the heart of epigenetic regulation of transcription is the dynamic nature of the histone modifications. For every enzyme that puts in place a posttranslational modification, there is usually an antagonizing enzyme that will act to remove the same mark (Aziz et al., 2010). In addition, there are epigenetic proteins that recognize and bind specific histone modifications (Ruthenburg, Allis, & Wysocka, 2007). Thus, the epigenetic machinery is often lumped into three groups of proteins: (1) Writers (enzymes that put in place epigenetic marks), (2) Erasers (enzymes that remove epigenetic marks), and (3) Readers (enzymes that specifically bind epigenetic marks). While many of these epigenetic enzymes are ubiquitously expressed, they play a key role in developmental patterning where they help establish tissue-specific gene expression programs (Ringrose, Ehret, & Paro, 2004). The tissue specificity is established by regulating how the Reader, Writers, and Erasers are targeted to specific sites in the genome. In this context, epigenetic enzymes that posttranslationally modify histone tails are typically targeted to specific loci by DNA-bound transcription factors to mediate gene activation and repression required for muscle cell lineage establishment. The contribution of methylation and acetylation of histones has been well studied in the context of myogenesis. However, much remains to be examined with respect to other histone modifications such as phosphorylation, sumoylation, and ubiquitination in the context of their role controlling cell fate changes during myogenesis.

In the following section, we discuss specific posttranslational modifications on histone tails that result in gene activation and repression during myogenesis. As will be shortly evident, these modifications play an important role in modulating the ability of quiescent satellite cell populations to become activated and committed toward terminal differentiation. Of particular focus will be methylation and acetylation events that control myogenesis along with the proteins that are responsible for reading, writing, or erasing these histone modifications.

2.1 Histone Methylation

2.1.1 Methylation Modifications That Cause Gene Activation During Myogenesis

Methylation status on histone tails plays a critical role in the epigenetic regulation of gene expression. Methylation events can occur on the ε-amino groups of either lysine or arginine residues. Importantly, distinct methylation patterns are correlated with chromatin in different cellular states. For instance, the trimethylation of histone 3 lysine 4 (H3K4me3) is thought to create a transcriptionally permissive chromatin environment that is suitable for gene expression, whereas methylation of histone H3 at lysine 27 (H3K27me3) or 9 (H3K9me2 or H3K9me3) is typically indicative of repressive chromatin environments. A series of regulatory pathways and epigenetic proteins used throughout myogenesis allow changes in histone methylation patterns to control repression and activation of specific genes during myogenic cellular state transitions. As such, methyltransferases and demethylases will be discussed together in this section. To date, most of the molecular epigenetic studies have been performed using C2 myoblasts (as opposed to isolated satellite cells or primary myoblasts) due to their ease of use. As C2 myoblasts represent an actively proliferating population of muscle progenitor cells, the transition from proliferation into terminal differentiation has been studied more intensely than activation of quiescent satellite cells which requires quiescent satellite cells from freshly isolated muscle. Therefore, we currently have a better understanding of epigenetic changes in methylation occurring during terminal differentiation compared to changes associated with activation of quiescent satellite cells.

2.1.1.1 Histone Methylation Patterns to Maintain Quiescence

Maintenance of satellite cells in a quiescent state requires that myogenic genes required for proliferation and differentiation remain repressed prior to satellite cell activation. Specific gene repression here is in part maintained by repressive histone marks that prevent myogenic gene expression. A key epigenetic complex implicated in myogenic gene repression during quiescence is the Polycomb Repressor Complex 2 (PRC2), of which the constituents are actively expressed in quiescent satellite cell populations. In this complex, the Ezh2 domain catalyzes trimethylation of histone H3K27, conferring the H3K27me3 silencing mark on the associated nucleosomes with which it interacts as shown in Fig. 2A (Bracken, Dietrich, Pasini, Hansen, & Helin, 2006; Caretti, Di Padova, Micales, Lyons, & Sartorelli, 2004; Margueron & Reinberg, 2011).

A second histone modification that has been studied in the context of quiescent satellite cells is the histone H4 lysine 20 dimethylation

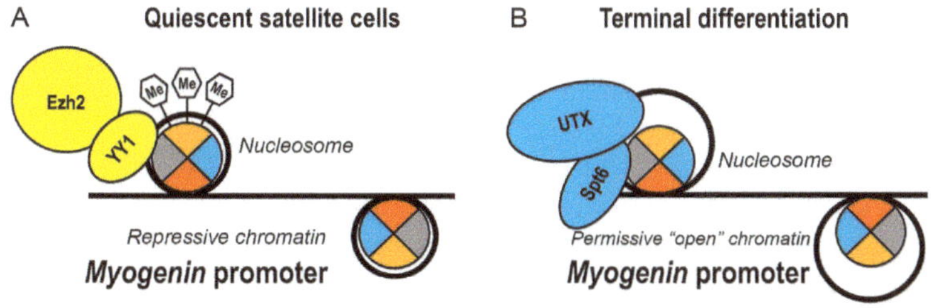

Fig. 2 Dynamic histone tail methylation allows control of muscle-specific genes during various stages of myogenesis. (A) During quiescent and proliferative stages, YY1 targets Ezh2 methyltransferase toward the promoter of muscle-specific genes such as *myogenin* and *CKM* to deposit repressive H3K27me3 marks. (B) Upon induced differentiation, Spt6 recruits the UTX demethylase to remove repressive methylation marks of H3K27 and form more permissive chromatin for easier gene expression.

(H4K20me2) (Jorgensen, Schotta, & Sorensen, 2013). The Suv4-20h1 methyltransferase has been implicated in maintaining quiescence in satellite cell populations by promoting heterochromatin formation to repress *MyoD* expression (Boonsanay et al., 2016). Gene expression studies in quiescent satellite cells showed high expression levels of the Suv4-20h1 methyltransferase, which is quickly reduced upon satellite cell activation. Showing the importance of Suv4-20h1 to quiescence, deletion of the enzyme in satellite cells induced satellite cell activation through the expression of MyoD. Electron microscopy also identified a role for Suv4-20h1 in maintaining the compacted heterochromatin state characteristic of quiescent satellite cells (Boonsanay et al., 2016). This suggests that compaction of the chromatin likely contributes to the maintenance of satellite cell quiescence.

The importance of Suv4-20h1 to satellite cell function was further confirmed in knockout mice where loss of the H4 lysine 20 methyltransferase led to defective regeneration upon repeated muscle injury (Boonsanay et al., 2016). These knockout mice also showed an increased presence of $Pax7^{+}MyoD^{+}$ expressing cells in the tibialis anterior and a reduced relative population of truly quiescent $Pax7^{+}MyoD^{-}$ satellite cells. This shows that Suv4-20h1 is required for establishing and maintaining satellite cells in a quiescent state under normal conditions. Based on these findings, it is understood that the H4K20 dimethylation activity of Suv4-20h1 is required to maintain specific heterochromatin states and prevent gene expression, to maintain quiescence. Of particular interest to this study was the observed decreased Suv4-20h1 expression in mdx mice (Boonsanay et al., 2016), a mouse model undergoing persistent muscle regeneration. This was attributed to endogenous downregulation of Suv4-20h1 arising from the continuous muscle regeneration (Boonsanay et al., 2016).

2.1.1.2 Changes in Methylation Patterns During Satellite Cell Activation and Proliferation

Activation of the quiescent satellite cell in response to extrinsic cues occurs through a series of gene activations that necessitate changes in histone methylation. A key regulator of satellite cell activation is the Ash2L/MLL2 methyltransferase complex that mediated H3K4 methylation of genes involved in establishing the muscle cell fate (Diao et al., 2012; Kawabe, Wang, McKinnell, Bedford, & Rudnicki, 2012; McKinnell et al., 2008). The targeting of the Ash2L/MLL2 complex to muscle genes is mediated by the satellite cell transcription factor Pax7 (McKinnell et al., 2008). Proteomic and biochemical analysis revealed that this targeting of Ash2L/MLL2 to muscle genes by Pax7 is regulated by the arginine methyltransferase CARM1 (Kawabe et al., 2012). As a means of controlling the timing of satellite cell activation and commitment to the muscle lineage, CARM1 methylates multiple arginine residues in the N-terminus of Pax7 during *asymmetric* satellite cell division (Kawabe et al., 2012). The CARM1-mediated methylation of Pax7 allows for its high-affinity interaction with the Ash2L/MLL2 methyltransferase complex. A key target of the Pax7-associated Ash2L/MLL2 methyltransferase complex is the *Myf5* promoter region where introduction of transcriptionally permissive H3K4me3 marks prime the *Myf5* promoter for subsequent activation (Diao et al., 2012; Kawabe et al., 2012; McKinnell et al., 2008) as reviewed in Fig. 3. This allows expression of Myf5 in one of the daughter cells during asymmetric division, allowing it to enter the proliferative state marked by expression of both Pax7 and Myf5. It is noteworthy to mention that in asymmetric division, specific activation of *Myf5* only occurs in the daughter cell that will become activated, whereas the other daughter cell that will return to a quiescent state will not activate *Myf5*.

The changes in methylation events leading to MyoD expression in activated satellite cells is not well established, though it is clear that a loss of the H4K20me2 mark is observed through a mechanism that remains to be determined (Boonsanay et al., 2016). An important question, once MyoD is

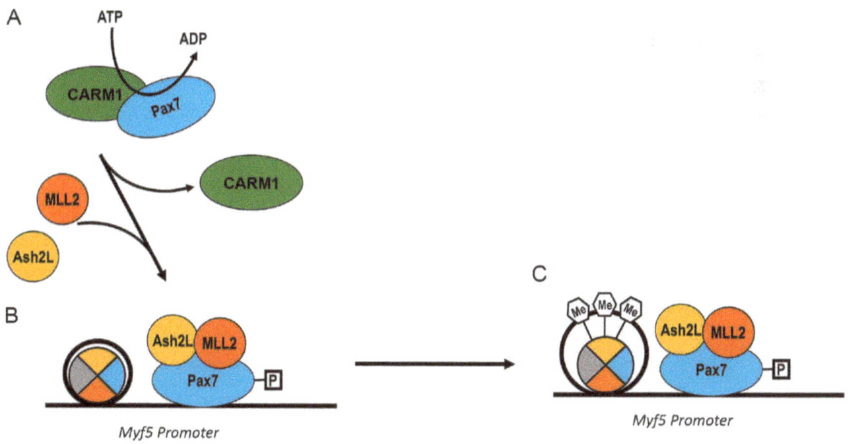

Fig. 3 During asymmetric division and satellite cell activation, (A) CARM1 targets Pax7 and phosphorylates N-terminal residues. (B) This allows Pax7 to bind MLL2 and Ash2L and targeted to the *Myf5* promoter. There, the methyltransferase activity of MLL2:Ash2L catalyzes activating trimethylation of H3K4. (C) This renders chromatin on the *Myf5* promoter more permissive to facilitate expression of the gene.

activated, is what prevents this master regulator of myogenesis from inducing terminal differentiation to allow expansion of the progenitor population? In part, the p38γ MAPK signaling cascade helps modulate MyoD activity by specifically phosphorylating Ser199 and Ser200 of MyoD to mediate active repression of myogenic genes such as *myogenin* (Gillespie et al., 2009). On MyoD-binding sites such as the *myogenin* promoter, phosphorylated MyoD protein recruits the H3K9 methyltransferase Suv39h1/KMT1A which marks the regional chromatin with repressive H3K9me2 and H3K9me3 methylation marks (Esteve et al., 2005; Fritsch et al., 2010). This prevents expression of *myogenin* so that myoblasts may remain in a proliferative state for myoblast population expansion. Interestingly, knockdown studies showed that Suv39h1 methyltransferase complex is also required for terminal differentiation as depletion of the enzyme caused a failure to repress key genes involved in S-phase transition during cell cycle progression (Ait-Si-Ali et al., 2004). Thus, the Suv39h1 complex plays key roles in facilitating both expansion of the muscle progenitor population and terminal differentiation.

2.1.1.3 Histone Tail Methylation in Terminal Differentiation

The ability of myogenic progenitor cells to undergo terminal differentiation requires a coordinated series of changes in gene expression that allow cell cycle exit in concert with the activation of muscle-specific genes such that allow the formation of functional multinucleated myofibers. The changes in gene expression observed throughout the differentiation process are heavily influenced through altered chromatin methylation patterns. These altered methylation patterns result from the erasing of repressive methylation marks and writing of new transcriptionally permissive marks at muscle-specific genes. In contrast, genes involved in cell cycle progression/proliferation will have their transcriptionally permissive marks erased, and their chromatin will become enriched in repressive histone marks.

One of the best-studied epigenetic marks regulating the transition between the proliferating muscle progenitor fate and that of the differentiated myotubes is the introduction of H3K27me3 marks at key muscle differentiation genes. In proliferating myoblasts, the presence of H3K27me3 methylation at the promoter of muscle differentiation genes such as myogenin ensure that cells continue to proliferate (Seenundun et al., 2010). This is a key regulatory step as myogenin expression represents a point of no return for progenitors to commit to muscle differentiation (Liu, Chakroun, et al., 2013; Singh & Dilworth, 2013).

The introduction of H3K27me3 at muscle functional genes is mediated by the PRC2/Ezh2 methyltransferase complex (Caretti et al., 2004). By association with the YY1 transcription factor, the PRC2/Ezh2 complex is recruited to myogenic gene promoters such as *myogenin* and *CKM* (creatine kinase, muscle), where the silencing histone marks prevent the expression of these myogenic genes in proliferating progenitor cell populations, as shown in Fig. 2A. Silencing of these myogenic genes is maintained throughout proliferation to allow proper temporal control of cellular state transitions. The observed downregulation of PRC2 in terminally differentiated muscle further supports its role to maintain repressive marks on myogenic genes in proliferating satellite cell populations (Bracken et al., 2006; Caretti et al., 2004; Levine, King, & Kingston, 2004), whereby its loss of expression upon differentiation allows a loss of repression on these specific genes to allow commitment toward terminal differentiation.

While decreased expression of PRC2/Ezh2 upon differentiation helps reduce H3K27me3 levels in differentiating myotubes, the loss of this repressive histone mark is also mediated through an active removal of the methyl groups. Important to terminal differentiation, the UTX histone demethylase (also known as KDM6A) can remove mono-, di-, and trimethylation marks of the repressive H3K27me2/3 histone mark (Agger et al., 2007; Hong et al., 2007; Lan et al., 2007). Preliminary work in cultured myoblasts showed that UTX knockdowns have an inability to form MyHC-positive multinucleated myotubes (Seenundun et al., 2010). Further work performed on mice has shown that satellite cell-specific UTX knockouts block the regenerative ability in damaged myofibers (Faralli et al., 2016), and an impeded ability to regenerate damaged muscle when UTX was rendered nonfunctional (Faralli et al., 2016). The role of UTX demethylation in myogenesis is explained through molecular analysis which has shown that the incorporation of UTX into the MLL4 complex (Cho et al., 2007; Hong et al., 2007; Lee et al., 2007) allows it to be recruited to the promoters of myogenic differentiation genes *myogenin* and *CKM*. There, the H3K27 demethylation activity of UTX allows subsequent activation of these genes so that complete differentiation of muscle progenitor cells may occur (Cho et al., 2007; Faralli et al., 2016; Hong et al., 2007; Lee et al., 2007). By removing H3K27me3 silencing marks on the myogenic differentiation genes *myogenin* and *CKM*, UTX is a key epigenetic enzyme that allows proper transitions between cellular fates of satellite cells and muscle progenitor cells during myogenesis, reviewed in Fig. 2. The specific removal of H3K27me3 repression marks within the gene body by UTX is directly facilitated by cooperative interactions with the histone chaperone Spt6 (Wang et al., 2013) (see Section 3 on nucleosome and histone chaperones), and recruitment of the UTX demethylase to specific sites for demethylation upon induced differentiation has been shown to be mediated by the transcription factor SIX4 (Seenundun et al., 2010).

In coordination with the removal of H3K27me3 marks at muscle-specific genes, the repression of genes involved in cell cycle progression must occur for muscle differentiation to proceed. Work from the Dynlacht group has clearly established a role for PRC2-mediated repression of cyclin D1 and other cell cycle-associated genes through marking of chromatin by H3K27me3 during differentiation (Blais, van Oevelen, Margueron, Acosta-Alvear, & Dynlacht, 2007). In the case of genes such as cyclin D1, the marking of chromatin by H3K27me3 was dependent on E2F-mediated recruitment of pRB which likely recruits PRC2 complexes to the target loci (Blais et al., 2007). However, it was noted that a subset of genes became repressed by PRC2 independent of pRB, suggesting that at least two mechanisms of PRC2 recruitment occur to repress genes involved in cell cycle—one that is pRB dependent and one that is pRB independent (Blais et al., 2007).

As stated in Section 2.1.1.2, MyoD plays an important role in maintaining muscle progenitors in a proliferative state by recruiting the repressive H3K9 methyltransferase Suv39h1 to myogenic genes such as myogenin. The activation of the myogenic program requires the removal of this active repression through the removal of H3K9me3 marks to allow for the establishment of H3K9 acetylation (Caretti et al., 2004; Hakelien, Delbarre, Gaustad, Buendia, & Collas, 2008; Jang, Choi, Kim, Cho, & Youn, 2007; Mal, 2006). The erasing of the H3K9me3 mark at muscle-specific genes is mediated by the histone demethylase LSD1 (Choi et al., 2010). Examining the role of LSD1 in myogenesis, it was revealed that MEF2 interacts directly with the demethylase to permit its recruitment to specific muscle promoters (Choi et al., 2010). Importantly, inhibition of LSD1 activity in progenitors impaired their ability to undergo differentiation, where H3K9me2 repressive histone marks continued to be observed on the promoters of muscle genes *MCK* and *myogenin* (Choi et al., 2010). It is noted that LSD1 levels rise only moderately during differentiation. This suggests that it is not the presence of LSD1 that regulated differentiation, but rather the ability of LSD1 to be targeted to muscle genes by MyoD and MEF2 to remove silencing H3K9me2/H3K9me3 methylation pattern to allow terminal differentiation.

The regulation of specific epigenetic-modifying proteins has further been shown to be regulated in part by secretion of TNF-α (tumor necrosis factor alpha), a cytokine that helps mediate the transition into terminal differentiation. Specifically, following damage to the muscle, an increased production of TNF-α by both the infiltration of macrophages in the damaged area and by myofibers themselves (De Bleecker, Meire, Declercq, & Van Aken, 1999; Tews & Goebel, 1996; Warren et al., 2002; Zador, Mendler, Takacs, de Bleecker, & Wuytack, 2001) allows activation of the p38 MAPK pathway. This activates p38α signaling, which finishes with phosphorylation of MEF2D to facilitate its interaction with the Ash2L/MLL2 methyltransferase complex (Rampalli et al., 2007). Through its interaction with MEF2D, the methyltransferase complex is recruited to the promoters of myogenic differentiation genes such as *myogenin* where subsequent deposition of transcriptionally permissive H3K4me3 marks renders the chromatin accessible for the subsequent expression of *myogenin* to commit the proliferating myoblasts to terminal differentiation, as reviewed in Fig. 4. Once expressed, myogenin activates its specific target genes for the proper progression of terminal differentiation and subsequent muscle repair.

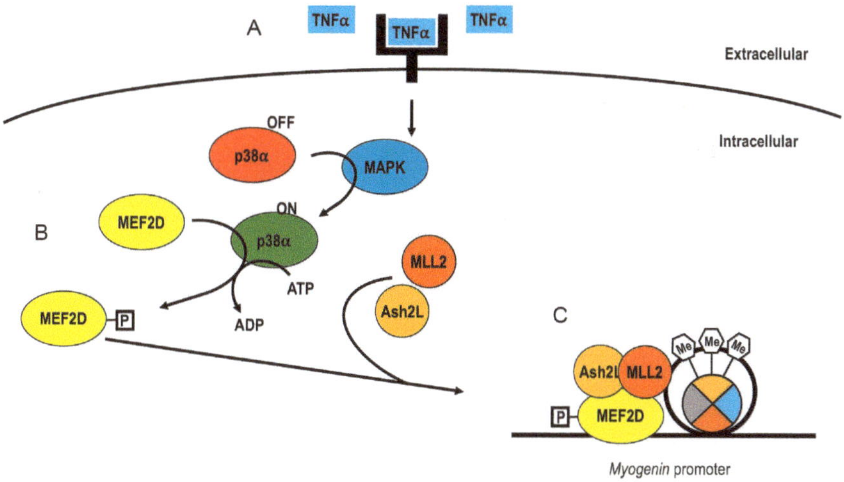

Fig. 4 (A) After muscle damage, TNFα is secreted by macrophages in the extracellular environment. These bind the TNFα receptor, which activate the MAPK pathway. (B) Activated p38α phosphorylates MEF2D, which (C) allows it to bind Ash2L and MLL2 and localize this methyltransferase complex to MEF2D-binding sites such as the *myogenin* promoter. There, the methyltransferase helps form permissive chromatin through H3K4 trimethylation.

The H3K4me3 mark in myogenesis for gene activation has also been studied in the context of Set7, a methyltransferase protein that specifically catalyzes H3K4 methylation. Studies in myoblasts have shown that an increased Set7 content is observed in differentiating myoblasts, and that a specific Set7 knockdown causes an overall reduced number of myotubes, and a loss in expression of *myogenin*, *MyHC*, and *MyoD* (Tao et al., 2011). It was proposed that upon differentiation, Set7 interacts with MyoD to recruits the complex to specific myogenic differentiation genes for their activation and expression. Although the specific mechanisms governing Set7-mediated epigenetic changes remain to be confirmed, it does appear to play a role in facilitating a transition toward terminal differentiation in myogenesis. The precise relationship between Set7-mediated and the aforementioned MLL-mediated methylation of H3K4 at muscle genes during myogenesis remains to be determined.

Lastly, the arginine methyltransferase CARM1 previously discussed in Pax7-mediated satellite cell activation is also implicated in direct histone tail methylation (Bedford, 2007). This epigenetic protein methylates histone H3 arginine 17 (H3R17) residues in specific promoters (Bedford & Richard, 2005). Specific to myogenesis, CARM1 interacts with the steroid receptor coactivator GRIP-1 where the two cooperatively stimulate MEF2C activity (Chen, Loffler, Chen, Stallcup, & Muscat, 2002). In vivo analysis has further supported the role where MEF2C targets CARM1 to the *CKm* promoter during differentiation (Chen et al., 2002), which strongly suggests that CARM1 and MEF2 cooperatively interact to stimulate muscle-specific genes during myogenic differentiation.

2.2 Histone Acetylation Modifications With Histone Acetyltransferases and Deacetylases

2.2.1 Dynamic Histone Tail Acetylation Helps Regulate Changes in Gene Expression

Histone tail acetylation is also a critical posttranslational modification for regulating dynamic chromatin structures during myogenesis. As a general guideline, histone acetylation reduces the nucleosome interaction with DNA and creates more permissive chromatin, whereas a loss of acetyl histone marks produces stronger interactions between nucleosomes and DNA which promotes formation of repressive chromatin. With respect to myogenesis, the transition between cellular states from quiescence to terminal differentiation is highly regulated by histone tail acetylation. Prior to terminal differentiation, the activity of many histone deacetylases (such as

SIRT1, Sir2, HDAC4, and HDAC5) promotes hypoacetylated nucleosomes at promoters and enhancers of muscle-specific genes to prevent their premature expression. As myogenic precursor cells proliferate and undergo terminal differentiation, a balance between a loss of deacetylases and a gain of activity in acetyltransferases promote acetylation of histone tails and more permissive chromatin at muscle-specific genes. In turn, this controls specific gene expression to allow proper transitions between myogenic cellular stages.

2.2.2 Histone Deacetylases and Their Importance in Myogenesis

Histone deacetylases (HDACs) are a family of proteins that can induce specific removal of acetyl groups from lysine residues of histones and other proteins. There are 19 deacetylases in mammals that are subcategorized into four families, known as Class I–IV HDACs, wherein Class II and Class III HDACs play important roles in the progression of myogenesis. Specifically, HDACs use their catalytic activity to remove acetylation marks from lysine residues to establish an unmodified state, while also priming the histone tails for potential subsequent modification such as methylation.

2.2.2.1 Class III HDACs Are Required for Satellite Cell Activation

The activity of class III HDACs, also called the sirtuins, is highly dependent on the availability of free NAD+ (Imai, Armstrong, Kaeberlein, & Guarente, 2000). An abundance of free NAD+ allows the maintenance of sirtuins in a catalytically active state so that they may deacetylate their specific target sites. On the contrary, metabolic shifts lowers levels of free NAD+, which causes sirtuins to be starved of their energy source and lose their deacetylation activity.

Satellite cells in a quiescent state generate ATP primarily through fatty acid oxidation. This generates NAD+ as a by-product which confers catalytic activity to the sirtuins. Upon activation, there is a change of cellular metabolism so that anaerobic glycolysis becomes the primary source of ATP production, and no longer produces NAD+ as a by-product. This results in a decreased availability of free NAD+ where reduced Class III HDACs activity can no longer maintain specific histone tails in a deacetylated, repressive state (Ryall et al., 2015).

Metabolic changes play an important role in controlling myogenic cellular state transitions by directly affecting the epigenetic landscape. In quiescent satellite cell populations, the HDAC Sirtuin 1 or SIRT1 removes acetylation marks on histone H4 lysine 16 (H4K16) to keep it in a hypoacetylated state. Upon activation and proliferation of satellite cells, the metabolic switch from fatty acid oxidation to glycolysis produces a drop in free NAD+ and SIRT1 loses its catalytic activity. This allows H4K16 to accumulate acetylated at muscle-specific genes which will be primed for expression upon induced differentiation (Ryall et al., 2015). When SIRT1 is knocked out from muscle populations, expression of muscle genes is increased though smaller myofiber size and impaired regenerative ability are observed in response to muscle injury (Ryall et al., 2015). Because H4K16ac helps maintain permissive chromatin states by preventing formation of the 30-nm fiber (Yao et al., 1998), it is proposed that SIRT1 maintains inactivation of muscle-specific genes during quiescence to prevent their premature expression. Upon activation and proliferation, the loss of function in SIRT1 allows loci to adopt a more transcriptionally permissive chromatin state such that these muscle-specific genes can be expressed under the proper conditions.

Sir2 is another class III HDAC that has been strongly implicated in regulating myogenic activity in skeletal muscle. Much like SIRT1, Sir2 activation is based on the [NAD+]:[NADH] ratio, where abundant levels of NAD+ are required for Sir2 deacetylase activity (Imai et al., 2000; Landry et al., 2000). Overexpression of Sir2 in myoblasts produces a delay in differentiation and myotube formation, whereas knockdown of Sir2 results in premature differentiation (Fulco et al., 2003), suggesting this deacetylase enzyme is required to maintain the muscle progenitor state. The mechanism of action by which Sir2 regulates a transition from proliferating myogenic precursors toward differentiation and myotube formation is through formation of a ternary complex between MyoD, pCAF, and Sir2 (Fulco et al., 2003). In proliferating myoblasts, this complex is directed to muscle-specific gene promoters such as *myogenin* and *MHC*, where the Sir2 deacetylase activity maintains H3K9 and H3K14 in a hypoacetylated state to prevent expression of the genes (Fulco et al., 2003). As the cells transition from a proliferating state into differentiation, a decrease in free NAD+ availability causes Sir2 to lose catalytic deacetylase activity. From there, hypoacetylation on the Sir2 target residues is lost, leading to more permissive chromatin and eventual expression of these myogenic genes.

2.2.2.2 Class I and II HDACs Help Prevent Differentiation in Myoblasts

HDAC4 and HDAC5, both class II HDACs, are also shown to regulate expression of muscle-specific genes prior to differentiation. When expressed, HDAC4 and HDAC5 have been shown to specifically interact with MEF2 and prevent the transcription factor from expressing its downstream muscle-specific genes prior to differentiation (Lu, McKinsey, Nicol, & Olson, 2000; Miska et al., 1999; Sparrow et al., 1999; Wang et al., 1999), but do not interact with MyoD (Lu, McKinsey, Zhang, & Olson, 2000). HDAC4 and HDAC5 can directly interact with promoter-bound MEF2 at muscle genes, where they act to decrease the level of acetylation on histone tails (Lu, McKinsey, Zhang, et al., 2000). When differentiation is induced, HDAC4 and HDAC5 are released from MEF2 by activation of calmodulin-dependent protein kinase signaling (Lu, McKinsey, Nicol, et al., 2000). Phosphorylation of HDAC4 and HDAC5 causes their relocalization to the cytoplasm where their deacetylase activity remains sequestered away from the target histones. This results in an increased accumulation of histone 4 acetylation (Lu, McKinsey, Zhang, et al., 2000). It must be noted that HDAC1 and HDAC3 do not interact with MEF2 (Lu, McKinsey, Nicol, et al., 2000; Miska et al., 1999; Sparrow et al., 1999; Wang et al., 1999), and their overexpression has no effect on cellular transitions of myoblasts from proliferation to terminal differentiation (Lu, McKinsey, Zhang, et al., 2000).

Class I HDACs are also implicated in regulating cellular state transitions in myogenesis, through their ability to interact with the zinc finger protein SNAI1. In proliferating myoblasts, SNAI1 preferentially interacts with both HDAC1 and HDAC2 (Class I HDACs) which bind G/C-rich E boxes of muscle-specific genes activated during differentiation (Soleimani et al., 2012). The binding of the SNAI1/HDAC complex at promoters competes with MyoD for E-box occupancy at muscle genes to prevent their activation, ultimately ensuring that progenitors maintain their proliferative state. Given the general role for HDAC-mediated histone hypoacetylation to promote repressive chromatin states, the recruitment of HDACs at these myogenic E-boxes likely leads to formation of repressive chromatin to further prevent expression of the muscle-specific genes. As progenitor cells transition to terminally differentiated myotubes, MRF-mediated expression of the microRNAs mir-30a and mir-206 is induced which targets the Snai1 and Snai2 mRNAs, respectively, for degradation (Soleimani et al., 2012). The loss of Snai transcripts frees myogenic E-boxes of the repressive Snai–HDAC complexes, which allows MyoD to bind for downstream muscle-specific gene expression. The association of HDACs to regulatory regions of *Myog, Myh10, CKM,* and *desmin* has also been identified in primary myoblasts (Caretti et al., 2004; Mal & Harter, 2003; Ohkawa, Marfella, & Imbalzano, 2006; Ohkawa et al., 2007) to represses their expression and maintain a proliferative state.

At the onset of differentiation, epigenetic modifications to produce more permissive chromatin for downstream gene expression require the removal of HDACs from many gene regulatory regions. This is achieved by a global reduction in expression of various HDACs (Puri et al., 2001). At the same time, remaining HDACs proteins can become phosphorylated which leads to their dissociation from chromatin and sequestration by 14-3-3 chaperone proteins for eventual export into the cytoplasm (McKinsey, Zhang, Lu, & Olson, 2000). This results in a reduced HDAC presence in the nucleus, and repressive chromatin may undergo further posttranslational modifications to facilitate downstream gene expression.

2.2.3 Histone Acetylation Marks in Myogenesis

Studies on histone tail acetylation in myogenesis have revealed the importance in their regulation during myogenesis. Here, acetylation is typically associated with more permissive chromatin states for enhanced transcriptional expression. Although our current understanding of acetylation processes in myogenesis has only recently begun to be understood with ChIP-seq mapping experiments, genome analysis show there is a global reduction in acetylation on H3K9, H3K18, and H4K12 in myotubes compared to conditions in proliferating myoblasts (Asp et al., 2011). As acetylation is typically associated with permissive chromatin to facilitate gene expression, this reduced acetylation is likely a result of global down-regulation of genes associated with proliferating conditions. Despite the global decrease in acetylation, an increase in acetylation is associated with both the *MyoD* and *Myf5* enhancer regions upon satellite cell activation. These have been attributed to specific activity of the p300 histone acetyltransferase protein to facilitate expression of these MRFs for the proper progression of the myogenic cycle, as discussed later.

2.2.3.1 The p300 Histone Acetyltransferase Activates Myogenic Genes

Two acetyltransferase proteins that play important roles in regulating global histone acetylation are p300 and the CREB-binding protein (CBP). Characterizing their role in myogenesis using mouse genetics revealed that these two related histone acetyltransferases play different roles in myogenesis, wherein myogenic defects are observed in the *p300* knockouts while the *cbp* knockouts appear phenotypically normal (Tanaka et al., 1997; Yao et al., 1998). A role for p300 in myogenesis was further established in studies using embryonic stem cells which showed that p300-null cells were unable to express MyoD and Myf5 (Roth et al., 2003). In contrast, *cbp* null ES cells were able to differentiate into multinucleated myotubes, suggesting this histone acetyltransferase is not critically required for modulating changes of cellular states in myogenesis (Roth et al., 2003). Collectively, these studies implicate the role of p300 in myogenesis by regulating expression of *MyoD* and *Myf5*.

Further research to probe how p300 might regulate *Myod* and *Myf5* expression revealed that it directly affects the acetylation state of histones associated with enhancer regions of each of these MRFs. The enhancer region regulating *MyoD* expression is composed of three main elements: the core enhancer region (CER) that is required for initial expression and activation of *MyoD* in embryogenesis (Goldhamer et al., 1995; Kablar et al., 1999) as well as the distal regulatory region (DRR) and proximal regulatory region (PRR) which are both implicated in maintenance of *MyoD* expression after initial activation (Asakura, Lyons, & Tapscott, 1995; Tapscott, Lassar, & Weintraub, 1992). The DRR and PRR enhancer regions are not sufficient for initiation of MyoD expression (Asp et al., 2011), so all three enhancer regions regulating *MyoD* must be active for regulated expression of *MyoD*. Comparing the interaction of p300 with each of these enhancer regions of *MyoD* through ChIP-qPCR experiments showed significant enrichment of p300 at the *MyoD* CER after differentiation has been induced in myoblasts (Hamed, Khilji, Chen, & Li, 2013). To confirm histone acetylation by p300 at the CER region was critical for activation of *MyoD*, the authors treated the myoblasts with curcumin to inhibit p300 histone acetyltransferase activity. These cells were unable to form myotubes, which was caused by an inability to express MyoD, showing acetylation plays critical roles in regulating MyoD expression for proper myogenic cellular transitions (Hamed et al., 2013). Although curcumin inhibits acetyltransferase ability in both p300 and CBP histone acetyltransferases (Balasubramanyam et al., 2004), the fact that CBP is not critically required for myogenesis (Tanaka et al., 1997; Yao et al., 1998) means treatment of cells with curcumin provides an effective way to study acetyltransferase defects in myogenesis specific to p300.

To understand which acetylation events at the *MyoD* enhancer region histones facilitated expression, chromosome immunoprecipitation (ChIP) experiments were performed on proliferating myoblasts and those induced for differentiation. This revealed an increase in H3K27ac at the CER of *MyoD* enhancer in differentiating myoblasts when compared to proliferating myoblasts (Hamed et al., 2013). When the experiments were repeated on a population of curcumin-treated myoblasts induced for differentiation, this

H3K27ac was not detected, showing that p300-mediated acetylation of H3K27 upon differentiation is required in myogenesis to express MyoD.

Other specific histone acetylation events that took place at the *MyoD* enhancer include (i) an increase of H3K18ac at all three enhancer regions (CER, DRR, PRR) upon induced differentiation, and (ii) an increase in H3K9ac most prominent at the *MyoD* PRR upon induced differentiation. Verification that these histone acetylation events were deposited by p300 was confirm by inhibiting the acetyltransferase with curcumin and assaying for the presence of these acetylated histone marks—all of which were not produced despite induced differentiation conditions (Hamed et al., 2013). Collectively, this shows that increased expression of *MyoD* upon induced differentiation requires activation of its enhancer region by direct histone acetylation events mediated by p300.

The activity of p300 was further explored to understand its effect on regulating *Myf5* expression during myogenesis. To do so, curcumin was used to inhibit p300 activity, while the results were verified using a p300 selective inhibitor C646 which does not inhibit CBP acetyltransferase activity (Bowers et al., 2010). In their analysis, myogenic differentiation was induced in P19 cells in the presence of the inhibitors of p300 acetyltransferase activity. Cells lacking p300 activity were unable to differentiate into multinucleated myotubes which was attributable to their inability to express myogenin. The lack of myogenin expression was a direct consequence of reduced *MyoD* and *Myf5* expression (Francetic et al., 2012). Characterization of the Myf5 enhancer showed enrichment of p300 binding at the early epaxial (EE) enhancer region that was lost upon p300 inhibition (Francetic et al., 2012). An increase in H3K27ac histone marks was also observed on the *Myf5* EE enhancer when differentiation was induced; however, this specific acetylation mark was not detected when p300 was inhibited through C646 treatment (Francetic et al., 2012). Collectively, these results indicate that p300-mediated acetylation of H3K27 during activation is required for the proper expression of *Myf5* (Francetic et al., 2012). It must be noted that P19 carcinoma cells (nonmyogenic) were used for this analysis, as they can be induced to differentiate into cardiac and skeletal muscle by direct treatment with DMSO. The specific implications of p300-mediated acetylation of H3K27 to activate the *Myf5* enhancer regions during activation remains to be verified in freshly cultured satellite cells.

ChIP-seq analysis to understand the effects of genomic MyoD binding has further implicated the role of acetylation in controlling changes in cellular states. Although a global decrease in acetylation is observed as proliferating myoblasts undergo differentiation to become multinucleated myotubes (Asp et al., 2011), ChIP-seq analysis found an overall increase in H4 acetylation at sites that were also bound by MyoD upon induced differentiation to form multinucleated myotubes (Cao et al., 2010). Because regions not bound by MyoD had no significant shifts in acetylation, this suggests that MyoD binding may allow for opening of chromatin structure through specific histone acetyltransferase recruitment. In fact, it was found that p300, pCAF and MyoD actively form complexes on promoters of MyoD target genes, and that when p300 and pCAF are prevented from interacting with MyoD, there will be no expression of the MyoD target genes despite MyoD still interacting with its target gene promoters (Puri et al., 1997). This mechanism was delineated by Dilworth and colleagues who showed sequential roles for the p300 and pCAF acetyltransferase proteins with MyoD. First, pCAF which acetylates MyoD to create an interaction surface that allows for recruitment of transcriptional coactivators (Dilworth, Seaver, Fishburn, Htet, & Tapscott, 2004), reviewed in Fig. 5. The acetylated MyoD can then establish a high-affinity interaction with p300 that leads to acetylation of histone H3 and H4 on target gene promoters (Dilworth et al., 2004). The specific targeting of p300 and pCAF to muscle genes by MyoD provides an explanation for the increased acetylation at a subset of genomic loci during terminal differentiation despite the observed global decrease in acetylation as myoblasts commit to terminal differentiation.

Many studies have implicated histone tail acetylation as an important modification that contributes to the regulation of myogenic progression.

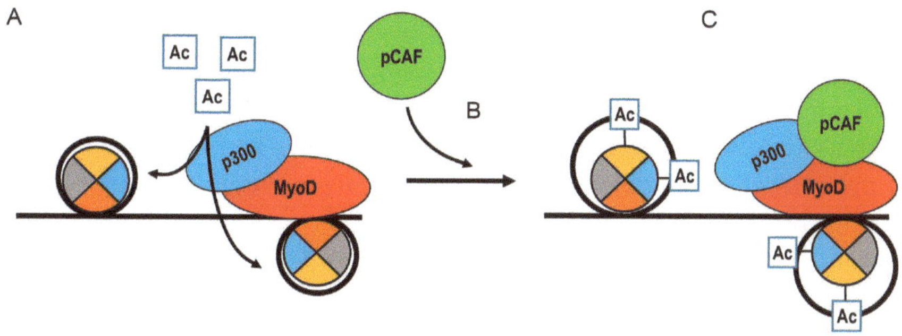

Fig. 5 Transactivation of MyoD target genes is facilitated by a two-step process involving p300 and pCAF-mediated acetylation. (A) MyoD and p300 interact at MyoD target gene promoters to acetylate H3 and H4 residues. (B) p300 recruits pCAF the complex, (C) which acetylates MyoD to facilitate its transactivation.

Specific acetylation marks are gained during myogenesis which allow the expression of the myogenic regulatory factors MyoD and Myf5. Although histone tail acetylation is typically associated with forming more permissive chromatin states, this often requires the prior activity of histone-modifying enzymes to remove previously established repressive methylation marks. Such is the case with H3K27me3 repressive marks associated with muscle-specific genes, where removal of repressive H3K27me3 mark by UTX is required prior to its acetylation (H3K27ac) to establish a permissive chromatin structure. This coordinated response for effective chromatin restructuring could in part be explained by formation of epigenetic reprogramming complexes that allow sequential chromatin remodeling events for a coordinated change in chromatin organization. The specific interactions between these histone-modifying enzymes and formation of complexes for regulated changes in posttranslational modifications remain an interesting area for future investigation.

3. CHAPERONES AND CHROMATIN REMODELING ENZYMES IN MYOGENESIS

Histone chaperones and ATP-dependent chromatin remodeling complexes play critical roles in regulating nucleosome assembly/disassembly and positioning in cells, which affects overall chromatin structure and DNA accessibility (Ransom, Dennehey, & Tyler, 2010). This helps establish a more permissive chromatin state to facilitate transcription factor binding and helps remove nucleosomes from DNA in transcription for easier transcriptional elongation. During cellular state transitions in myogenesis, it is understood that histone chaperones and other nucleosome remodeling proteins facilitate transitions between chromatin states. Although histone chaperones and remodeling enzymes are expressed ubiquitously, their implication in myogenic cellular transitions makes them a crucial component in muscle regeneration.

In the following section, we will discuss histone chaperones and nucleosome remodeling enzymes implicated in myogenesis and some of the specific bridging proteins that are required to link posttranslational histone-modifying proteins and their histone modifications to nucleosomes, as discussed in Section 2.

3.1 The SWI/SNF Complex Is Required for Myogenic Differentiation

SWI/SNF is an ATP-dependent chromatin remodeling complex present in many different cell lineages and tissue types where it is used to alter nucleosome positioning, control chromatin accessibility, and therefore regulate gene expression (de la Serna, Ohkawa, & Imbalzano, 2006; Flowers, Nagl, Beck, & Moran, 2009; Wilson & Roberts, 2011; Young et al., 2005). The composition of SWI/SNF complexes varies between cell types based on alternate incorporation of noncore subunits (Lessard & Crabtree, 2010). The SWI/SNF complex is generally composed of 9–12 subunits that invariably contain a core ATPase of either Brg1 or Brm (Brahma) which provides the complex the necessary energy for nucleosome remodeling activity (Phelan, Sif, Narlikar, & Kingston, 1999; Wang et al., 1996). When the ATPase activity of Brg1 was rendered inactive, the SWI/SNF complex could still bind its target genes (such as *myogenin*), but it could no longer form permissive chromatin and its muscle target genes were not expressed (de La Serna et al., 2000; de la Serna, Carlson, & Imbalzano, 2001). The SWI/SNF complex also contains some preserved core subunits between different cell types which include the Ini1, BAF155, and BAF170 proteins (Phelan et al., 1999). The variability in the remaining subunits of SWI/SNF is present to control complex targeting and lineage-specific regulators for functions of the complex in different tissue types (Lessard et al., 2007; Lickert et al., 2004; Wu, Lessard, & Crabtree, 2009; Yan et al., 2008).

The nucleosome remodeling ability of this complex is in part mediated by direct association of either Brg1 or Brm ATPase subunits with the complex, which produces the necessary energy requirements for the catalytic activity of the complex through direct ATP hydrolysis (Smith, Horowitz-Scherer, Flanagan, Woodcock, & Peterson, 2003). In myogenic differentiation, SWI/SNF complexes containing either Brg1 or Brm ATPase subunits have been shown to interact with different genes: SWI/SNF with Brg1 regulates early differentiation genes such as *myogenin*, and SWI/SNF associated with Brm regulates genes required for cell cycle arrest such as *Ccnd1*, and those in late differentiation such as *MyHC* (Albini et al., 2015; Flowers et al., 2009; Toto, Puri, & Albini, 2016). These findings are largely based on knockdown studies, where Brg1 depletion in myoblasts undergoing differentiation could withdraw from the cell cycle, but did not form multinucleated myotubes. When Brm was depleted in myoblasts undergoing differentiation, myoblasts either remained in a proliferative state or underwent partial differentiation to form truncated myotubes characterized by a limited number of fusion events (Albini et al., 2015). The different phenotypes arising from Brm and Brg1 depletion in myoblasts is a direct consequence of these alternate remodeling enzymes targeting different subsets of genes during terminal differentiation. Under normal conditions, induction of differentiation leads to an increased presence of Brm-containing

SWI/SNF complexes at the *Ccnd1* promoter to mediate repression. This repression of cyclin D1 causes an arrest in the cell cycle that allows for the proper transitions into terminal differentiation. Thus, in the absence of Brm, cyclin D1 continues to be expressed which prevents a cell cycle arrest. As such the Brm depleted myoblasts to continue to proliferate despite being exposed to differentiation medium. In contrast, Brg1 associates with early differentiation genes such as *myogenin* to stimulate expression which commits myoblasts toward terminal differentiation. Upon depletion of Brg1, the activation of early differentiation genes no longer occurs, which explains why Brg1 depletion fails to form myotubes despite their withdrawal from the cell cycle (Albini et al., 2015). It must be mentioned that Brm also plays a role in regulating late myogenic differentiation genes. Indeed, the depletion of cyclin D1 in Brm null myoblasts leads to cell cycle exit and activation of early myogenic genes. However, late muscle differentiation genes such as *MyHC* are not expressed (Albini et al., 2015).

The ability of Brg1 to activate early myogenic genes depends upon specific targeting of the SWI/SNF complexes to the target loci. This targeting of SWI/SNF to the myogenic loci is mediated through a direct interaction between MyoD and the BAF60c subunit of the complex. Already in the proliferating progenitor cell, BAF60c can be found associated with MyoD on muscle-specific gene promoters (Forcales et al., 2012). However, the remaining components of the SWI/SNF complex are not recruited to these promoters in muscle progenitor cells undergoing proliferation (Forcales et al., 2012). When committed toward terminal differentiation, activation of the p38α/β kinases phosphorylates a highly conserved Thr-229 residue on BAF60c, which allows BAF60c incorporation into the Brg1–SWI/SNF complex at the MyoD-directed muscle-specific gene promoter (Forcales et al., 2012; Simone et al., 2004). In turn, the Brg1 ATPase catalyzes chromosome remodeling to effectively remove nucleosomes on the muscle gene promoters to drive the differentiation process, reviewed in Fig. 6.

The histone arginine methyltransferase PRMT5 is also critical to regulate the Brg1 ATPase during myogenic differentiation. Through a series of coimmunoprecipitation, chromatin immunoprecipitation, and depletion experiments in NIH3T3 cells (with ectopic MyoD expression to commit them toward myogenic progression), it was shown that PRMT5 is recruited to the *myogenin* promoter prior to differentiation where it mediates dimethylation of histone 3 arginine 8 (H3R8me2). This serves as a recruiting platform for the Brg1 ATPase at the *myogenin* promoter to further facilitate

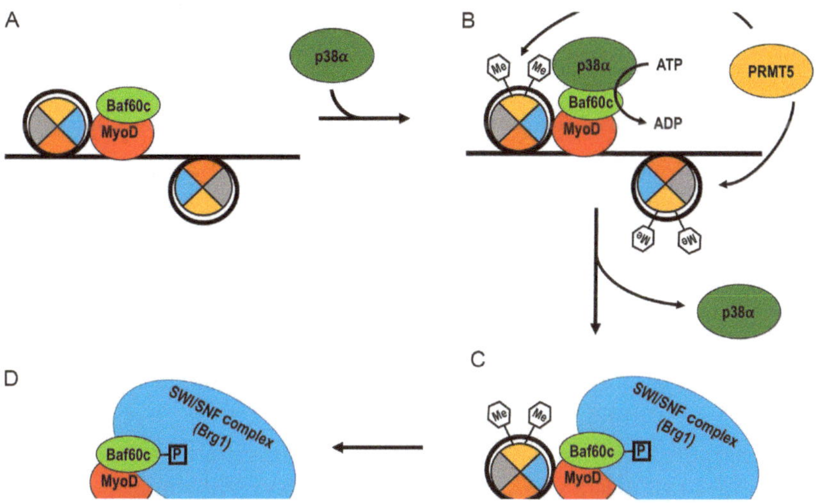

Fig. 6 The SWI/SNF histone chaperone plays an important role in removing nucleosomes to facilitate downstream gene expression. (A) During proliferation, Baf60c interacts with MyoD and is recruited to certain MyoD target genes. (B) When terminal differentiation begins, (i) PRMT5 dimethylates H3R8 residues, and (ii) p38α/β phosphorylates Baf60c. (C) These both work to recruit SWI/SNF and Brg1 ATPase to remove nucleosomes from DNA and (D) form DNA that may readily be transcribed. The interaction between PRMT5 and the SWI/SNF complex is not currently known.

Brg1 binding (Dacwag, Ohkawa, Pal, Sif, & Imbalzano, 2007). A reduction of PRMT5 results in a loss of H3R8me2 at the *myogenin* promoter which prevents Brg1 ATPase binding and no SWI/SNF binding at the *myogenin* promoter. As a result, there is no chromatin remodeling to form a permissive chromatin state surrounding the *myogenin* promoter, and *myogenin* is no longer expressed. This was further observed in primary satellite cells and myofibers isolated from mice (Dacwag et al., 2007), where the enriched presence of Prmt5, Brg1, and H3R8me2 marks at the *myogenin* promoter at different times during differentiation suggest similar *myogenin* activation as studied in cell culture (Dacwag et al., 2007). Recruitment of Brm arising from specific H3R8 dimethylation was not found, as this is likely one method that allows differential recruitment of Brg1 vs Brm at specific target sites during myogenic differentiation.

In addition to previously discussed epigenetic changes required for activation of *myogenin* during differentiation, MEF2D is also required in initial and maintained expression of *myogenin*. After its initial production, in addition to binding muscle-specific gene promoters for their expression, myogenin also binds its own promoter to maintain its expression through a feed-forward loop. Maintenance of this feed-forward loop through myogenin and MEF2D binding also requires interaction with the Brg1 ATPase as differentiation proceeds (Ohkawa et al., 2007). This assures permissive chromatin is maintained around *myogenin* so that both MEF2D and myogenin may continue to bind its promoter for continued expression of the gene and progression of differentiation. The requirement of Brg1 to maintain *myogenin* expression in late differentiation was validated in mouse by in vivo electroporation of the ATPase-deficient Brg1 subunit, where satellite cells expressing the dominant-negative SWI/SNF complex were not able to activate *myogenin expression* in regenerating satellite cells (Ohkawa et al., 2007).

Collectively, these experiments show that initiation of differentiation by p38α/β and H3R8me2 marks deposited by PRMT5 during induced differentiation is required to recruit the SWI/SNF enzyme Brg1 to the *myogenin* promoter. Here, Brg1 remains during early and late differentiation to maintain a permissive chromatin structure and therefore continued expression of myogenin. The Brm ATPase is also critical for proper myogenic differentiation, as it allows myoblasts to exit the cell cycle through repressed *Ccnd1*, and proceed in late myogenic differentiation (de La Serna et al., 2000, 2001).

3.2 The HIRA Complex Is a Histone Chaperone Required During Myogenesis

The HIRA complex is a replication-independent histone chaperone, that in humans participates in both transcriptional activation (Dutta et al., 2010; Yang, Song, et al., 2011) and repression (Zhang et al., 2005), contributing to both euchromatic and heterochromatic states (Elsaesser, Goldberg, & Allis, 2010; Ray-Gallet et al., 2002; Tagami, Ray-Gallet, Almouzni, & Nakatani, 2004). The specific activity of this complex lies in its ability to recruit histone-modifying proteins such as histone acetyltransferases and HDACs for histone tail modifications and to help regulate specific histone variant depositions through a replication-independent pathway (De Koning, Corpet, Haber, & Almouzni, 2007; Ray-Gallet et al., 2002; Tagami et al., 2004).

The HIRA histone chaperone plays several critical roles in regulating myogenic progression. This chaperone functions to deliver the histone variant H3.3 through a replication-independent pathway (De Koning et al., 2007; Ray-Gallet et al., 2002; Tagami et al., 2004), which has been associated with transcriptionally active genes given its enrichment in permissive posttranslational histone modifications and similar genome-wide occupancy profiles to that of RNA Polymerase II (Hake et al., 2006; Loyola, Bonaldi, Roche, Imhof, & Almouzni, 2006; McKittrick, Gafken, Ahmad, & Henikoff, 2004; Mito, Henikoff, & Henikoff, 2005). In myoblasts, it was shown that association between HIRA and Asf1a allows specific deposition of histone variant H3.3 at the *MyoD* promoter among other genomic regions (Yang, Song, et al., 2011). It is thought that the incorporation of Histone H3.3 into chromatin makes the individual H3.3-containing nucleosomes more dynamic for their turnover, facilitating access to the gene-regulatory elements. This primes *MyoD* for quicker and easier expression for an rapid increase in MyoD expression (Yang, Song, et al., 2011). Although the specific mechanism that relate H3.3 deposition to permissive chromatin is not fully understood, an increase in H3.3 deposition has been observed to correlate with a decrease in histone linker H1 binding to chromatin regions (Yang, Song, et al., 2011). Loss of histone H1 would allow for less tightly packed chromatin and more readily accessible DNA for transcription factor binding to activate downstream gene expression. The association between HIRA and Asf1 has further been implicated in myogenic terminal differentiation for transcriptional coactivation, when a R460A mutation was generated in HIRA to prevent its association with Asf1. This prevented coactivation of MEF2, indicating the interaction of HIRA and Asf1 is required for proper myogenic progression (Yang, Choi, et al., 2011).

Immunoprecipitation analysis of the HIRA complex further showed its implications in regulating myogenic differentiation. It was found that HIRA interacts with MEF2D with both its B-domain and C-terminal domains (Yang, Choi, et al., 2011), allowing recruitment of the HIRA complex to MEF2 target sites, such as the *myogenin* promoter (Yang, Choi, et al., 2011). Here, the HIRA complex regulates gene expression by interacting with histone-modifying enzymes to either repress or activate genes, depending on the specific myogenic cellular state. During proliferation, the MEF2D–HIRA complex recruits Cabin1 (a calcium-dependent MEF2 repressor in myogenesis) through direct interaction of the Cabin1 N-terminal domain with the HIRA C-terminal domain. Once recruited, Cabin1 recruits the methyltransferase Suv39h1 (Jang et al., 2007; Youn & Liu, 2000) which catalyzes H3 methylation, and Cabin1 also recruits HDAC1 and HDAC2 with a mSin3 intermediate, which works to deacetylate histone residues at the MEF2 target sites. Association of HIRA

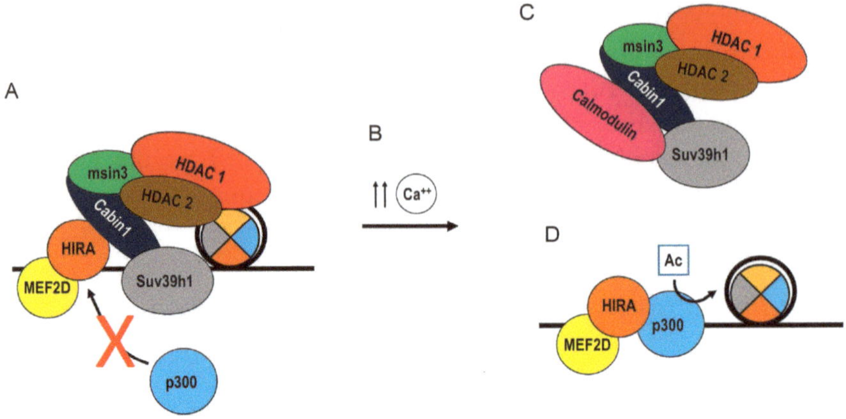

Fig. 7 The HIRA histone chaperone plays many roles in regulating chromatin structure in myogenesis. (A) During proliferation, HIRA interacts with MEF2D and brought to its target sites. Here, HIRA recruits Cabin1:mSin3:HDAC1:HDAC2:Suv39h1 where HDAC1 and HDAC2 remove histone acetylation marks and Suv39h1 adds repressive methylation on histone tails to collectively favor repressed chromatin. (B) During differentiation, increased calcium concentrations cause (C) calmodulin to bind Cabin1 and release the Cabin1:HDAC1:HDAC2:Suv39h1 complex from HIRA. (D) This leaves the MEF2D:HIRA complex open for p300 binding which acetylates histone tail residues and forms permissive chromatin.

with Cabin1:HDAC1:HDAC2:Suv39h1 also creates a block that prevents binding of the p300 coactivator to the MEF2–HIRA complex. Collectively, the Suv39h1-mediated methylation and HDAC1/2-mediated deacetylation favor a repressed chromatin state at the MEF2 target sites which prevents subsequent gene expression during myoblast proliferation (Jang et al., 2007; Yang, Choi, et al., 2011; Youn & Liu, 2000). Upon differentiation, increased intracellular calcium concentrations (Porter, Makuck, & Rivkees, 2002) allow calmodulin to displace the Cabin1:HDAC1:HDAC2:Suv39h1 complex from MEF2D:HIRA to permit gene activation (Jang et al., 2007). Indeed, the departure of the repressive methyltransferase and HDAC epigenetic enzymes provides the p300 acetyltransferase coactivator access to its binding site on HIRA complex. This allows the newly formed MEF2D:HIRA:p300 complex to catalyze a change in the epigenetic landscape and favor a more permissive chromatin state to ultimately allow downstream expression of the MEF target genes, as reviewed in Fig. 7. The HIRA complex also recruits Asf1 where it acts as a coactivator to help express MEF2 target genes (Yang, Choi, et al., 2011).

As HIRA can both repress myogenin expression through Cabin1–Suv39 recruitment and activate *myogenin* by chromatin remodeling to render nucleosomes more permissive, it may act as a regulatory switch to facilitate transitions from proliferation into terminal differentiation.

3.3 FACT Is Another Histone Chaperone Required for Myogenesis

The histone chaperone complex FACT (facilitates chromatin transcription) is implicated in many roles in various cellular lineages, such as nucleosome organization during transcription (Orphanides, LeRoy, Chang, Luse, & Reinberg, 1998; Orphanides, Wu, Lane, Hampsey, & Reinberg, 1999), replication (Wittmeyer & Formosa, 1997; Wittmeyer, Joss, & Formosa, 1999), and repair (Bruhn, Pil, Essigmann, Housman, & Lippard, 1992; Yarnell, Oh, Reinberg, & Lippard, 2001). It is composed of both Spt16 and SSRP1 subunits to form a heterodimeric complex. After transcription initiation has occurred, RNA Polymerase II must navigate nucleosome-induced impairment to active transcriptional elongation. Here, FACT participates in nucleosome disassembly by binding and removing the H2A—H2B dimers from a nucleosome complex. The dismantled nucleosomes allow RNA Polymerase II to overcome the impairment to elongation and proceed with high level gene expression (Kemble, McCullough, Whitby, Formosa, & Hill, 2015). This histone chaperone contributes to many nucleosome remodeling events (reviewed in Formosa, 2012) and has been implicated in regulating terminal myogenesis through its role as a coactivator of myogenin (Lolis et al., 2013).

Interaction studies established that FACT associates with the C-terminal domain of myogenin but cannot bind MyoD (Lolis et al., 2013). The interaction between FACT and myogenin occurs immediately upon induction of differentiation, implicating a role in establishing gene expression needed for early differentiation (Lolis et al., 2013). In contrast, FACT is not required for late myogenic differentiation as a downregulation of the Spt16 and SSRP1 subunits is observed after 3 days of differentiation. Consistent with this, the association of FACT with the myogenin target genes *Tnni* and *lmod2* is lost as differentiation progresses (Lolis et al., 2013). Thus, it is thought that through its specific interactions with myogenin, FACT is recruited to muscle target genes during early differentiation to establish a transcriptionally permissive state. Once established, chromatin surrounding these muscle genes would remain in a permissive state for their continued expression even after the departure of the FACT complex (Lolis et al., 2013).

3.4 Spt6 Helps Bridge UTX Demethylase to Its Target Histones

The histone chaperone Spt6 is also implicated in regulating transitions in myogenic cellular states. Broadly, this histone chaperone interacts with FACT to facilitate transcriptional elongation by nucleosome disassembly (Belotserkovskaya & Reinberg, 2004) and in nucleosome reassembly on transcribed DNA (Belotserkovskaya et al., 2003; Bortvin & Winston, 1996; Saunders et al., 2003; Youdell et al., 2008). Of particular interest to myogenesis is its role in tethering UTX to the elongating RNA Polymerase II for demethylation of the repressive H3K27me3 mark in gene bodies.

Spt6 is recruited to muscle-specific genes such as *myogenin* and *myh3* that are activated during differentiation, and knockdowns of Spt6 in myoblasts prevent differentiation (Wang et al., 2013). Abnormal muscle differentiation in the absence of Spt6 was also observed in zebrafish embryogenesis (Kok et al., 2007), suggesting a role in myogenesis that is evolutionarily conserved. Further investigation showed that when Spt6 is depleted, UTX cannot associate to chromatin, which is consistent with the increased levels of H3K27me3 observed (Wang et al., 2013). These observations led to the discovery that during the transition from proliferation toward terminal differentiation in myogenesis, Spt6 acts as a direct linker to allow UTX to associate with the elongating RNA Polymerase II (Wang et al., 2013), reviewed in Fig. 2B. Once associated to the RNA Polymerase II, UTX erases the repressive H3K27me3 histone marks within gene bodies, allowing a more permissive chromatin state and enhanced gene expression.

Histone chaperones are inevitably important epigenetic regulators of cellular state transitions during myogenesis. Their specific recruitment at muscle genes by myogenic regulatory factors MyoD and myogenin results in an altered chromatin state that facilitates subsequent gene expression. Although HIRA, the SWI/SNF enzyme Brg1 ATPase, and FACT are all recruited at the *myogenin* promoter to facilitate chromatin remodeling, the interactions between these complexes are not yet understood. There may be collective chromatin remodeling to facilitate differentiation, such as an initial deposition of histone variant H3.3 by HIRA to decrease histone linker H1 for more permissive chromatin structure followed by Brg1 and FACT recruitment to render chromatin sufficiently permissive for expression of *muscle genes*. However, this remains to be further explored.

4. DNA METHYLATION AS AN EPIGENETIC REGULATOR IN MYOGENESIS

4.1 DNA Methylation Controls Gene Expression to Regulate Cell Lineage and Cellular Fate Transitions During Myogenesis

The direct modification of DNA via methylation is yet another important epigenetic mechanism to control gene expression, and changes in myogenic cell fate. Here, the cytosine in CpG dinucleotides may be methylated on the carbon at position 5 to form 5-methyl-cytosine (5mC). These methylation events can occur both within regions of high (also called CpG islands) and low CpG dinucleotide repeat content. The introduction of the methyl group to the cytosine introduces steric hindrance in the DNA major groove and can effectively prevent binding of specific DNA-binding proteins (Bogdanovic & Veenstra, 2009), or may recruit repressive methylated DNA-binding proteins (MBD, MeCP2) to maintain silencing of specific genes (Hansen, Ghosh, & Woodcock, 2010). DNA methylation occurs in both intragenic and intergenic regions, where intragenic methylation has been implicated with regulating alternate splicing (Maunakea, Chepelev, Cui, & Zhao, 2013; Sati et al., 2012; Shukla et al., 2011) and promoting gene expression (Yang et al., 2014), whereas intergenic DNA hypermethylation on promoters and enhancers directly affects downstream gene expression (Baylin, 2005). Although DNA methylation has been classically assigned as a gene silencing method to prevent expression of nonlineage genes, recent work suggests that direct methylation of DNA may be more dynamic and directly contribute to epigenetic regulation of cellular fate transitions. As will be discussed in the following section, recent research has shown potent changes in DNA methylation occur during myogenesis as adult quiescent satellite cells become activated, undergo proliferation, and finally undergo terminal differentiation.

Two protein families play a central role in regulating the methylation status of DNA. The first family represents the DNA methyltransferases (DNMTs) where DNMT1 plays a role in maintaining methylation patterns after DNA synthesis (Trasler et al., 2003), and DNMT3a and DNMT3b which catalyze de novo DNA methylation for dynamic DNA silencing (Bird, 2002). Antagonizing the DNMTs are the ten-eleven translocation methylcytosine dioxygenase (TET) family of proteins, which act to demethylate DNA. There are three known TET family members that have been termed TET1, TET2, and TET3. These directly convert 5mC into 5-hydroxymethyl cytosine (5hmC) which undergoes further changes for its excision and replacement with unmethylated cytosine (Ito et al., 2010; Tahiliani et al., 2009), through different possible pathways as reviewed by Wu and Zhang (2010). This allows methylated CpG dinucleotides to lose repressive DNA methylation marks and become accessible to DNA-binding proteins for their activation upon recruitment of the appropriate transcription factors and coactivators.

4.2 Global Analysis of DNA Methylation Patterns

Comparisons between DNA methylation pattern in embryonic stem cells and that of somatic cells of various lineages have shown substantial changes in methylation as cells acquire specialization. Generally, a gain of methylation is observed between pluripotent stem cells and lineage specified cell lineages in order to suppress nonspecific genes. At the same time, a loss of methylation is observed at promoters of genes required for specification of the newly acquired cell lineage (Calvanese et al., 2012; Fouse et al., 2008; Isagawa et al., 2011; Meissner et al., 2008; Nagae et al., 2011; Nazor et al., 2012). These dynamic changes in DNA methylation allow (i) specific cell lineages to be produced by downregulation of global genes required during embryogenesis and (ii) specific demethylation of cell-specific genes allows their activation to drive differentiation toward the different lineages.

Specific to myogenesis, early work has shown differential expression of proteins directly implicated in DNA methylation and demethylation. During activation of quiescent satellite cells, a reduction of *TET* isoforms and *Dnmt3a* along with an increased expression of *Dnmt1* isoforms has been observed (Aguirre-Arteta, Grunewald, Cardoso, & Leonhardt, 2000; Liu, Cheung, et al., 2013; Liu, Sun, & Jost, 1996; Pallafacchina et al., 2010; Robertson et al., 1999; Ryall et al., 2015). Analysis of DNA methylation profiles in proliferating and differentiating muscle precursor cells compared to that of mature muscle fibers revealed a significant loss of hypermethylated regions in mature skeletal muscle when compared to the proliferating myogenic progenitors (Carrio et al., 2015; Tsumagari et al., 2013).

Global gene analysis using gene ontology has further supported the role of a dynamic change in DNA methylation during cellular fates of myogenesis in human myoblasts (Miyata et al., 2015). Specifically, hypermethylation was found to occur predominantly at two conserved

8-base pair motif repeats within the genome (Motif 1 C[*AT*]GCTG[CGT][CTG] and Motif 2 [CG][*AT*]G[*GC*]C[*TA*]G[GC]), of which many of the genes downstream of these sites were muscle specific (Miyata et al., 2015). Many of these eight nucleotide sequences were found in the Id protein promoter, suggesting that inhibition of Id protein expression through direct methylation of their promoters could play a role in decreasing available Id protein (Miyata et al., 2015). The reduced amount of Id proteins would effectively remove their inhibitory activity so that MyoD and E proteins are liberated for target binding and downstream gene expression.

4.3 Regulation of Specific Myogenic Regulatory Factors and Other Transcription Factors by DNA Methylation

A specific focus on the methylation status of the MRFs showed the CpG sites at the *MyoD* distal enhancer (20 kb upstream of its transcription start site) maintain a hypomethylated status in muscle progenitors, whereas these sites are hypermethylated in nonmyogenic somatic cells (Brunk, Goldhamer, & Emerson, 1996). These findings suggest DNA methylation status as a key requirement to control cellular fate transitions. Analysis of myoblast differentiation revealed that as quickly as 2 h after serum depletion there is demethylation of the *myogenin* promoter and that this same site maintains a demethylated status in vivo in mature mouse myofibers, but not nonmyogenic lineages (Lucarelli, Fuso, Strom, & Scarpa, 2001).

The *Myf5/Myf6* gene regulation region has also been shown to undergo variable methylation. In embryonic stem cells, all five *Myf5/Myf6* enhancer regions analyzed were seen to be hypermethylated, which keeps them in a repressed state and prevents activation of the enhancer region. Upon satellite cell activation, a partial demethylation was observed in the −111 and −17 kb regions of the gene regulation region, and a complete demethylation observed in the EE enhancer, proximal arch element, and minimal promoter region of *Myf5* (Carrio et al., 2015). Upon differentiation of proliferating progenitors, all the Myf5/Myf6 enhancer regions undergo methylation which is maintained in mature myofibers (Carrio et al., 2015). These alterations in methylation during changes in satellite cell fates are highly specific, as regions upstream and downstream of the enhancer regions show no changes in DNA methylation throughout the cell cycle (Carrio et al., 2015), and other nonmyogenic tissue analyzed retained methylation throughout the *Myf5/Myf6* gene regulation region. This work shows the specific epigenetic modifications through direct DNA methylation are required to control cellular fate transitions during myogenesis.

The role of DNMT1 in myogenesis was investigated using 5-azacytindine (AZA), a cytidine analog that cannot undergo methylation, creating a hypomethylated genome. In a classic experiment, it was shown that CH310T1/2 cells (fibroblasts) treated with AZA lead to myogenic conversion of the lineage to form multinucleated myotubes (Constantinides, Jones, & Gevers, 1977; Szyf, Rouleau, Theberge, & Bozovic, 1992). More relevant to myogenesis, treatment of proliferating progenitors with AZA was sufficient to induce the cells to undergo differentiation as evidenced by an increase in myogenin protein levels and myotube formation (Hupkes, Jonsson, et al., 2011; Hupkes, van Someren, et al., 2011; Scarpa, Lucarelli, Palitti, Carotti, & Strom, 1996). Temporal analysis of AZA-treated cells showed that upon treatment, proliferating progenitors have a reduced proliferative ability and an increased accumulation of MyoD. During induced differentiation, the AZA-treated progenitors also displayed increased expression of *Myf5, MyoD,* and *p21* (Montesano, Luzi, Senesi, & Terruzzi, 2013) supporting the notion that DNA demethylation of myogenic genes is crucial in cellular fate transitions during myogenesis. Although the specific methylated regions were not investigated, it is highly likely that the incorporation of AZA in proliferating myoblasts resulted in a loss of methylation at sites involved in differentiation such as the *myogenin* promoter, which resulted in the premature expression of differentiation genes and the arising phenotype observed.

The role of DNA methylation in myogenesis was further explored using muscle-specific DNMT3a knockout mice. This study showed that when DNMT3a was knocked out from satellite cells in developing mice, they showed greatly reduced muscle mass. The myogenic regeneration ability in adult populations of these knockout mice was also unable to undergo proper repair of damaged muscle, showing that de novo DNA methylation by DNMT3a is required for the proper myogenic development (Naito et al., 2016). When these knockout primary myoblasts were cultured ex vivo, they displayed a proliferation defect (Naito et al., 2016). Molecular analysis showed that direct methylation of the *p57Kip2* promoter by DNMT3a upon activation of satellite cells is required to suppress the expression of the $p57^{kip2}$ cyclin-dependent kinase. In the absence of DNMT3a, $p57^{kip2}$ confers antiproliferative properties to the progenitors by preventing the progression of the cell cycle and results in satellite cells returning to a quiescent state.

Typically, methylated DNA regions are bound by the methylated DNA-binding proteins 1–3 (MBD1–3) and methyl CpG-binding protein

(MeCP2), where they prevent transcription factor binding and RNA Polymerase II association with genes. In myogenesis, analysis has shown that these methylated DNA-binding proteins do not play any significant role in regulating cellular fate transitions (Sasai & Defossez, 2009). Rather the CtBP-interacting BTB zinc finger protein (CIBZ), another methylated CpG DNA-binding protein, is directly implicated in myogenesis (Oikawa et al., 2011). Specifically, CIBZ binds directly to methylated CpG dinucleotides where it acts as a transcriptional repressor of the downstream regions (Filion et al., 2006). During myogenesis, depletion of CIBZ protein content was seen to correlate with an increased expression of *myogenin* and premature formation of myotubes (Filion et al., 2006). Conversely, when proliferating progenitors are induced to undergo differentiation by low-serum starvation, expression of exogenous CIBZ inhibits the expression of *myogenin*. In accordance with the previously described AZA studies, this shows that DNA methylation helps to repress the expression of *myogenin*, likely among other myogenic genes, by direct recruitment of transcriptional repressors such as CIBZ. Further work has shown that specific methylation of the *myogenin* promoter is a likely regulatory step that prevents its premature expression during myogenesis, where it can only undergo expression when sufficient levels of MEF2A and SIX1 are present for the subsequent expression of *myogenin* (Liu et al., 2010; Niro et al., 2010). Although it has not been experimentally verified, it is likely that SIX1 is recruited to the methylated *myogenin* promoter where it recruits "*helper*" proteins for DNA demethylation and subsequent chromatin accessibility of the *myogenin* promoter (Palacios, Summerbell, Rigby, & Boyes, 2010). Although hypothetical interactions between methylated CpG dinucleotides of the *myogenin* promoter and specific proteins such as CIBZ, SIX1, and MEF2A have been proposed, the specific interaction of these proteins or their organization in a larger complex has yet to be studied and remains an interesting area for future studies to further understand the epigenetic regulations of myogenesis from a DNA methylation perspective.

4.4 Future Prospects in Myogenesis and DNA Methylation

The discovery of TET proteins as mediators of the first step in the demethylation of DNA has revealed the truly dynamic nature of DNA methylation, and highlighting the importance of this epigenetic modification in allowing specific gene activation and silencing to regulate specific cellular fate transitions. A global change in DNA methylation during myogenic cell fate transitions plays key roles in regulating cell fate changes from an epigenetic perspective. By blocking direct access of transcription factors, coactivators, and RNA Polymerase II, DNA methylation can be used to regulate DNA accessibility and therefore gene expression. Global methylation analysis showed dynamic changes in DNA methylation between embryonic stem cells and mature muscle fibers, and highly specific differences in methylation of promoter and enhancer regions of MRFs have started to be identified. Because DNMTs and TETs have no specific DNA-binding domains, their ability to catalyze methylation of DNA must be through the interaction with specific transcription factors. This has yet to be investigated and presents a potentially prosperous area of research in the coming years.

The direct implication of myogenic-specific transcription factors such as Pax7, MyoD, Myf5, Myogenin, and the MEF2 family of proteins in regulating DNA methylation has yet to be established and remains an interesting area for future investigation to further comprehend the role of DNA methylation in epigenetic regulation of cellular fate changes in myogenesis. The development of therapeutics that alter DNMT and TET function at specific genes could 1 day prove as successful epigenetic-based treatment for muscle regeneration of persons affected by muscle-wasting diseases.

5. CONCLUSION AND OPEN QUESTIONS

Gene regulation through direct epigenetic control plays a critical role in cell state transitions during myogenesis. An interplay between the three levels of epigenetic regulation discussed allow the repression and activation of specific genes involved in cell cycle and commitment to the muscle lineage which in turn allow coordinated and effective transition between multipotency, proliferation, and terminal differentiation.

Although we have discussed changes in the epigenetic landscape during myogenesis, we must emphasize that these are not specific to myogenesis. That is to say the enzymes coordinating epigenetic changes are expressed in various cell types to mediate similar epigenetic changes. The specificity of these enzymes acting on muscle gene promoters to regulate gene expression is an inherent property of these epigenetic enzymes to interact with the myogenic-specific transcription factors MRFs, Pax7, SIX1/SIX4, and MEF2. In turn, this recruits epigenetic-modifying enzymes to gene-specific promoters and catalyze different chromatin organization to regulate dynamic changes in expression and repression of the genes.

6. OPEN-ENDED QUESTIONS

Despite the discussed epigenetic regulation to control myogenic transcription networks during cellular state changes, there remains many areas for further exploration to understand the full epigenetic implications in myogenesis. First, the mechanisms discussed that regulate histone modifications, histone chaperone activity, and DNA methylation are actively being studied as separate phenomena and are rarely studied simultaneously. However, given the meticulous organization and gene expression networks in myogenic cell progression, there is a likely interplay between these epigenetic changes that govern ordered epigenetic changes for effective gene expression during cellular state changes. This remains to be fully understood and is an interesting area for further investigation.

As shown, histone tail modifications readily allow the change in nucleosome–DNA interaction to control DNA accessibility. Because there are many lysine residues within histone tails that can be modified in any given nucleosome, this leaves the opportunity for multiple posttranslational modifications on a given nucleosome. To our knowledge, the temporal interaction of histone-modifying enzymes with one another during histone tail modifications is not known. As such it is unsure if histone-modifying enzymes access and modify histone tails simultaneously, if there is an order in which histone tails are preferentially modified, and if there are interactions among histone tail modifiers to form complexes to promote multiple simultaneous changes in histone tails. One hypothetical example of histone-modifying complexes could be through interactions of HDACss and methyltransferases to maintain repressive marks on specific histone residues to confer an overall repressed DNA–nucleosome assembly. This is purely speculative and remains an area for future investigation. A collective understanding of the complete epigenetic processes regulating cell fate transitions in myogenesis could lead to developments of therapeutics to delay the progression or revert the effect of muscle-wasting diseases.

In this chapter, the epigenetic mechanisms discussed to regulate cellular state transitions during myogenesis are all at the most basic level of DNA compaction, namely, the interaction of DNA with nucleosomes. Given the recent discovery that heterochromatin compaction is maintained in quiescent satellite cells with the histone methyltransferase Suv4-20h1 (Boonsanay et al., 2016) implicates the likeliness that these base levels of epigenetic regulation could in fact harbor implications in higher order chromatin structure. Although higher levels of chromatin structure are not yet fully understood in terms of epigenetic regulation, understanding chromatin looping and organization to regulate transcriptional networks remains a fascinating area for further investigations. The implication of promoter enhancer looping to regulate gene expression proves to be a very recent method of epigenetic control of gene expression. This could play critical roles to regulate myogenic cell fate transitions where grouping of enhancer promoter regions for a specific gene can actively regulate gene expression, or different genes from different chromosomes become localized within a specific area for easier regulation. This is purely speculative, and the implication of higher order chromatin organization to regulate myogenic cellular fate transitions will be an interesting area for further investigation.

7. FINAL REMARKS

A complete understanding of the epigenetic regulations governing cellular fate transitions will be critical for a complete understanding of muscle regeneration. This could also lead to new therapeutics to treat muscle-wasting diseases by directly targeting epigenetic regulating mechanisms to promote efficient muscle regeneration in persons affected by muscle-wasting diseases. Epigenetic therapies for various diseases are currently being investigated (Brueckner et al., 2005; Cheng et al., 2003; Heerboth et al., 2014). HDAC inhibitors have been shown to promote differentiation of myoblasts, which can contribute to increased muscle regeneration (Consalvi et al., 2013; Consalvi, Saccone, & Mozzetta, 2014; Iezzi, Cossu, Nervi, Sartorelli, & Puri, 2002), which could provide a way to delay muscle wasting. A further understanding of the epigenetic regulatory mechanisms governing cellular state transitions during myogenesis could provide new innovative epigenetic therapies to treat muscle-wasting diseases.

REFERENCES

Agger, K., Cloos, P. A., Christensen, J., Pasini, D., Rose, S., Rappsilber, J., et al. (2007). UTX and JMJD3 are histone H3K27 demethylases involved in HOX gene regulation and development. *Nature, 449*, 731–734.

Aguirre-Arteta, A. M., Grunewald, I., Cardoso, M. C., & Leonhardt, H. (2000). Expression of an alternative Dnmt1 isoform during muscle differentiation. *Cell Growth & Differentiation: The Molecular Biology Journal of the American Association for Cancer Research, 11*, 551–559.

Ait-Si-Ali, S., Guasconi, V., Fritsch, L., Yahi, H., Sekhri, R., Naguibneva, I., et al. (2004). A Suv39h-dependent mechanism for silencing S-phase genes in differentiating but not in cycling cells. *The EMBO Journal, 23*, 605–615.

Albini, S., Coutinho Toto, P., Dall'Agnese, A., Malecova, B., Cenciarelli, C., Felsani, A., et al. (2015). Brahma is required for cell cycle arrest and late muscle gene expression during skeletal myogenesis. *EMBO Reports*, *16*, 1037–1050.

Asakura, A., Lyons, G. E., & Tapscott, S. J. (1995). The regulation of MyoD gene expression: Conserved elements mediate expression in embryonic axial muscle. *Developmental Biology*, *171*, 386–398.

Asp, P., Blum, R., Vethantham, V., Parisi, F., Micsinai, M., Cheng, J., et al. (2011). Genome-wide remodeling of the epigenetic landscape during myogenic differentiation. *Proceedings of the National Academy of Sciences of the United States of America*, *108*, E149–158.

Aziz, A., Liu, Q. C., & Dilworth, F. J. (2010). Regulating a master regulator: Establishing tissue-specific gene expression in skeletal muscle. *Epigenetics*, *5*, 692–696.

Balasubramanyam, K., Varier, R. A., Altaf, M., Swaminathan, V., Siddappa, N. B., Ranga, U., et al. (2004). Curcumin, a novel p300/CREB-binding protein-specific inhibitor of acetyltransferase, represses the acetylation of histone/nonhistone proteins and histone acetyltransferase-dependent chromatin transcription. *The Journal of Biological Chemistry*, *279*, 51163–51171.

Baylin, S. B. (2005). DNA methylation and gene silencing in cancer. *Nature Clinical Practice Oncology*, *2*(Suppl. 1), S4–11.

Bedford, M. T. (2007). Arginine methylation at a glance. *Journal of Cell Science*, *120*, 4243–4246.

Bedford, M. T., & Richard, S. (2005). Arginine methylation an emerging regulator of protein function. *Molecular Cell*, *18*, 263–272.

Belotserkovskaya, R., Oh, S., Bondarenko, V. A., Orphanides, G., Studitsky, V. M., & Reinberg, D. (2003). FACT facilitates transcription-dependent nucleosome alteration. *Science*, *301*, 1090–1093.

Belotserkovskaya, R., & Reinberg, D. (2004). Facts about FACT and transcript elongation through chromatin. *Current Opinion in Genetics & Development*, *14*, 139–146.

Bird, A. (2002). DNA methylation patterns and epigenetic memory. *Genes & Development*, *16*, 6–21.

Black, B. L., & Olson, E. N. (1998). Transcriptional control of muscle development by myocyte enhancer factor-2 (MEF2) proteins. *Annual Review of Cell and Developmental Biology*, *14*, 167–196.

Blais, A., van Oevelen, C. J., Margueron, R., Acosta-Alvear, D., & Dynlacht, B. D. (2007). Retinoblastoma tumor suppressor protein-dependent methylation of histone H3 lysine 27 is associated with irreversible cell cycle exit. *The Journal of Cell Biology*, *179*, 1399–1412.

Bogdanovic, O., & Veenstra, G. J. (2009). DNA methylation and methyl-CpG binding proteins: Developmental requirements and function. *Chromosoma*, *118*, 549–565.

Boonsanay, V., Zhang, T., Georgieva, A., Kostin, S., Qi, H., Yuan, X., et al. (2016). Regulation of skeletal muscle stem cell quiescence by Suv4-20h1-dependent facultative heterochromatin formation. *Cell Stem Cell*, *18*, 229–242.

Bortvin, A., & Winston, F. (1996). Evidence that Spt6p controls chromatin structure by a direct interaction with histones. *Science*, *272*, 1473–1476.

Bowers, E. M., Yan, G., Mukherjee, C., Orry, A., Wang, L., Holbert, M. A., et al. (2010). Virtual ligand screening of the p300/CBP histone acetyltransferase: Identification of a selective small molecule inhibitor. *Chemistry & Biology*, *17*, 471–482.

Bracken, A. P., Dietrich, N., Pasini, D., Hansen, K. H., & Helin, K. (2006). Genome-wide mapping of Polycomb target genes unravels their roles in cell fate transitions. *Genes & Development*, *20*, 1123–1136.

Brueckner, B., Garcia Boy, R., Siedlecki, P., Musch, T., Kliem, H. C., Zielenkiewicz, P., et al. (2005). Epigenetic reactivation of tumor suppressor genes by a novel small-molecule inhibitor of human DNA methyltransferases. *Cancer Research*, *65*, 6305–6311.

Bruhn, S. L., Pil, P. M., Essigmann, J. M., Housman, D. E., & Lippard, S. J. (1992). Isolation and characterization of human cDNA clones encoding a high mobility group box protein that recognizes structural distortions to DNA caused by binding of the anticancer agent cisplatin. *Proceedings of the National Academy of Sciences of the United States of America, 89*, 2307–2311.

Brunk, B. P., Goldhamer, D. J., & Emerson, C. P., Jr. (1996). Regulated demethylation of the myoD distal enhancer during skeletal myogenesis. *Developmental Biology, 177*, 490–503.

Calvanese, V., Fernandez, A. F., Urdinguio, R. G., Suarez-Alvarez, B., Mangas, C., Perez-Garcia, V., et al. (2012). A promoter DNA demethylation landscape of human hematopoietic differentiation. *Nucleic Acids Research, 40*, 116–131.

Cao, Y., Yao, Z., Sarkar, D., Lawrence, M., Sanchez, G. J., Parker, M. H., et al. (2010). Genome-wide MyoD binding in skeletal muscle cells: A potential for broad cellular reprogramming. *Developmental Cell, 18*, 662–674.

Caretti, G., Di Padova, M., Micales, B., Lyons, G. E., & Sartorelli, V. (2004). The Polycomb Ezh2 methyltransferase regulates muscle gene expression and skeletal muscle differentiation. *Genes & Development, 18*, 2627–2638.

Carrio, E., Diez-Villanueva, A., Lois, S., Mallona, I., Cases, I., Forn, M., et al. (2015). Deconstruction of DNA methylation patterns during myogenesis reveals specific epigenetic events in the establishment of the skeletal muscle lineage. *Stem Cells, 33*, 2025–2036.

Charge, S. B., & Rudnicki, M. A. (2004). Cellular and molecular regulation of muscle regeneration. *Physiological Reviews, 84*, 209–238.

Chen, S. L., Loffler, K. A., Chen, D., Stallcup, M. R., & Muscat, G. E. (2002). The coactivator-associated arginine methyltransferase is necessary for muscle differentiation: CARM1 coactivates myocyte enhancer factor-2. *The Journal of Biological Chemistry, 277*, 4324–4333.

Cheng, J. C., Matsen, C. B., Gonzales, F. A., Ye, W., Greer, S., Marquez, V. E., et al. (2003). Inhibition of DNA methylation and reactivation of silenced genes by zebularine. *Journal of the National Cancer Institute, 95*, 399–409.

Cho, Y. W., Hong, T., Hong, S., Guo, H., Yu, H., Kim, D., et al. (2007). PTIP associates with MLL3- and MLL4-containing histone H3 lysine 4 methyltransferase complex. *The Journal of Biological Chemistry, 282*, 20395–20406.

Choi, J., Jang, H., Kim, H., Kim, S. T., Cho, E. J., & Youn, H. D. (2010). Histone demethylase LSD1 is required to induce skeletal muscle differentiation by regulating myogenic factors. *Biochemical and Biophysical Research Communications, 401*, 327–332.

Consalvi, S., Mozzetta, C., Bettica, P., Germani, M., Fiorentini, F., Del Bene, F., et al. (2013). Preclinical studies in the mdx mouse model of duchenne muscular dystrophy with the histone deacetylase inhibitor givinostat. *Molecular Medicine, 19*, 79–87.

Consalvi, S., Saccone, V., & Mozzetta, C. (2014). Histone deacetylase inhibitors: A potential epigenetic treatment for Duchenne muscular dystrophy. *Epigenomics, 6*, 547–560.

Constantinides, P. G., Jones, P. A., & Gevers, W. (1977). Functional striated muscle cells from non-myoblast precursors following 5-azacytidine treatment. *Nature, 267*, 364–366.

Dacwag, C. S., Ohkawa, Y., Pal, S., Sif, S., & Imbalzano, A. N. (2007). The protein arginine methyltransferase Prmt5 is required for myogenesis because it facilitates ATP-dependent chromatin remodeling. *Molecular and Cellular Biology, 27*, 384–394.

De Bleecker, J. L., Meire, V. I., Declercq, W., & Van Aken, E. H. (1999). Immunolocalization of tumor necrosis factor-alpha and its receptors in inflammatory myopathies. *Neuromuscular Disorders, 9*, 239–246.

De Koning, L., Corpet, A., Haber, J. E., & Almouzni, G. (2007). Histone chaperones: An escort network regulating histone traffic. *Nature Structural & Molecular Biology, 14*, 997–1007.

de La Serna, I. L., Carlson, K. A., Hill, D. A., Guidi, C. J., Stephenson, R. O., Sif, S., et al. (2000). Mammalian SWI-SNF complexes contribute to activation of the hsp70 gene. *Molecular and Cellular Biology*, *20*, 2839–2851.

de la Serna, I. L., Carlson, K. A., & Imbalzano, A. N. (2001). Mammalian SWI/SNF complexes promote MyoD-mediated muscle differentiation. *Nature Genetics*, *27*, 187–190.

de la Serna, I. L., Ohkawa, Y., & Imbalzano, A. N. (2006). Chromatin remodelling in mammalian differentiation: Lessons from ATP-dependent remodellers. *Nature Reviews. Genetics*, *7*, 461–473.

Diao, Y., Guo, X., Li, Y., Sun, K., Lu, L., Jiang, L., et al. (2012). Pax3/7BP is a Pax7- and Pax3-binding protein that regulates the proliferation of muscle precursor cells by an epigenetic mechanism. *Cell Stem Cell*, *11*, 231–241.

Dilworth, F. J., Seaver, K. J., Fishburn, A. L., Htet, S. L., & Tapscott, S. J. (2004). In vitro transcription system delineates the distinct roles of the coactivators pCAF and p300 during MyoD/E47-dependent transactivation. *Proceedings of the National Academy of Sciences of the United States of America*, *101*, 11593–11598.

Dostie, J., & Dekker, J. (2007). Mapping networks of physical interactions between genomic elements using 5C technology. *Nature Protocols*, *2*, 988–1002.

Dutta, D., Ray, S., Home, P., Saha, B., Wang, S., Sheibani, N., et al. (2010). Regulation of angiogenesis by histone chaperone HIRA-mediated incorporation of lysine 56-acetylated histone H3.3 at chromatin domains of endothelial genes. *The Journal of Biological Chemistry*, *285*, 41567–41577.

Elsaesser, S. J., Goldberg, A. D., & Allis, C. D. (2010). New functions for an old variant: No substitute for histone H3.3. *Current Opinion in Genetics & Development*, *20*, 110–117.

Esteve, P. O., Patnaik, D., Chin, H. G., Benner, J., Teitell, M. A., & Pradhan, S. (2005). Functional analysis of the N- and C-terminus of mammalian G9a histone H3 methyltransferase. *Nucleic Acids Research*, *33*, 3211–3223.

Faralli, H., Wang, C., Nakka, K., Benyoucef, A., Sebastian, S., Zhuang, L., et al. (2016). UTX demethylase activity is required for satellite cell-mediated muscle regeneration. *The Journal of Clinical Investigation*, *126*, 1555–1565.

Filion, G. J., Zhenilo, S., Salozhin, S., Yamada, D., Prokhortchouk, E., & Defossez, P. A. (2006). A family of human zinc finger proteins that bind methylated DNA and repress transcription. *Molecular and Cellular Biology*, *26*, 169–181.

Flowers, S., Nagl, N. G., Jr., Beck, G. R., Jr., & Moran, E. (2009). Antagonistic roles for BRM and BRG1 SWI/SNF complexes in differentiation. *The Journal of Biological Chemistry*, *284*, 10067–10075.

Forcales, S. V., Albini, S., Giordani, L., Malecova, B., Cignolo, L., Chernov, A., et al. (2012). Signal-dependent incorporation of MyoD-BAF60c into Brg1-based SWI/SNF chromatin-remodelling complex. *The EMBO Journal*, *31*, 301–316.

Formosa, T. (2012). The role of FACT in making and breaking nucleosomes. *Biochimica et Biophysica Acta*, *1819*, 247–255.

Fouse, S. D., Shen, Y., Pellegrini, M., Cole, S., Meissner, A., Van Neste, L., et al. (2008). Promoter CpG methylation contributes to ES cell gene regulation in parallel with Oct4/Nanog, PcG complex, and histone H3 K4/K27 trimethylation. *Cell Stem Cell*, *2*, 160–169.

Francetic, T., Le May, M., Hamed, M., Mach, H., Meyers, D., Cole, P. A., et al. (2012). Regulation of Myf5 early enhancer by histone acetyltransferase p300 during stem cell differentiation. *Molecular Biology*, *1*, 5019.

Fritsch, L., Robin, P., Mathieu, J. R., Souidi, M., Hinaux, H., Rougeulle, C., et al. (2010). A subset of the histone H3 lysine 9 methyltransferases Suv39h1, G9a, GLP, and SETDB1 participate in a multimeric complex. *Molecular Cell*, *37*, 46–56.

Fulco, M., Schiltz, R. L., Iezzi, S., King, M. T., Zhao, P., Kashiwaya, Y., et al. (2003). Sir2 regulates skeletal muscle differentiation as a potential sensor of the redox state. *Molecular Cell, 12*, 51–62.

Gillespie, M. A., Le Grand, F., Scime, A., Kuang, S., von Maltzahn, J., Seale, V., et al. (2009). p38-{gamma}-dependent gene silencing restricts entry into the myogenic differentiation program. *The Journal of Cell Biology, 187*, 991–1005.

Goldhamer, D. J., Brunk, B. P., Faerman, A., King, A., Shani, M., & Emerson, C. P., Jr. (1995). Embryonic activation of the myoD gene is regulated by a highly conserved distal control element. *Development, 121*, 637–649.

Hackett, J. A., & Surani, M. A. (2013). DNA methylation dynamics during the mammalian life cycle. *Philosophical Transactions of the Royal Society of London Series B, Biological Sciences, 368*. 20110328.

Hake, S. B., Garcia, B. A., Duncan, E. M., Kauer, M., Dellaire, G., Shabanowitz, J., et al. (2006). Expression patterns and post-translational modifications associated with mammalian histone H3 variants. *The Journal of Biological Chemistry, 281*, 559–568.

Hakelien, A. M., Delbarre, E., Gaustad, K. G., Buendia, B., & Collas, P. (2008). Expression of the myodystrophic R453W mutation of lamin A in C2C12 myoblasts causes promoter-specific and global epigenetic defects. *Experimental Cell Research, 314*, 1869–1880.

Hamed, M., Khilji, S., Chen, J., & Li, Q. (2013). Stepwise acetyltransferase association and histone acetylation at the Myod1 locus during myogenic differentiation. *Scientific Reports, 3*, 2390.

Hansen, J. C., Ghosh, R. P., & Woodcock, C. L. (2010). Binding of the Rett syndrome protein, MeCP2, to methylated and unmethylated DNA and chromatin. *IUBMB Life, 62*, 732–738.

Heerboth, S., Lapinska, K., Snyder, N., Leary, M., Rollinson, S., & Sarkar, S. (2014). Use of epigenetic drugs in disease: An overview. *Genetics & Epigenetics, 6*, 9–19.

Holterman, C. E., & Rudnicki, M. A. (2005). Molecular regulation of satellite cell function. *Seminars in Cell & Developmental Biology, 16*, 575–584.

Hong, S., Cho, Y. W., Yu, L. R., Yu, H., Veenstra, T. D., & Ge, K. (2007). Identification of JmjC domain-containing UTX and JMJD3 as histone H3 lysine 27 demethylases. *Proceedings of the National Academy of Sciences of the United States of America, 104*, 18439–18444.

Hupkes, M., Jonsson, M. K., Scheenen, W. J., van Rotterdam, W., Sotoca, A. M., van Someren, E. P., et al. (2011). Epigenetics: DNA demethylation promotes skeletal myotube maturation. *The FASEB Journal, 25*, 3861–3872.

Hupkes, M., van Someren, E. P., Middelkamp, S. H., Piek, E., van Zoelen, E. J., & Dechering, K. J. (2011). DNA methylation restricts spontaneous multi-lineage differentiation of mesenchymal progenitor cells, but is stable during growth factor-induced terminal differentiation. *Biochimica et Biophysica Acta, 1813*, 839–849.

Iezzi, S., Cossu, G., Nervi, C., Sartorelli, V., & Puri, P. L. (2002). Stage-specific modulation of skeletal myogenesis by inhibitors of nuclear deacetylases. *Proceedings of the National Academy of Sciences of the United States of America, 99*, 7757–7762.

Imai, S., Armstrong, C. M., Kaeberlein, M., & Guarente, L. (2000). Transcriptional silencing and longevity protein Sir2 is an NAD-dependent histone deacetylase. *Nature, 403*, 795–800.

Isagawa, T., Nagae, G., Shiraki, N., Fujita, T., Sato, N., Ishikawa, S., et al. (2011). DNA methylation profiling of embryonic stem cell differentiation into the three germ layers. *PLoS One, 6*. e26052.

Ito, S., D'Alessio, A. C., Taranova, O. V., Hong, K., Sowers, L. C., & Zhang, Y. (2010). Role of Tet proteins in 5mC to 5hmC conversion, ES-cell self-renewal and inner cell mass specification. *Nature, 466*, 1129–1133.

Jang, H., Choi, D. E., Kim, H., Cho, E. J., & Youn, H. D. (2007). Cabin1 represses MEF2 transcriptional activity by association with a methyltransferase, SUV39H1. *The Journal of Biological Chemistry*, *282*, 11172–11179.
Jenuwein, T., & Allis, C. D. (2001). Translating the histone code. *Science*, *293*, 1074–1080.
Jeon, Y., & Lee, J. T. (2011). YY1 tethers Xist RNA to the inactive X nucleation center. *Cell*, *146*, 119–133.
Jorgensen, S., Schotta, G., & Sorensen, C. S. (2013). Histone H4 lysine 20 methylation: Key player in epigenetic regulation of genomic integrity. *Nucleic Acids Research*, *41*, 2797–2806.
Kablar, B., Krastel, K., Ying, C., Tapscott, S. J., Goldhamer, D. J., & Rudnicki, M. A. (1999). Myogenic determination occurs independently in somites and limb buds. *Developmental Biology*, *206*, 219–231.
Kadauke, S., & Blobel, G. A. (2009). Chromatin loops in gene regulation. *Biochimica et Biophysica Acta*, *1789*, 17–25.
Kaimori, J. Y., Maehara, K., Hayashi-Takanaka, Y., Harada, A., Fukuda, M., Yamamoto, S., et al. (2016). Histone H4 lysine 20 acetylation is associated with gene repression in human cells. *Scientific Reports*, *6*. 24318.
Kawabe, Y., Wang, Y. X., McKinnell, I. W., Bedford, M. T., & Rudnicki, M. A. (2012). Carm1 regulates Pax7 transcriptional activity through MLL1/2 recruitment during asymmetric satellite stem cell divisions. *Cell Stem Cell*, *11*, 333–345.
Kemble, D. J., McCullough, L. L., Whitby, F. G., Formosa, T., & Hill, C. P. (2015). FACT disrupts nucleosome structure by binding H2A-H2B with conserved peptide motifs. *Molecular Cell*, *60*, 294–306.
Kok, F. O., Oster, E., Mentzer, L., Hsieh, J. C., Henry, C. A., & Sirotkin, H. I. (2007). The role of the SPT6 chromatin remodeling factor in zebrafish embryogenesis. *Developmental Biology*, *307*, 214–226.
Lafontaine, D. L., & Tollervey, D. (1998). Birth of the snoRNPs: The evolution of the modification-guide snoRNAs. *Trends in Biochemical Sciences*, *23*, 383–388.
Lan, F., Bayliss, P. E., Rinn, J. L., Whetstine, J. R., Wang, J. K., Chen, S., et al. (2007). A histone H3 lysine 27 demethylase regulates animal posterior development. *Nature*, *449*, 689–694.
Landry, J., Sutton, A., Tafrov, S. T., Heller, R. C., Stebbins, J., Pillus, L., et al. (2000). The silencing protein SIR2 and its homologs are NAD-dependent protein deacetylases. *Proceedings of the National Academy of Sciences of the United States of America*, *97*, 5807–5811.
Lee, M. G., Villa, R., Trojer, P., Norman, J., Yan, K. P., Reinberg, D., et al. (2007). Demethylation of H3K27 regulates polycomb recruitment and H2A ubiquitination. *Science*, *318*, 447–450.
Lessard, J. A., & Crabtree, G. R. (2010). Chromatin regulatory mechanisms in pluripotency. *Annual Review of Cell and Developmental Biology*, *26*, 503–532.
Lessard, J., Wu, J. I., Ranish, J. A., Wan, M., Winslow, M. M., Staahl, B. T., et al. (2007). An essential switch in subunit composition of a chromatin remodeling complex during neural development. *Neuron*, *55*, 201–215.
Levine, S. S., King, I. F., & Kingston, R. E. (2004). Division of labor in polycomb group repression. *Trends in Biochemical Sciences*, *29*, 478–485.
Lickert, H., Takeuchi, J. K., Von Both, I., Walls, J. R., McAuliffe, F., Adamson, S. L., et al. (2004). Baf60c is essential for function of BAF chromatin remodelling complexes in heart development. *Nature*, *432*, 107–112.
Liu, Y., Chakroun, I., Yang, D., Horner, E., Liang, J., Aziz, A., et al. (2013). Six1 regulates MyoD expression in adult muscle progenitor cells. *PLoS One*, *8*. e67762.
Liu, L., Cheung, T. H., Charville, G. W., Hurgo, B. M., Leavitt, T., Shih, J., et al. (2013). Chromatin modifications as determinants of muscle stem cell quiescence and chronological aging. *Cell Reports*, *4*, 189–204.

Liu, Y., Chu, A., Chakroun, I., Islam, U., & Blais, A. (2010). Cooperation between myogenic regulatory factors and SIX family transcription factors is important for myoblast differentiation. *Nucleic Acids Research*, *38*, 6857–6871.

Liu, Y., Sun, L., & Jost, J. P. (1996). In differentiating mouse myoblasts DNA methyltransferase is posttranscriptionally and posttranslationally regulated. *Nucleic Acids Research*, *24*, 2718–2722.

Lolis, A. A., Londhe, P., Beggs, B. C., Byrum, S. D., Tackett, A. J., & Davie, J. K. (2013). Myogenin recruits the histone chaperone facilitates chromatin transcription (FACT) to promote nucleosome disassembly at muscle-specific genes. *The Journal of Biological Chemistry*, *288*, 7676–7687.

Loyola, A., Bonaldi, T., Roche, D., Imhof, A., & Almouzni, G. (2006). PTMs on H3 variants before chromatin assembly potentiate their final epigenetic state. *Molecular Cell*, *24*, 309–316.

Lu, J., McKinsey, T. A., Nicol, R. L., & Olson, E. N. (2000). Signal-dependent activation of the MEF2 transcription factor by dissociation from histone deacetylases. *Proceedings of the National Academy of Sciences of the United States of America*, *97*, 4070–4075.

Lu, J., McKinsey, T. A., Zhang, C. L., & Olson, E. N. (2000). Regulation of skeletal myogenesis by association of the MEF2 transcription factor with class II histone deacetylases. *Molecular Cell*, *6*, 233–244.

Lucarelli, M., Fuso, A., Strom, R., & Scarpa, S. (2001). The dynamics of myogenin site-specific demethylation is strongly correlated with its expression and with muscle differentiation. *The Journal of Biological Chemistry*, *276*, 7500–7506.

Majid, S., Dar, A. A., Saini, S., Yamamura, S., Hirata, H., Tanaka, Y., et al. (2010). MicroRNA-205-directed transcriptional activation of tumor suppressor genes in prostate cancer. *Cancer*, *116*, 5637–5649.

Mal, A. K. (2006). Histone methyltransferase Suv39h1 represses MyoD-stimulated myogenic differentiation. *The EMBO Journal*, *25*, 3323–3334.

Mal, A., & Harter, M. L. (2003). MyoD is functionally linked to the silencing of a muscle-specific regulatory gene prior to skeletal myogenesis. *Proceedings of the National Academy of Sciences of the United States of America*, *100*, 1735–1739.

Margueron, R., & Reinberg, D. (2011). The Polycomb complex PRC2 and its mark in life. *Nature*, *469*, 343–349.

Maunakea, A. K., Chepelev, I., Cui, K., & Zhao, K. (2013). Intragenic DNA methylation modulates alternative splicing by recruiting MeCP2 to promote exon recognition. *Cell Research*, *23*, 1256–1269.

Mauro, A. (1961). Satellite cell of skeletal muscle fibers. *The Journal of Biophysical and Biochemical Cytology*, *9*, 493–495.

McKinnell, I. W., Ishibashi, J., Le Grand, F., Punch, V. G., Addicks, G. C., Greenblatt, J. F., et al. (2008). Pax7 activates myogenic genes by recruitment of a histone methyltransferase complex. *Nature Cell Biology*, *10*, 77–84.

McKinsey, T. A., Zhang, C. L., Lu, J., & Olson, E. N. (2000). Signal-dependent nuclear export of a histone deacetylase regulates muscle differentiation. *Nature*, *408*, 106–111.

McKittrick, E., Gafken, P. R., Ahmad, K., & Henikoff, S. (2004). Histone H3.3 is enriched in covalent modifications associated with active chromatin. *Proceedings of the National Academy of Sciences of the United States of America*, *101*, 1525–1530.

Meissner, A., Mikkelsen, T. S., Gu, H., Wernig, M., Hanna, J., Sivachenko, A., et al. (2008). Genome-scale DNA methylation maps of pluripotent and differentiated cells. *Nature*, *454*, 766–770.

Miska, E. A., Karlsson, C., Langley, E., Nielsen, S. J., Pines, J., & Kouzarides, T. (1999). HDAC4 deacetylase associates with and represses the MEF2 transcription factor. *The EMBO Journal*, *18*, 5099–5107.

Mito, Y., Henikoff, J. G., & Henikoff, S. (2005). Genome-scale profiling of histone H3.3 replacement patterns. *Nature Genetics*, *37*, 1090–1097.

Miyata, K., Miyata, T., Nakabayashi, K., Okamura, K., Naito, M., Kawai, T., et al. (2015). DNA methylation analysis of human myoblasts during in vitro myogenic differentiation: De novo methylation of promoters of muscle-related genes and its involvement in transcriptional down-regulation. *Human Molecular Genetics*, *24*, 410–423.

Montesano, A., Luzi, L., Senesi, P., & Terruzzi, I. (2013). Modulation of cell cycle progression by 5-azacytidine is associated with early myogenesis induction in murine myoblasts. *International Journal of Biological Sciences*, *9*, 391–402.

Nagae, G., Isagawa, T., Shiraki, N., Fujita, T., Yamamoto, S., Tsutsumi, S., et al. (2011). Tissue-specific demethylation in CpG-poor promoters during cellular differentiation. *Human Molecular Genetics*, *20*, 2710–2721.

Naito, M., Mori, M., Inagawa, M., Miyata, K., Hashimoto, N., Tanaka, S., et al. (2016). Dnmt3a regulates proliferation of muscle satellite cells via p57Kip2. *PLoS Genetics*, *12*. e1006167.

Nazor, K. L., Altun, G., Lynch, C., Tran, H., Harness, J. V., Slavin, I., et al. (2012). Recurrent variations in DNA methylation in human pluripotent stem cells and their differentiated derivatives. *Cell Stem Cell*, *10*, 620–634.

Niro, C., Demignon, J., Vincent, S., Liu, Y., Giordani, J., Sgarioto, N., et al. (2010). Six1 and Six4 gene expression is necessary to activate the fast-type muscle gene program in the mouse primary myotome. *Developmental Biology*, *338*, 168–182.

Ohkawa, Y., Marfella, C. G., & Imbalzano, A. N. (2006). Skeletal muscle specification by myogenin and Mef2D via the SWI/SNF ATPase Brg1. *The EMBO Journal*, *25*, 490–501.

Ohkawa, Y., Yoshimura, S., Higashi, C., Marfella, C. G., Dacwag, C. S., Tachibana, T., et al. (2007). Myogenin and the SWI/SNF ATPase Brg1 maintain myogenic gene expression at different stages of skeletal myogenesis. *The Journal of Biological Chemistry*, *282*, 6564–6570.

Oikawa, Y., Omori, R., Nishii, T., Ishida, Y., Kawaichi, M., & Matsuda, E. (2011). The methyl-CpG-binding protein CIBZ suppresses myogenic differentiation by directly inhibiting myogenin expression. *Cell Research*, *21*, 1578–1590.

Orphanides, G., LeRoy, G., Chang, C. H., Luse, D. S., & Reinberg, D. (1998). FACT, a factor that facilitates transcript elongation through nucleosomes. *Cell*, *92*, 105–116.

Orphanides, G., Wu, W. H., Lane, W. S., Hampsey, M., & Reinberg, D. (1999). The chromatin-specific transcription elongation factor FACT comprises human SPT16 and SSRP1 proteins. *Nature*, *400*, 284–288.

Palacios, D., Summerbell, D., Rigby, P. W., & Boyes, J. (2010). Interplay between DNA methylation and transcription factor availability: Implications for developmental activation of the mouse Myogenin gene. *Molecular and Cellular Biology*, *30*, 3805–3815.

Pallafacchina, G., Francois, S., Regnault, B., Czarny, B., Dive, V., Cumano, A., et al. (2010). An adult tissue-specific stem cell in its niche: A gene profiling analysis of in vivo quiescent and activated muscle satellite cells. *Stem Cell Research*, *4*, 77–91.

Pawlikowski, B., Pulliam, C., Betta, N. D., Kardon, G., & Olwin, B. B. (2015). Pervasive satellite cell contribution to uninjured adult muscle fibers. *Skeletal Muscle*, *5*, 42.

Phelan, M. L., Sif, S., Narlikar, G. J., & Kingston, R. E. (1999). Reconstitution of a core chromatin remodeling complex from SWI/SNF subunits. *Molecular Cell*, *3*, 247–253.

Porter, G. A., Jr., Makuck, R. F., & Rivkees, S. A. (2002). Reduction in intracellular calcium levels inhibits myoblast differentiation. *The Journal of Biological Chemistry*, *277*, 28942–28947.

Puri, P. L., Iezzi, S., Stiegler, P., Chen, T. T., Schiltz, R. L., Muscat, G. E., et al. (2001). Class I histone deacetylases sequentially interact with MyoD and pRb during skeletal myogenesis. *Molecular Cell*, *8*, 885–897.

Puri, P. L., Sartorelli, V., Yang, X. J., Hamamori, Y., Ogryzko, V. V., Howard, B. H., et al. (1997). Differential roles of p300 and PCAF acetyltransferases in muscle differentiation. *Molecular Cell*, *1*, 35–45.

Rampalli, S., Li, L., Mak, E., Ge, K., Brand, M., Tapscott, S. J., et al. (2007). p38 MAPK signaling regulates recruitment of Ash2L-containing methyltransferase complexes to specific genes during differentiation. *Nature Structural & Molecular Biology, 14,* 1150–1156.

Ransom, M., Dennehey, B. K., & Tyler, J. K. (2010). Chaperoning histones during DNA replication and repair. *Cell, 140,* 183–195.

Ray-Gallet, D., Quivy, J. P., Scamps, C., Martini, E. M., Lipinski, M., & Almouzni, G. (2002). HIRA is critical for a nucleosome assembly pathway independent of DNA synthesis. *Molecular Cell, 9,* 1091–1100.

Ringrose, L., Ehret, H., & Paro, R. (2004). Distinct contributions of histone H3 lysine 9 and 27 methylation to locus-specific stability of polycomb complexes. *Molecular Cell, 16,* 641–653.

Robertson, K. D., Uzvolgyi, E., Liang, G., Talmadge, C., Sumegi, J., Gonzales, F. A., et al. (1999). The human DNA methyltransferases (DNMTs) 1, 3a and 3b: Coordinate mRNA expression in normal tissues and overexpression in tumors. *Nucleic Acids Research, 27,* 2291–2298.

Roth, J. F., Shikama, N., Henzen, C., Desbaillets, I., Lutz, W., Marino, S., et al. (2003). Differential role of p300 and CBP acetyltransferase during myogenesis: p300 acts upstream of MyoD and Myf5. *The EMBO Journal, 22,* 5186–5196.

Ruthenburg, A. J., Allis, C. D., & Wysocka, J. (2007). Methylation of lysine 4 on histone H3: Intricacy of writing and reading a single epigenetic mark. *Molecular Cell, 25,* 15–30.

Ryall, J. G., Dell'Orso, S., Derfoul, A., Juan, A., Zare, H., Feng, X., et al. (2015). The NAD (+)-dependent SIRT1 deacetylase translates a metabolic switch into regulatory epigenetics in skeletal muscle stem cells. *Cell Stem Cell, 16,* 171–183.

Sasai, N., & Defossez, P. A. (2009). Many paths to one goal? The proteins that recognize methylated DNA in eukaryotes. *The International Journal of Developmental Biology, 53,* 323–334.

Sati, S., Tanwar, V. S., Kumar, K. A., Patowary, A., Jain, V., Ghosh, S., et al. (2012). High resolution methylome map of rat indicates role of intragenic DNA methylation in identification of coding region. *PLoS One,* 7. e31621.

Saunders, A., Werner, J., Andrulis, E. D., Nakayama, T., Hirose, S., Reinberg, D., et al. (2003). Tracking FACT and the RNA polymerase II elongation complex through chromatin in vivo. *Science, 301,* 1094–1096.

Scarpa, S., Lucarelli, M., Palitti, F., Carotti, D., & Strom, R. (1996). Simultaneous myogenin expression and overall DNA hypomethylation promote in vitro myoblast differentiation. *Cell Growth & Differentiation: The Molecular Biology Journal of the American Association for Cancer Research,* 7, 1051–1058.

Seale, P., Sabourin, L. A., Girgis-Gabardo, A., Mansouri, A., Gruss, P., & Rudnicki, M. A. (2000). Pax7 is required for the specification of myogenic satellite cells. *Cell, 102,* 777–786.

Seenundun, S., Rampalli, S., Liu, Q. C., Aziz, A., Palii, C., Hong, S., et al. (2010). UTX mediates demethylation of H3K27me3 at muscle-specific genes during myogenesis. *The EMBO Journal, 29,* 1401–1411.

Shukla, S., Kavak, E., Gregory, M., Imashimizu, M., Shutinoski, B., Kashlev, M., et al. (2011). CTCF-promoted RNA polymerase II pausing links DNA methylation to splicing. *Nature, 479,* 74–79.

Simone, C., Forcales, S. V., Hill, D. A., Imbalzano, A. N., Latella, L., & Puri, P. L. (2004). p38 pathway targets SWI-SNF chromatin-remodeling complex to muscle-specific loci. *Nature Genetics, 36,* 738–743.

Singh, K., & Dilworth, F. J. (2013). Differential modulation of cell cycle progression distinguishes members of the myogenic regulatory factor family of transcription factors. *The FEBS Journal, 280,* 3991–4003.

Smith, C. L., Horowitz-Scherer, R., Flanagan, J. F., Woodcock, C. L., & Peterson, C. L. (2003). Structural analysis of the yeast SWI/SNF chromatin remodeling complex. *Nature Structural Biology, 10*, 141–145.

Soleimani, V. D., Yin, H., Jahani-Asl, A., Ming, H., Kockx, C. E., van Ijcken, W. F., et al. (2012). Snail regulates MyoD binding-site occupancy to direct enhancer switching and differentiation-specific transcription in myogenesis. *Molecular Cell, 47*, 457–468.

Sparrow, D. B., Miska, E. A., Langley, E., Reynaud-Deonauth, S., Kotecha, S., Towers, N., et al. (1999). MEF-2 function is modified by a novel co-repressor, MITR. *The EMBO Journal, 18*, 5085–5098.

Szyf, M., Rouleau, J., Theberge, J., & Bozovic, V. (1992). Induction of myogenic differentiation by an expression vector encoding the DNA methyltransferase cDNA sequence in the antisense orientation. *The Journal of Biological Chemistry, 267*, 12831–12836.

Tagami, H., Ray-Gallet, D., Almouzni, G., & Nakatani, Y. (2004). Histone H3.1 and H3.3 complexes mediate nucleosome assembly pathways dependent or independent of DNA synthesis. *Cell, 116*, 51–61.

Tahiliani, M., Koh, K. P., Shen, Y., Pastor, W. A., Bandukwala, H., Brudno, Y., et al. (2009). Conversion of 5-methylcytosine to 5-hydroxymethylcytosine in mammalian DNA by MLL partner TET1. *Science, 324*, 930–935.

Tanaka, Y., Naruse, I., Maekawa, T., Masuya, H., Shiroishi, T., & Ishii, S. (1997). Abnormal skeletal patterning in embryos lacking a single Cbp allele: A partial similarity with Rubinstein-Taybi syndrome. *Proceedings of the National Academy of Sciences of the United States of America, 94*, 10215–10220.

Tao, Y., Neppl, R. L., Huang, Z. P., Chen, J., Tang, R. H., Cao, R., et al. (2011). The histone methyltransferase Set7/9 promotes myoblast differentiation and myofibril assembly. *The Journal of Cell Biology, 194*, 551–565.

Tapscott, S. J., Lassar, A. B., & Weintraub, H. (1992). A novel myoblast enhancer element mediates MyoD transcription. *Molecular and Cellular Biology, 12*, 4994–5003.

Tews, D. S., & Goebel, H. H. (1996). Cytokine expression profile in idiopathic inflammatory myopathies. *Journal of Neuropathology and Experimental Neurology, 55*, 342–347.

Toto, P. C., Puri, P. L., & Albini, S. (2016). SWI/SNF-directed stem cell lineage specification: Dynamic composition regulates specific stages of skeletal myogenesis. *Cellular and Molecular Life Sciences, 73*, 3887–3896.

Trasler, J., Deng, L., Melnyk, S., Pogribny, I., Hiou-Tim, F., Sibani, S., et al. (2003). Impact of Dnmt1 deficiency, with and without low folate diets, on tumor numbers and DNA methylation in Min mice. *Carcinogenesis, 24*, 39–45.

Trojer, P., & Reinberg, D. (2007). Facultative heterochromatin: Is there a distinctive molecular signature? *Molecular Cell, 28*, 1–13.

Tsumagari, K., Baribault, C., Terragni, J., Varley, K. E., Gertz, J., Pradhan, S., et al. (2013). Early de novo DNA methylation and prolonged demethylation in the muscle lineage. *Epigenetics, 8*, 317–332.

Wang, A. H., Bertos, N. R., Vezmar, M., Pelletier, N., Crosato, M., Heng, H. H., et al. (1999). HDAC4, a human histone deacetylase related to yeast HDA1, is a transcriptional corepressor. *Molecular and Cellular Biology, 19*, 7816–7827.

Wang, W., Cote, J., Xue, Y., Zhou, S., Khavari, P. A., Biggar, S. R., et al. (1996). Purification and biochemical heterogeneity of the mammalian SWI-SNF complex. *The EMBO Journal, 15*, 5370–5382.

Wang, A. H., Zare, H., Mousavi, K., Wang, C., Moravec, C. E., Sirotkin, H. I., et al. (2013). The histone chaperone Spt6 coordinates histone H3K27 demethylation and myogenesis. *The EMBO Journal*.

Warren, G. L., Hulderman, T., Jensen, N., McKinstry, M., Mishra, M., Luster, M. I., et al. (2002). Physiological role of tumor necrosis factor alpha in traumatic muscle injury. *The FASEB Journal, 16*, 1630–1632.

Wilson, B. G., & Roberts, C. W. (2011). SWI/SNF nucleosome remodellers and cancer. *Nature Reviews. Cancer, 11*, 481–492.
Wittmeyer, J., & Formosa, T. (1997). The Saccharomyces cerevisiae DNA polymerase alpha catalytic subunit interacts with Cdc68/Spt16 and with Pob3, a protein similar to an HMG1-like protein. *Molecular and Cellular Biology, 17*, 4178–4190.
Wittmeyer, J., Joss, L., & Formosa, T. (1999). Spt16 and Pob3 of Saccharomyces cerevisiae form an essential, abundant heterodimer that is nuclear, chromatin-associated, and copurifies with DNA polymerase alpha. *Biochemistry, 38*, 8961–8971.
Wu, J. I., Lessard, J., & Crabtree, G. R. (2009). Understanding the words of chromatin regulation. *Cell, 136*, 200–206.
Wu, S. C., & Zhang, Y. (2010). Active DNA demethylation: Many roads lead to Rome. *Nature Reviews. Molecular Cell Biology, 11*, 607–620.
Yan, Z., Wang, Z., Sharova, L., Sharov, A. A., Ling, C., Piao, Y., et al. (2008). BAF250B-associated SWI/SNF chromatin-remodeling complex is required to maintain undifferentiated mouse embryonic stem cells. *Stem Cells, 26*, 1155–1165.
Yang, J. H., Choi, J. H., Jang, H., Park, J. Y., Han, J. W., Youn, H. D., et al. (2011). Histone chaperones cooperate to mediate Mef2-targeted transcriptional regulation during skeletal myogenesis. *Biochemical and Biophysical Research Communications, 407*, 541–547.
Yang, X., Han, H., De Carvalho, D. D., Lay, F. D., Jones, P. A., & Liang, G. (2014). Gene body methylation can alter gene expression and is a therapeutic target in cancer. *Cancer Cell, 26*, 577–590.
Yang, J. H., Song, Y., Seol, J. H., Park, J. Y., Yang, Y. J., Han, J. W., et al. (2011). Myogenic transcriptional activation of MyoD mediated by replication-independent histone deposition. *Proceedings of the National Academy of Sciences of the United States of America, 108*, 85–90.
Yao, T. P., Oh, S. P., Fuchs, M., Zhou, N. D., Ch'ng, L. E., Newsome, D., et al. (1998). Gene dosage-dependent embryonic development and proliferation defects in mice lacking the transcriptional integrator p300. *Cell, 93*, 361–372.
Yarnell, A. T., Oh, S., Reinberg, D., & Lippard, S. J. (2001). Interaction of FACT, SSRP1, and the high mobility group (HMG) domain of SSRP1 with DNA damaged by the anticancer drug cisplatin. *The Journal of Biological Chemistry, 276*, 25736–25741.
Youdell, M. L., Kizer, K. O., Kisseleva-Romanova, E., Fuchs, S. M., Duro, E., Strahl, B. D., et al. (2008). Roles for Ctk1 and Spt6 in regulating the different methylation states of histone H3 lysine 36. *Molecular and Cellular Biology, 28*, 4915–4926.
Youn, H. D., & Liu, J. O. (2000). Cabin1 represses MEF2-dependent Nur77 expression and T cell apoptosis by controlling association of histone deacetylases and acetylases with MEF2. *Immunity, 13*, 85–94.
Young, D. W., Pratap, J., Javed, A., Weiner, B., Ohkawa, Y., van Wijnen, A., et al. (2005). SWI/SNF chromatin remodeling complex is obligatory for BMP2-induced, Runx2-dependent skeletal gene expression that controls osteoblast differentiation. *Journal of Cellular Biochemistry, 94*, 720–730.
Zador, E., Mendler, L., Takacs, V., de Bleecker, J., & Wuytack, F. (2001). Regenerating soleus and extensor digitorum longus muscles of the rat show elevated levels of TNF-alpha and its receptors, TNFR-60 and TNFR-80. *Muscle & Nerve, 24*, 1058–1067.
Zhang, R., Poustovoitov, M. V., Ye, X., Santos, H. A., Chen, W., Daganzo, S. M., et al. (2005). Formation of MacroH2A-containing senescence-associated heterochromatin foci and senescence driven by ASF1a and HIRA. *Developmental Cell, 8*, 19–30.
Zhang, G., & Pradhan, S. (2014). Mammalian epigenetic mechanisms. *IUBMB Life, 66*, 240–256.
Zhu, P., & Li, G. (2016). Structural insights of nucleosome and the 30-nm chromatin fiber. *Current Opinion in Structural Biology, 36*, 106–115.
Ziller, M. J., Gu, H., Muller, F., Donaghey, J., Tsai, L. T., Kohlbacher, O., et al. (2013). Charting a dynamic DNA methylation landscape of the human genome. *Nature, 500*, 477–481.

CHAPTER NINE

Dysregulated Myogenesis in Rhabdomyosarcoma

Peter Y. Yu[*,†], **Denis C. Guttridge**[*,‡,1]
[*]Arthur G. James Comprehensive Cancer Center, The Ohio State University, Columbus, OH, United States
[†]College of Medicine, The Ohio State University, Columbus, OH, United States
[‡]The Ohio State University, Columbus, OH, United States
[1]Corresponding author: e-mail address: denis.guttridge@osumc.edu

Contents

Abstract

Rhabdomyosarcoma is a mesenchymal malignancy associated with the skeletal muscle lineage and is also the most common pediatric soft tissue cancer. Between the two pediatric subtypes, embryonal and alveolar rhabdomyosarcoma, the alveolar subtype is generally more aggressive and high-risk. Despite intensive multimodal therapy, patients with high-risk rhabdomyosarcoma continue to have poor prognosis. In this chapter we address the mechanisms underlying the dysregulation of myogenesis in rhabdomyosarcoma. We specifically focus on recently identified signaling pathways that function to inhibit myogenesis and how similar functions have been shown to overlap in rhabdomyosarcoma, potentially contributing to the disease.

1. INTRODUCTION

Rhabdomyosarcoma (RMS) is the most common soft tissue sarcoma in children and young adults (Loeb, Thornton, & Shokek, 2008). While the cell of origin is unknown, RMSs are defined by myogenic features as they

Current Topics in Developmental Biology, Volume 126
ISSN 0070-2153
https://doi.org/10.1016/bs.ctdb.2017.10.007
© 2018 Elsevier Inc.
All rights reserved.

are skeletal muscle-like tumors that typically arise in muscle and express myogenic regulatory factors (MRFs). Pediatric RMS includes embryonal (ERMS) and alveolar (ARMS) subtypes, and adult rhabdomyosarcoma is pleomorphic rhabdomyosarcoma. ERMS is associated with better outcomes than ARMS, and the relative 5-year survival rate improved from 60.9% in 1976–1980 to 73.4% in 1996–2000. In contrast, 5-year survival rate for ARMS has not improved significantly as this rate has only increased from 40.1% to 47.8% during the same period (Ognjanovic et al., 2010). The higher mortality rate in patients with ARMS reflects its more aggressive and metastatic nature (Wang, 2012). ARMS are characterized genetically by PAX3–FOXO1 or PAX7–FOXO1 gene fusions that correlate with their poor clinical outcome. Reports on the global landscape of mutations on RMS have varied, with recent larger studies suggesting that mutations are rare. Analysis of TP53 mutation and MDM2 amplification in 37 ARMS and 38 ERMS frozen tumor samples revealed only 1.3% mutations in *TP53* and no *MDM2* amplification (Ognjanovic et al., 2012).

All body muscles derive initially from the somite mesoderm (Christ & Ordahl, 1995). As somites migrate, they express Pax3 and Pax7, two transcription factors that control the proliferation and survival of progenitor muscle cells, and are downregulated during progression of myogenesis (Buckingham, 2001). Progenitor cells which continue to express Pax3 and Pax7 are called satellite cells and have the ability to regenerate postnatal muscle (Relaix, Rocancourt, Mansouri, & Buckingham, 2005). During myogenesis, pluripotent mesodermal precursor cells commit to myoblast lineage, proliferate, differentiate, fuse into multinucleated myotubes, and mature to form myofibers. This process is regulated by a family of conserved basic helix–loop–helix (bHLH) transcription factors called MRFs that include MyoD, Myf5, Myogenin, and MRF4 (Weintraub et al., 1991). These MRFs heterodimerize with other bHLH proteins including E-proteins in vitro and in vivo (Lassar et al., 1991). Among the MRFs, MyoD is considered a master myogenic transcription factor, as its activity can convert nonmuscle cells into the muscle lineage. Myf5 cooperates with MyoD to establish and maintain muscle myoblasts in a proliferative state (Emerson, 1990). During differentiation, the expression of Myogenin and MRF4 is induced, and their activities are coordinated with MyoD and Myf5 to complete the differentiation process (Edmondson & Olson, 1989). Myogenin and MRF4 are also differentially activated during differentiation, leading to muscle specialization (Pownall, Gustafsson, & Emerson, 2002).

In this review, we discuss recent genome-wide analyses that have led to insights into the various oncogenic signaling factors in RMS. These oncogenic signals contribute to repressing the activities of MRFs and their ability to differentiate RMS cells. Attempts to push the myogenesis cascade in RMS tumor cells have been discussed as a current potential therapeutic avenue.

2. EXPRESSION AND FUNCTION OF MYOGENIC FACTORS IN RMS

MyoD and Myogenin are expressed in RMS and thus have been used as diagnostic markers to identify the disease (Kumar, Perlman, Harris, Raffeld, & Tsokos, 2000). However, because tumor cells are impaired in differentiation and MyoD and Myogenin function as differentiation transcription factors, there has been significant interest in investigating how the function of these factors are perturbed during the development of RMS. In normal myogenesis, MyoD promotes histone acetylation on a genome-wide scale (Cao et al., 2010). This acetylation is regulated through the MyoD-dependent recruitment of histone-modifying enzymes, such as CREB-binding protein (CBP), p300, and p300/CBP-associated factor, that induce chromatin remodeling and in turn activate genes coding for proteins involved in cell cycle arrest, myoblast fusion, and myotube formation (Sartorelli, Huang, Hamamori, & Kedes, 1997; Sartorelli et al., 1999).

Further, MyoD regulates myogenesis through the interaction of other transcription factors, including MEF2C, RUNX, NFIC, and JDP2. Although the levels of MyoD in RMS cells per se do not significantly change from normal skeletal muscle cells, gene profiling combined with chromatin immunoprecipitation-coupled high-throughput sequencing (ChIP-seq) results has shown that the differentiation defect in RMS cells relates more to the suppression of MyoD cofactors. The decrease in the expression of these cofactors is significant since increasing the expression of these cofactors in RMS is capable of restoring differentiation of RMS cells (MacQuarrie et al., 2013). Recent studies have also shown that the inability of RMS to properly differentiate might be linked to the improper binding of MyoD to its target genes (MacQuarrie et al., 2013; Tapscott, Thayer, & Weintraub, 1993). This was revealed by whole-exome and whole-transcriptome sequencing of ARMS and ERMS that led to the discovery of a recurrent somatic mutation encoding p.Leu122Arg in MyoD in a subset (approximately 10%) of ERMS tumors that caused MyoD to preferentially bind to *MYC* consensus sequences, switching cells to proliferate rather than differentiate (Kohsaka et al., 2014; Van Antwerp, Chen, Chang, & Prochownik, 1992).

3. MYOGENESIS IS BLOCKED IN RMS THROUGH VARIOUS SIGNALING PATHWAYS

RMS cells show histologic and gene expression patterns resembling developmentally arrested skeletal muscle. While RMS tumors are heterogeneous in their molecular alterations, some common pathways have been implicated including nuclear factor-kappa B (NF-κB), yes-associated protein (YAP)/transcriptional coactivator with PDZ-binding motif (TAZ), and Hippo, Notch, Wnt, and MEK/ERK signaling (Fig. 1).

NF-κB functions as a negative regulator of myogenic differentiation (Guttridge, Albanese, Reuther, Pestell, & Baldwin, 1999; Guttridge, Mayo, Madrid, Wang, & Baldwin, 2000; Peterson, Bakkar, & Guttridge, 2011). NF-κB expression is increased in RMS, and it suppresses differentiation through decreased expression of prodifferentiator miR-29 through Ying Yang 1 and the PRC2 polycomb complex (Wang et al., 2008). In ARMS cells with PAX3/7–FHKR translocation, NF-κB activity is upregulated as a function of PAX3/7–FKHR expression due to upregulation of the PI3K/AKT pathway. In this context, NF-κB was shown to inhibit myogenesis via activation of cyclinD1/cdk4 complexes, which sequester MyoD (Charytonowicz et al., 2012).

YAP/TAZ are inactivated by the Hippo pathway, which consists of components MST1/2, SAV1, and LATS1/2 (Wackerhage, Del Re,

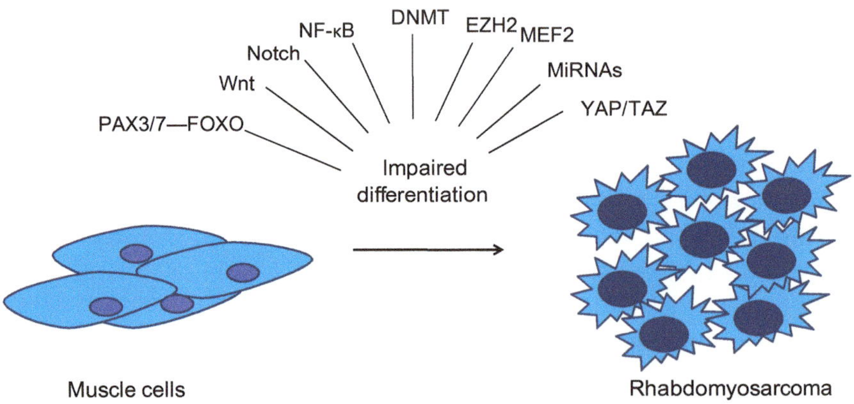

Fig. 1 Modulators of several signaling pathways that are discussed in the chapter as negative regulators of myogenesis and potential contributors to the development of rhabdomyosarcoma.

Judson, Sudol, & Sadoshima, 2014). The proliferation of activated Pax7+ and MyoD+ muscle progenitors cells is promoted by YAP activity, and this activity must be inactivated for muscle differentiation to occur (Judson et al., 2012; Watt et al., 2010). YAP overexpression has been shown to reduce MyoD expression in *Xenopus laevis* embryos in vivo (Gee, Milgram, Kramer, Conlon, & Moody, 2011). In RMS, YAP protein levels are elevated in RMS cells, and YAP nuclear staining is increased in RMS tumors. Overexpression of YAP in activated satellite cells generates tumors similar to ERMS, and knockdown of YAP rescues tumorigenicity (Tremblay et al., 2014). In ARMS containing the PAX3–FOXO1 fusion, PAX3–FOXO1 directly upregulates RASSF4, which associates with MST1 and inhibits its tumor suppressor function (Crose et al., 2014). Thus, YAP and the Hippo signaling pathways are implicated in the maintenance of proproliferative, antidifferentiation phenotypes in ERMS (Svalina & Keller, 2014).

In myoblasts, Notch signaling is another important regulator of myogenesis as Notch1 activation sustains myoblast proliferation, and Notch3 activation inhibits myoblast fusion and differentiation (De Salvo et al., 2014). In RMS, Notch signaling is upregulated and cross talks with Hippo, Hedgehog, Wnt, and TGF-β pathways (Conti, Slemmons, Rota, & Linardic, 2016). Specifically in ERMS, Notch1 signaling is activated and functions in these tumors by regulating SNAIL1 and blocking MEF2C-regulated differentiation, resulting in dedifferentiation of ERMS cells into self-renewing tumor-propagating cells (Ignatius et al., 2017). Another target of Notch signaling is Hairy/enhancer-of-split (Hes) related with YRPW motif 1 (Hey1), which is a bHLH transcription factor known to block myogenin transcription by binding to the myogenin promoter during myogenesis (Buas, Kabak, & Kadesch, 2010). In ERMS, Notch inhibition blocked tumorigenesis in vivo, but did not exhibit evidence of terminal differentiation (Belyea, Naini, Bentley, & Linardic, 2011). Inhibition of the Notch pathway showed a significant decrease in RMS cell mobility and invasiveness in vitro (Roma, Masia, Reventos, Sanchez de Toledo, & Gallego, 2011). Notch3 in these tumors is also required as depletion of Notch3-reduced PAX3–FOXO1+ ARMS tumor growth in vivo (De Salvo et al., 2014). Notch3 can also be activated in ARMS where it functions to inhibit the differentiation of tumor cells by activating the expression of Hes1 (Raimondi et al., 2012). Inhibition of Notch3 resulted in differentiation of RMS cells as evidenced by formation of multinucleated myotubes expressing myosin heavy chain (MYH) and reduced the tumorigenic potential of RMS cells both in vitro and in vivo (Raimondi et al., 2012).

Wnt signaling is another major pathway impacting the development of skeletal muscle at various levels including cell proliferation, differentiation, and hemeostasis (Chen, 2016; von Maltzahn, Chang, Bentzinger, & Rudnicki, 2012). Wnt signaling performs these functions, in part through the positive regulation of MRFs, Myf5, and MyoD (Munsterberg, Kitajewski, Bumcrot, McMahon, & Lassar, 1995; Tajbakhsh et al., 1998). In keeping with its function to promote myogenesis, activation of Wnt via its natural agonist Wnt3a was shown to decrease proliferation and induce myogenic differentiation in ARMS cells, suggesting that Wnt signaling in RMS functions as a tumor suppressor (Annavarapu et al., 2013).

Although not necessarily a signaling mediator, DNA methyltransferase (DNMT) enzymes are responsible for de novo DNA methylation during DNA replication that also impinge on gene expression (Kim, Samaranayake, & Pradhan, 2009). DNMT3B is a specific family member that has been found to be upregulated in patient RMS tumors and cell lines compared to normal skeletal muscle (Megiorni et al., 2016). Functionally, DNMT3B appears to also function as a negative regulator of myogenesis in RMS as knockdown of this gene in ERMS cells promotes myogenic terminal differentiation by increasing the expression of MyoD, Myogenein, and MYH. DNMT3B silencing also reversed the oncogenic phenotype by decreasing migratory capacity and clonogenic potential, and with regards to cell proliferation causes RMS cells to undergo G1 arrest by reducing expression of cell cycle regulators, Cyclin B1, Cyclin D1, and Cyclin E2, while increasing expression of p21 and p27. Inhibition of MEK/ERK signaling resulted in reduced DNMT3B protein, suggesting that DNMT3B functions downstream of MEK/ERK signaling (Megiorni et al., 2016).

4. PATHOGENIC REGULATORS OF DIFFERENTIATION IN RMS

4.1 PAX3/7–FOXO1 Fusion in ARMS

The PAX3/7–FOXO1 fusions, which are the hallmark feature of ARMS, also serve as inhibitors of myogenic differentiation of RMS tumor cells. As described earlier, PAX3–FOXO1 promotes tumorigenesis via Hippo pathway suppression (Crose et al., 2014). PAX3–FOXO1 and PAX7–FOXO1 products can also prevent myogenic differentiation by suppressing the transcriptional activation of MyoD-target genes (Calhabeu, Hayashi, Morgan, Relaix, & Zammit, 2013), even while increasing MyoD expression. P-cadherin was found to be a direct target of PAX3–FOXO1 in ARMS cell lines (Thuault et al., 2013). P-cadherin is expressed in muscle progenitor cells in both the dermatomyotome and primary myotome in vivo, and its expression decreases once muscle progenitor cells begin to express MRFs. In metastatic ARMS, P-cadherin expression is increased, which promotes a cadherin switch where N- and M-cadherins are downregulated. Cadherins play a critical role in tumor progression and metastasis (Christofori, 2003), as P-cadherin is involved in ARMS cell proliferation, transformation, migration, and invasion (Thuault et al., 2013). Another direct target of the PAX3–FOXO1 fusion protein is *JARID2*, which encodes a protein known to recruit various histone-methylating complexes to their target genes. *JARID2* is associated with metastases at diagnosis of RMS, and silencing *JARID2* results in reduced cell proliferation and increased expression of myogenin and myosin light chain in both ERMS and ARMS. JARID2 was shown to bind to and alter methylation of histone H3 lysine 27 (H3K27) in the promoter regions of MRFs, and this interaction is dependent on embryonic ectoderm development (EED), a core component of the polycomb repressive complex 2. Through this mechanism, JARID2 maintains the undifferentiated myogenic phenotype of RMS (Walters et al., 2014).

4.2 EZH2

The Polycomb group proteins are epigenetic silencers of transcription that have key roles in differentiation by functioning within complexes called Polycomb repressive complexes (PRC) (Di Croce & Helin, 2013). PRC2 complex catalyzes H3K27 trimethylation and contains several components including Suppressor of zeste 12 homolog (Suz12), EED, and the catalytic subunit Enhancer of zeste homolog 2 (EZH2). In skeletal muscle progenitors, EZH2 represses muscle-specific gene expression and is downregulated during differentiation (Caretti, Di Padova, Micales, Lyons, & Sartorelli, 2004; Ciarapica et al., 2009; Marchesi, Fiorentino, Rizzolio, Giordano, & Bagella, 2012). In ERMS cells, EZH2 directly binds muscle-specific gene promoters including myogenein, MYH, and MyoD. Ablation of EZH2 in ERMS cells decreases the trimethylation of H3K27 and increases expression of Myogenin, MYH, and MyoD. In addition, MyoD binding to muscle-specific promoters was increased, resulting in partial reversion of the tumor phenotype by the formation of myotubes in culture (Marchesi et al., 2012). In vivo studies have shown that ARMS tumor growth is impaired when EZH2 function is blocked through the use of two different types of inhibitors, leading to tumor cell apoptosis (Ciarapica et al., 2009). Thus, reducing the activity of EZH2 has been proposed as a novel adjuvant strategy against high-risk ARMS (Ciarapica et al., 2014; Ramaglia et al., 2016).

4.3 MEF2 family

The myocyte enhancer factor 2 (MEF2) family of proteins regulate many developmental programs including myogenesis (Potthoff & Olson, 2007). Both MEF2C and MEF2D regulate myogenesis (Penn, Bergstrom, Dilworth, Bengal, & Tapscott, 2004; Potthoff & Olson, 2007) by working in combination with MRFs (Molkentin, Black, Martin, & Olson, 1995; Penn et al., 2004; Potthoff & Olson, 2007). In RMS cell lines, MEF2D levels are downregulated and there is reduced association with muscle-specific promoters (RNNI2, LMOD2, DES, and CKM), while MRFs and E proteins remain present (Zhang, Truscott, & Davie, 2013). Similar to other myogenic regulators, reexpression of MEF2D in ARMS cells promotes their differentiation as determined by expression of MYH. This phenotype is associated with a reduction in proliferation, motility, and anchorage-independent growth in vitro, and strong inhibition of tumor growth in vivo (Zhang et al., 2013). Regulation of MEF2C in ARMS cells follows a similar pattern to MEF2D (Nishijo et al., 2009), and functionally, exogenous addition of MEF2C in EMRS promotes the expression of differentiation-specific genes and increases MyoD binding at sites of its target genes, suggesting that MEF2C might stabilize binding of MyoD or increase accessibility of these sites to MyoD (MacQuarrie et al., 2013). Activation of MEF2C during myogenesis is also maintained through the alternative splicing of its mRNA. The MEF2Cα1 isoform lacks myogenic activity, while the MEF2Cα2 isoform is required for efficient differentiation of human myoblast cells and RMS cells (Zhang, Zhu, & Davie, 2015). The switch between MEF2Cα1 to MEF2Cα2 is regulated by SRSF protein kinase 3 (SRPK3), which itself is downregulated in RMS cells. Further, the MEF2Cα1 isoform preferentially interacts with HDAC5, and as a complex functions in transcriptional suppression of myogenic genes. In contrast, addition of the MEF2Cα2 isoform inhibits recruitment of HDAC4 and HDAC5 to target promoters, leading to a block in both proliferation- and anchorage-dependent growth in RMS cells (Zhang et al., 2015).

4.4 MicroRNAs

A group of miRNAs have been referred to as myomiRs (miR-1, -133a/b, -206) for their contribution to promoting skeletal muscle differentiation

(Ciesla, Dulak, & Jozkowicz, 2014; Ju, Yang, Sheng, & Jiang, 2015; Novak, Vinklarek, Bienertova-Vasku, & Slaby, 2013; Romania et al., 2012). They subsequently been shown to possess anticancer functions in numerous malignancies including RMS. Among these, miR-206 has been the most studied in the context of myogenesis in RMS. While miR-206 increases during myogenesis, its levels are downregulated in RMS (Macquarrie, Yao, Young, Cao, & Tapscott, 2012). Functionally, miR-206 targets antidifferentiation mediators, PAX7, PAX3, NOTCH3, and CCND2 for degradation, thus promoting differentiation (Hanna et al., 2016). In addition, miR-206 directly targets the BAF53a subunit of the SWI/SNF complex. BAF53a interferes with differentiation in myogenic cells and silencing BAF53a in RMS increases the expression of myogenic markers and inhibits proliferation and anchorage-independent growth (Taulli et al., 2014).

5. FUTURE DIRECTIONS

As this review points out, there are numerous strategies to arrest the growth of RMS cells. Our understanding of how MRFs are regulated during sarcomagenesis and how signaling pathways are modulated has provided pivotal insight in how RMS tumors develop. As the mechanisms are elucidated and potential therapeutic options are considered, the potential of differentiation therapy will need to be thoroughly tested. While some evidence points toward differentiation as a potential therapeutic alternative to chemoradiotherapy options (Cruz & Matushansky, 2012), recent evidence also points to oncogenic roles for MyoD and Myf5 (Tenente et al., 2017). Further investigation is warranted in understanding the arrest of myogenesis in RMS and the potential therapeutic implications.

REFERENCES

Annavarapu, S. R., Cialfi, S., Dominici, C., Kokai, G. K., Uccini, S., Ceccarelli, S., et al. (2013). Characterization of Wnt/beta-catenin signaling in rhabdomyosarcoma. *Laboratory Investigation: A Journal of Technical Methods and Pathology*, *93*, 1090–1099.

Belyea, B. C., Naini, S., Bentley, R. C., & Linardic, C. M. (2011). Inhibition of the Notch-Hey1 axis blocks embryonal rhabdomyosarcoma tumorigenesis. *Clinical Cancer Research: An Official Journal of the American Association for Cancer Research*, *17*, 7324–7336.

Buas, M. F., Kabak, S., & Kadesch, T. (2010). The Notch effector Hey1 associates with myogenic target genes to repress myogenesis. *The Journal of Biological Chemistry*, *285*, 1249–1258.

Buckingham, M. (2001). Skeletal muscle formation in vertebrates. *Current Opinion in Genetics & Development*, *11*, 440–448.

Calhabeu, F., Hayashi, S., Morgan, J. E., Relaix, F., & Zammit, P. S. (2013). Alveolar rhabdomyosarcoma-associated proteins PAX3/FOXO1A and PAX7/FOXO1A suppress the transcriptional activity of MyoD-target genes in muscle stem cells. *Oncogene, 32*, 651–662.

Cao, Y., Yao, Z., Sarkar, D., Lawrence, M., Sanchez, G. J., Parker, M. H., et al. (2010). Genome-wide MyoD binding in skeletal muscle cells: A potential for broad cellular reprogramming. *Developmental Cell, 18*, 662–674.

Caretti, G., Di Padova, M., Micales, B., Lyons, G. E., & Sartorelli, V. (2004). The Polycomb Ezh2 methyltransferase regulates muscle gene expression and skeletal muscle differentiation. *Genes & Development, 18*, 2627–2638.

Charytonowicz, E., Matushansky, I., Domenech, J. D., Castillo-Martin, M., Ladanyi, M., Cordon-Cardo, C., et al. (2012). PAX7-FKHR fusion gene inhibits myogenic differentiation via NF-kappaB upregulation. *Clinical & Translational Oncology: Official Publication of the Federation of Spanish Oncology Societies and of the National Cancer Institute of Mexico, 14*, 197–206.

Chen, E. (2016). Wnt signaling in rhabdomyosarcoma—A potential targeted therapy option. *Current Drug Targets, 17*, 1245–1251.

Christ, B., & Ordahl, C. P. (1995). Early stages of chick somite development. *Anatomy and Embryology, 191*, 381–396.

Christofori, G. (2003). Changing neighbours, changing behaviour: Cell adhesion molecule-mediated signalling during tumour progression. *The EMBO Journal, 22*, 2318–2323.

Ciarapica, R., De Salvo, M., Carcarino, E., Bracaglia, G., Adesso, L., Leoncini, P. P., et al. (2014). The Polycomb group (PcG) protein EZH2 supports the survival of PAX3-FOXO1 alveolar rhabdomyosarcoma by repressing FBXO32 (Atrogin1/MAFbx). *Oncogene, 33*, 4173–4184.

Ciarapica, R., Russo, G., Verginelli, F., Raimondi, L., Donfrancesco, A., Rota, R., et al. (2009). Deregulated expression of miR-26a and Ezh2 in rhabdomyosarcoma. *Cell Cycle, 8*, 172–175.

Ciesla, M., Dulak, J., & Jozkowicz, A. (2014). MicroRNAs and epigenetic mechanisms of rhabdomyosarcoma development. *The International Journal of Biochemistry & Cell Biology, 53*, 482–492.

Conti, B., Slemmons, K. K., Rota, R., & Linardic, C. M. (2016). Recent insights into Notch signaling in embryonal rhabdomyosarcoma. *Current Drug Targets, 17*, 1235–1244.

Crose, L. E., Galindo, K. A., Kephart, J. G., Chen, C., Fitamant, J., Bardeesy, N., et al. (2014). Alveolar rhabdomyosarcoma-associated PAX3-FOXO1 promotes tumorigenesis via Hippo pathway suppression. *The Journal of Clinical Investigation, 124*, 285–296.

Cruz, F. D., & Matushansky, I. (2012). Solid tumor differentiation therapy—Is it possible? *Oncotarget, 3*, 559–567.

De Salvo, M., Raimondi, L., Vella, S., Adesso, L., Ciarapica, R., Verginelli, F., et al. (2014). Hyper-activation of Notch3 amplifies the proliferative potential of rhabdomyosarcoma cells. *PLoS One, 9*, e96238.

Di Croce, L., & Helin, K. (2013). Transcriptional regulation by Polycomb group proteins. *Nature Structural & Molecular Biology, 20*, 1147–1155.

Edmondson, D. G., & Olson, E. N. (1989). A gene with homology to the myc similarity region of MyoD1 is expressed during myogenesis and is sufficient to activate the muscle differentiation program. *Genes & Development, 3*, 628–640.

Emerson, C. P. (1990). Myogenesis and developmental control genes. *Current Opinion in Cell Biology, 2*, 1065–1075.

Gee, S. T., Milgram, S. L., Kramer, K. L., Conlon, F. L., & Moody, S. A. (2011). Yes-associated protein 65 (YAP) expands neural progenitors and regulates Pax3 expression in the neural plate border zone. *PLoS One, 6*, e20309.

Guttridge, D. C., Albanese, C., Reuther, J. Y., Pestell, R. G., & Baldwin, A. S., Jr. (1999). NF-kappaB controls cell growth and differentiation through transcriptional regulation of cyclin D1. *Molecular and Cellular Biology, 19*, 5785–5799.

Guttridge, D. C., Mayo, M. W., Madrid, L. V., Wang, C. Y., & Baldwin, A. S., Jr. (2000). NF-kappaB-induced loss of MyoD messenger RNA: Possible role in muscle decay and cachexia. *Science, 289*, 2363–2366.

Hanna, J. A., Garcia, M. R., Go, J. C., Finkelstein, D., Kodali, K., Pagala, V., et al. (2016). PAX7 is a required target for microRNA-206-induced differentiation of fusion-negative rhabdomyosarcoma. *Cell Death & Disease, 7*, e2256.

Ignatius, M. S., Hayes, M. N., Lobbardi, R., Chen, E. Y., McCarthy, K. M., Sreenivas, P., et al. (2017). The NOTCH1/SNAIL1/MEF2C pathway regulates growth and self-renewal in embryonal rhabdomyosarcoma. *Cell Reports, 19*, 2304–2318.

Ju, H., Yang, Y., Sheng, A., & Jiang, X. (2015). Role of microRNAs in skeletal muscle development and rhabdomyosarcoma (review). *Molecular Medicine Reports, 11*, 4019–4024.

Judson, R. N., Tremblay, A. M., Knopp, P., White, R. B., Urcia, R., De Bari, C., et al. (2012). The Hippo pathway member Yap plays a key role in influencing fate decisions in muscle satellite cells. *Journal of Cell Science, 125*, 6009–6019.

Kim, J. K., Samaranayake, M., & Pradhan, S. (2009). Epigenetic mechanisms in mammals. *Cellular and Molecular Life Sciences: CMLS, 66*, 596–612.

Kohsaka, S., Shukla, N., Ameur, N., Ito, T., Ng, C. K., Wang, L., et al. (2014). A recurrent neomorphic mutation in MYOD1 defines a clinically aggressive subset of embryonal rhabdomyosarcoma associated with PI3K-AKT pathway mutations. *Nature Genetics, 46*, 595–600.

Kumar, S., Perlman, E., Harris, C. A., Raffeld, M., & Tsokos, M. (2000). Myogenin is a specific marker for rhabdomyosarcoma: An immunohistochemical study in paraffin-embedded tissues. *Modern Pathology: An Official Journal of the United States and Canadian Academy of Pathology, Inc, 13*, 988–993.

Lassar, A. B., Davis, R. L., Wright, W. E., Kadesch, T., Murre, C., Voronova, A., et al. (1991). Functional activity of myogenic HLH proteins requires hetero-oligomerization with E12/E47-like proteins in vivo. *Cell, 66*, 305–315.

Loeb, D. M., Thornton, K., & Shokek, O. (2008). Pediatric soft tissue sarcomas. *The Surgical Clinics of North America, 88*, 615–627. vii.

MacQuarrie, K. L., Yao, Z., Fong, A. P., Diede, S. J., Rudzinski, E. R., Hawkins, D. S., et al. (2013). Comparison of genome-wide binding of MyoD in normal human myogenic cells and rhabdomyosarcomas identifies regional and local suppression of promyogenic transcription factors. *Molecular and Cellular Biology, 33*, 773–784.

Macquarrie, K. L., Yao, Z., Young, J. M., Cao, Y., & Tapscott, S. J. (2012). miR-206 integrates multiple components of differentiation pathways to control the transition from growth to differentiation in rhabdomyosarcoma cells. *Skeletal Muscle, 2*, 7.

Marchesi, I., Fiorentino, F. P., Rizzolio, F., Giordano, A., & Bagella, L. (2012). The ablation of EZH2 uncovers its crucial role in rhabdomyosarcoma formation. *Cell Cycle, 11*, 3828–3836.

Megiorni, F., Camero, S., Ceccarelli, S., McDowell, H. P., Mannarino, O., Marampon, F., et al. (2016). DNMT3B in vitro knocking-down is able to reverse embryonal rhabdomyosarcoma cell phenotype through inhibition of proliferation and induction of myogenic differentiation. *Oncotarget, 7*, 79342–79356.

Molkentin, J. D., Black, B. L., Martin, J. F., & Olson, E. N. (1995). Cooperative activation of muscle gene expression by MEF2 and myogenic bHLH proteins. *Cell, 83*, 1125–1136.

Munsterberg, A. E., Kitajewski, J., Bumcrot, D. A., McMahon, A. P., & Lassar, A. B. (1995). Combinatorial signaling by Sonic hedgehog and Wnt family members induces myogenic bHLH gene expression in the somite. *Genes & Development, 9*, 2911–2922.

Nishijo, K., Chen, Q. R., Zhang, L., McCleish, A. T., Rodriguez, A., Cho, M. J., et al. (2009). Credentialing a preclinical mouse model of alveolar rhabdomyosarcoma. *Cancer Research, 69*, 2902–2911.

Novak, J., Vinklarek, J., Bienertova-Vasku, J., & Slaby, O. (2013). MicroRNAs involved in skeletal muscle development and their roles in rhabdomyosarcoma pathogenesis. *Pediatric Blood & Cancer*, *60*, 1739–1746.

Ognjanovic, S., Carozza, S. E., Chow, E. J., Fox, E. E., Horel, S., McLaughlin, C. C., et al. (2010). Birth characteristics and the risk of childhood rhabdomyosarcoma based on histological subtype. *British Journal of Cancer*, *102*, 227–231.

Ognjanovic, S., Martel, G., Manivel, C., Olivier, M., Langer, E., & Hainaut, P. (2012). Low prevalence of TP53 mutations and MDM2 amplifications in pediatric rhabdomyosarcoma. *Sarcoma*, *2012*. 492086.

Penn, B. H., Bergstrom, D. A., Dilworth, F. J., Bengal, E., & Tapscott, S. J. (2004). A MyoD-generated feed-forward circuit temporally patterns gene expression during skeletal muscle differentiation. *Genes & Development*, *18*, 2348–2353.

Peterson, J. M., Bakkar, N., & Guttridge, D. C. (2011). NF-kappaB signaling in skeletal muscle health and disease. *Current Topics in Developmental Biology*, *96*, 85–119.

Potthoff, M. J., & Olson, E. N. (2007). MEF2: A central regulator of diverse developmental programs. *Development*, *134*, 4131–4140.

Pownall, M. E., Gustafsson, M. K., & Emerson, C. P., Jr. (2002). Myogenic regulatory factors and the specification of muscle progenitors in vertebrate embryos. *Annual Review of Cell and Developmental Biology*, *18*, 747–783.

Raimondi, L., Ciarapica, R., De Salvo, M., Verginelli, F., Gueguen, M., Martini, C., et al. (2012). Inhibition of Notch3 signalling induces rhabdomyosarcoma cell differentiation promoting p38 phosphorylation and p21(Cip1) expression and hampers tumour cell growth in vitro and in vivo. *Cell Death and Differentiation*, *19*, 871–881.

Ramaglia, M., D'Angelo, V., Iannotta, A., Di Pinto, D., Pota, E., Affinita, M. C., et al. (2016). High EZH2 expression is correlated to metastatic disease in pediatric soft tissue sarcomas. *Cancer Cell International*, *16*, 59.

Relaix, F., Rocancourt, D., Mansouri, A., & Buckingham, M. (2005). A Pax3/Pax7-dependent population of skeletal muscle progenitor cells. *Nature*, *435*, 948–953.

Roma, J., Masia, A., Reventos, J., Sanchez de Toledo, J., & Gallego, S. (2011). Notch pathway inhibition significantly reduces rhabdomyosarcoma invasiveness and mobility in vitro. *Clinical Cancer Research: An Official Journal of the American Association for Cancer Research*, *17*, 505–513.

Romania, P., Bertaina, A., Bracaglia, G., Locatelli, F., Fruci, D., & Rota, R. (2012). Epigenetic deregulation of microRNAs in rhabdomyosarcoma and neuroblastoma and translational perspectives. *International Journal of Molecular Sciences*, *13*, 16554–16579.

Sartorelli, V., Huang, J., Hamamori, Y., & Kedes, L. (1997). Molecular mechanisms of myogenic coactivation by p300: Direct interaction with the activation domain of MyoD and with the MADS box of MEF2C. *Molecular and Cellular Biology*, *17*, 1010–1026.

Sartorelli, V., Puri, P. L., Hamamori, Y., Ogryzko, V., Chung, G., Nakatani, Y., et al. (1999). Acetylation of MyoD directed by PCAF is necessary for the execution of the muscle program. *Molecular Cell*, *4*, 725–734.

Svalina, M. N., & Keller, C. (2014). YAPping about differentiation therapy in muscle cancer. *Cancer Cell*, *26*, 154–155.

Tajbakhsh, S., Borello, U., Vivarelli, E., Kelly, R., Papkoff, J., Duprez, D., et al. (1998). Differential activation of Myf5 and MyoD by different Wnts in explants of mouse paraxial mesoderm and the later activation of myogenesis in the absence of Myf5. *Development*, *125*, 4155–4162.

Tapscott, S. J., Thayer, M. J., & Weintraub, H. (1993). Deficiency in rhabdomyosarcomas of a factor required for MyoD activity and myogenesis. *Science*, *259*, 1450–1453.

Taulli, R., Foglizzo, V., Morena, D., Coda, D. M., Ala, U., Bersani, F., et al. (2014). Failure to downregulate the BAF53a subunit of the SWI/SNF chromatin remodeling complex contributes to the differentiation block in rhabdomyosarcoma. *Oncogene*, *33*, 2354–2362.

Tenente, I. M., Hayes, M. N., Ignatius, M. S., McCarthy, K., Yohe, M., Sindiri, S., et al. (2017). Myogenic regulatory transcription factors regulate growth in rhabdomyosarcoma. *eLife*, *6*, e19214.

Thuault, S., Hayashi, S., Lagirand-Cantaloube, J., Plutoni, C., Comunale, F., Delattre, O., et al. (2013). P-cadherin is a direct PAX3-FOXO1A target involved in alveolar rhabdomyosarcoma aggressiveness. *Oncogene*, *32*, 1876–1887.

Tremblay, A. M., Missiaglia, E., Galli, G. G., Hettmer, S., Urcia, R., Carrara, M., et al. (2014). The Hippo transducer YAP1 transforms activated satellite cells and is a potent effector of embryonal rhabdomyosarcoma formation. *Cancer Cell*, *26*, 273–287.

Van Antwerp, M. E., Chen, D. G., Chang, C., & Prochownik, E. V. (1992). A point mutation in the MyoD basic domain imparts c-Myc-like properties. *Proceedings of the National Academy of Sciences of the United States of America*, *89*, 9010–9014.

von Maltzahn, J., Chang, N. C., Bentzinger, C. F., & Rudnicki, M. A. (2012). Wnt signaling in myogenesis. *Trends in Cell Biology*, *22*, 602–609.

Wackerhage, H., Del Re, D. P., Judson, R. N., Sudol, M., & Sadoshima, J. (2014). The Hippo signal transduction network in skeletal and cardiac muscle. *Science Signaling*, 7, re4.

Walters, Z. S., Villarejo-Balcells, B., Olmos, D., Buist, T. W., Missiaglia, E., Allen, R., et al. (2014). JARID2 is a direct target of the PAX3-FOXO1 fusion protein and inhibits myogenic differentiation of rhabdomyosarcoma cells. *Oncogene*, *33*, 1148–1157.

Wang, C. (2012). Childhood rhabdomyosarcoma: Recent advances and prospective views. *Journal of Dental Research*, *91*, 341–350.

Wang, H., Garzon, R., Sun, H., Ladner, K. J., Singh, R., Dahlman, J., et al. (2008). NF-kappaB-YY1-miR-29 regulatory circuitry in skeletal myogenesis and rhabdomyosarcoma. *Cancer Cell*, *14*, 369–381.

Watt, K. I., Judson, R., Medlow, P., Reid, K., Kurth, T. B., Burniston, J. G., et al. (2010). Yap is a novel regulator of C2C12 myogenesis. *Biochemical and Biophysical Research Communications*, *393*, 619–624.

Weintraub, H., Davis, R., Tapscott, S., Thayer, M., Krause, M., Benezra, R., et al. (1991). The myoD gene family: Nodal point during specification of the muscle cell lineage. *Science*, *251*, 761–766.

Zhang, M., Truscott, J., & Davie, J. (2013). Loss of MEF2D expression inhibits differentiation and contributes to oncogenesis in rhabdomyosarcoma cells. *Molecular Cancer*, *12*, 150.

Zhang, M., Zhu, B., & Davie, J. (2015). Alternative splicing of MEF2C pre-mRNA controls its activity in normal myogenesis and promotes tumorigenicity in rhabdomyosarcoma cells. *The Journal of Biological Chemistry*, *290*, 310–324.

CHAPTER TEN

Muscle Stem Cells and Aging

Ara B. Hwang, Andrew S. Brack[1]
Broad Center of Regeneration Medicine and Stem Cell Research, University of California, San Francisco, San Francisco, CA, USA
[1]Corresponding author: e-mail address: andrew.brack@ucsf.edu

Contents

Abstract

Satellite cells (SCs) are a population of muscle-resident stem cells that are essential for efficient tissue repair. SCs reside in a relatively quiescent state during normal tissue turnover, but are activated in response to injury through the microenvironment and

Current Topics in Developmental Biology, Volume 126
ISSN 0070-2153
https://doi.org/10.1016/bs.ctdb.2017.08.008
© 2018 Elsevier Inc.
All rights reserved.

cell-intrinsic signals. During aging, SC dysfunction is a major contributor to the decline in regenerative potential of muscle tissue. Recent studies have demonstrated that both cell-intrinsic and cell-extrinsic factors are deregulated during aging. Interventions that reverse age-associated changes in SCs or the niche have shown to partially rejuvenate the regenerative capacity of aged muscle SCs. In this review, we discuss recent advances in SC biology as it pertains to the deleterious effects of aging. A better understanding of how age-dependent changes in the SC and its environment niche impact muscle regeneration could lead to interventions to ameliorate the effects of aging in humans.

1. INTRODUCTION

Aging leads to a decline in the regenerative capacity of tissues across all mammalian species. This is driven primarily by the functional decline in tissue-resident stem cell populations. In skeletal muscle, the resident stem cell population, SCs are numerically and functionally compromised, and the microenvironment (the niche) of the SC becomes less supportive (Blau, Cosgrove, & Ho, 2015). In this chapter, we will review the topic of age-related changes in SC function and muscle regeneration. We will focus on the regulatory changes of the aged SC that are imposed by the niche and are driven cell autonomously. We specifically pay attention to more current developments in the SC-aging field such as the instructive role of the niche of SC function, and the contribution of senescence, metabolism, and epigenetics to the decline in SC function (Fig. 1).

2. AGE-DEPENDENT CHANGES IN MUSCLE REGENERATION AND SATELLITE CELL FUNCTION

In the absence of overt muscle injury, the adult SC pool resides in a quiescent state. In response to muscle injury, SCs activate, proliferate, and fuse to form new myofibers, a subset of proliferating SCs return back to a quiescent state to replenish the SC pool. Upon injury to aged muscle, the regeneration process is impaired, as indicated by limited expansion of SCs and their descendants, delayed differentiation of progenitors into new myofibers, and an increase in extracellular matrix (fibrosis), together suggesting a less efficient muscle regeneration program (Brack & Munoz-Canoves, 2016). Aged and adult mice surgically paired reveal a role of circulatory factors in the muscle regeneration process (Brack et al., 2007; Conboy et al., 2005), while cell transplantation of aged SCs engrafted into

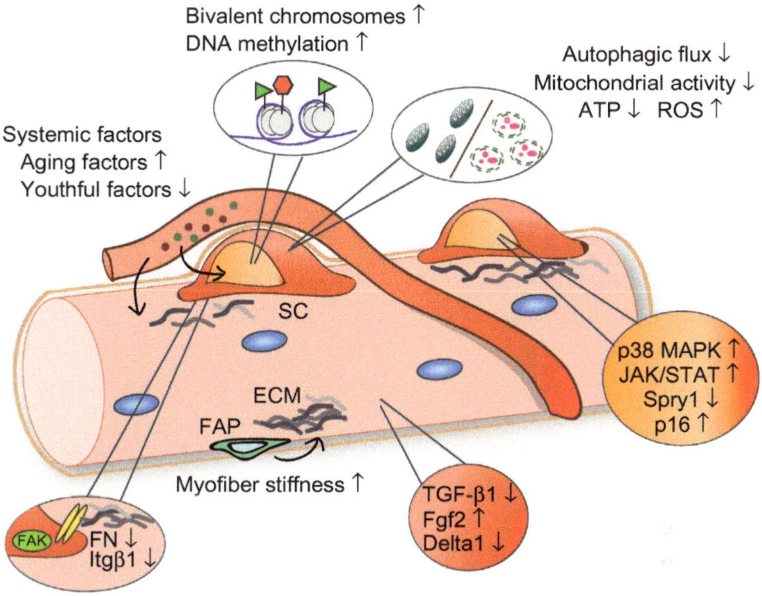

Fig. 1 Age-dependent changes that occur in the muscle satellite cell and niche. The muscle goes through systemic changes in both muscle satellite cells (SCs), which lays under the basal lamina (indicated in *orange*), and muscle niche (circulatory system, myofiber, fibro-/adipogenic progenitors (FAPs)) that limit the regenerative capacity of SCs. Myofibers become stiffer due to increase in ECM (extracellular matrix; indicated in *gray*) deposition from FAPs and other sources. Fibronectin (FN) and its interacting factors, integrin-β1 (Itgβ1) and FAK, decrease, leading to a less adhesive niche. Growth factor ligands including TGF-β1, Fgf2, and Delta1 change in the aged muscle fiber, becoming a less supportive niche. A global change in the epigenome shows that the chromatin becomes more bivalent (permissive marks: *green*; repressive marks: *red*). p38 MAPK, JAK/STAT, Spry1, and p16, which are major regulators of SC function, undergo age-dependent changes. Reduced autophagic activity and mitochondrial activity lead to a decreased ATP production and increased ROS generation in aged SCs.

adult preinjured muscle shows a reduced functionality of aged SCs, suggesting a cell-intrinsic element to SC functional demise during aging (Chakkalakal, Jones, Basson, & Brack, 2012; Cosgrove et al., 2014; Price et al., 2014). The defects in aged muscle regeneration can be caused by many potential cellular defects, from insufficient numbers of SCs, inappropriate regulation of cell fate decisions in response to injury cues, and a loss of functional subsets based on their heterogeneous nature. Below we discuss recent advances in each of these aspects of SC biology.

2.1 Satellite Cell Behavior in the Absence of Injury

The cell cycle state of quiescence correlates positively with stem cell potential across different stem cell compartments (Cheung & Rando, 2013). For many years, the SC pool was thought to be in a quiescent state and only activated in response to increased demand such as injury or exercise. Based on long-term label retention assays, it is estimated that the SC pool goes through approximately 4–8 divisions between 3 and 26 months of age (Chakkalakal et al., 2014, 2012). SCs that go through four divisions retain label (LR-SCs), while those that go through eight divisions lose label (nLR-SCs) (Chakkalakal et al., 2014). Therefore, the adult SC pool undergoes modest turnover throughout life. Moreover, aged SCs are depleted over time (Brack, Bildsoe, & Hughes, 2005; Shefer, Van de Mark, Richardson, & Yablonka-Reuveni, 2006). This suggests a defect in self-renewal potential of aged SCs under normal homeostatic conditions.

2.2 Cell Fate Decisions in Response to Activating Stimuli

Under appropriate stimulatory cues, SCs proliferate and either self-renew or differentiate to replenish the SC pool and remake muscle, respectively (Brack & Rando, 2012). In low mitogen conditions in vitro, aged SCs are biased to either commit to differentiation or die, at the expense of self-renewal (Chakkalakal et al., 2012). This could explain why SC pool depletion is coincident with proliferation in aged muscle. In geriatric mice, a subset ($\sim$10%) of SCs are susceptible to senescence upon mitogen exposure, thus further limiting the ability of aged SCs to contribute to tissue homeostasis after injury (Garcia-Prat et al., 2016).

2.3 A Loss of Satellite Cell Heterogeneity During Aging

Analysis of SC function under transplantation assays reveals that LR-SCs function as bone fide stem cells, and nLR-SCs as committed progenitors (Chakkalakal et al., 2014). During aging, stem cells (LR-SCs) are lost and committed progenitors (nLR-SCs) are retained (Chakkalakal et al., 2012), suggesting that the overall decline in transplantation potential of aged SCs is due to the loss of cells harboring stem cell potential. In support, aged muscle SCs engrafted into preinjured adult hosts under limiting dilution reveal that the number of functional SCs is diminished by approximately two-thirds (Cosgrove et al., 2014; Price et al., 2014). Moreover, cell-intrinsic defects can be overcome by transplanting a larger population (Collins, Zammit, Ruiz, Morgan, & Partridge, 2007; Cosgrove et al., 2014).

Therefore, aging promotes the loss of the more competent SCs. This likely explains the decrease in self-renewal and SC output of aged SCs under transplantation.

3. THE AGING NICHE AND ITS EFFECT ON SC FUNCTION

Tight regulation of muscle SC function from the local microenvironment (niche) and circulatory factors in the bloodstream are crucial for efficient muscle tissue regeneration. The niche consists of various cell types that reside in the muscle including the muscle fibers where the SC resides, fibro-/adipogenic progenitors (FAPs), endothelial cells, fibroblasts, and macrophages (Bentzinger, Wang, Dumont, & Rudnicki, 2013). Among these different cell types, age-associated changes in the interactions between endothelial cells or macrophages with muscle SCs are not well studied. Therefore, in the following section, we will discuss the role of the circulatory factors and cellular niche components (muscle fibers and FAPs) on SC function in muscle aging.

3.1 Circulatory Factors

Heterochronic parabiosis has been used to interrogate the role of circulatory factors on muscle SC function. Pairing adult and aged mice to create anastomosis revealed an inhibitory effect of aged circulatory factors on adult mice, and stimulatory effect of young serum on aged mice (Conboy, Conboy, & Rando, 2013). Exposing aged mice to a more youthful systemic environment promoted regeneration, as indicated by increased SC proliferation and differentiation and reduced fibrotic deposition. In vitro studies showed that SC proliferation and cell fate are under direct control of serum factors (Brack et al., 2007; Carlson, Hsu, & Conboy, 2008; Conboy et al., 2005). Changes in RTK/ERK, Notch, Wnt, C1q, and TGF-β signaling, oxytocin hormones, and growth differentiation factor 11 have been proposed as mechanisms by which serum factors alter the regenerative potential of skeletal muscle (Brack et al., 2007; Carlson et al., 2008; Conboy, Conboy, Smythe, & Rando, 2003; Egerman et al., 2015; Elabd et al., 2014; Naito et al., 2012; Paliwal, Pishesha, Wijaya, & Conboy, 2012; Sinha et al., 2014). The paradigms used in the parabiotic studies to date have maintained the shared circulation prior to and during regeneration; therefore, it remains unknown whether the effect of serum factors impacts SCs at distinct or all stages of lineage progression. In addition, the impact of circulatory factors on other muscle-resident niche and progenitor cells warrants examination.

3.2 Extracellular Matrix

The extracellular matrix (ECM) plays multiple roles in tissue function, providing a mechanically stiff substrate for structural integrity and cellular organization. Muscle fibers and tissue-resident FAPs are the primary source of matrix deposition (Joe et al., 2010; Lemos et al., 2015; Uezumi, Fukada, Yamamoto, Takeda, & Tsuchida, 2010; Uezumi et al., 2011). During aging, the abundance of ECM within skeletal muscle increases, leading to fibrosis and increased tissue stiffness (Brack et al., 2007; Lukjanenko et al., 2016; Mann et al., 2011). The physical stiffness of the cellular microenvironment plays a crucial role in stem cell fate determination including SCs (Gilbert et al., 2010; Lv et al., 2015; Wood et al., 2014). Single myofibers isolated from aged mice have increased physical stiffness and this correlates with increase of cross-linked collagen levels (Lacraz et al., 2015; Wood et al., 2014). Adult primary myoblasts cultured on hydrogels that mimic the stiffness of aged muscle promoted differentiation at the expense of proliferation (Lacraz et al., 2015). On the contrary, adult SCs cultured on hydrogel that mimics the softness of adult muscle, compared to ECM-coated plastic dish, showed an increase in proliferation and self-renewal in vitro, and improved engraftment in cell transplantation assays (Gilbert et al., 2010). Thus, maintaining the physical properties of adult muscle fibers is crucial to preserve SC function and muscle regeneration in aged animals.

FAPs are rapidly activated during muscle injury and by reducing proliferation correlates with reduction of ECM deposition (Fiore et al., 2016; Joe et al., 2010). Inhibiting proliferation of FAPs in aged muscle through suppression of PDGFRα signaling prevents much of the fibrotic deposition during regeneration (Mueller, van Velthoven, Fukumoto, Cheung, & Rando, 2016). Increased fibrotic deposition negatively correlates with myogenesis (Brack et al., 2007). Therefore, interventions that target fibrosis directly may lead to a niche where its physical properties support SC proliferation and myogenesis during regeneration.

ECM plays an important role in providing a scaffold to regenerate nascent muscle. Although the overall expression of ECM components is increased in aged uninjured muscle, in response to injury, ECM deposition is blunted (Lukjanenko et al., 2016). Injury-induced induction of fibronectin (Fn1) is contributed mainly from cells of immune, hematopoietic, and endothelial lineages (Lin+). Aged Lin+ cells express lower levels of Fn1 (Lukjanenko et al., 2016). Using a temporally controllable allele to target fibronectin expression in a ubiquitous manner led to a reduction in the number of Pax7+ cells during adult muscle repair, mimicking muscle regeneration of aged mice. In addition, injection of fibronectin in a regenerating muscle restored the regenerative capacity of an aged muscle (Lukjanenko et al., 2016). Thus, muscle stiffness and the proper deposition and clearance of ECM are all important to maintain a functioning SC pool during aging.

While accumulation of ECM, derived predominantly from muscle fibers and FAPs, reveals a nonautonomous role of ECM on SCs, there is also evidence that SCs can regulate ECM deposition from fibroblasts. A lifelong ablation of SCs using a diphtheria toxin-inducible allele targeted to adult Pax7 + cells led to the accumulation of ECM (Fry et al., 2015). This suggests that the SC pool provides feedback signals to ECM-producing cells in the stroma. In vitro experiments revealed that myoblasts carrying exosomes containing mir-206 were sufficient to repress a fibrogenic program in fibroblasts (Fry, Kirby, Kosmac, McCarthy, & Peterson, 2017). The expression of mir-206 in aged muscle SC has not been characterized yet, but this study suggests that the reduction in the number of SCs during aging dysregulates exosome signaling in the niche. Determining whether exosomes are able to pass through the basal lamina in vivo to impact a sufficient number of fibrogenic cells is worthy of further investigation.

3.3 Muscle Fiber-Derived Growth Factors

Growth factor signaling from the niche plays a key role in stem cell function (Gattazzo, Urciuolo, & Bonaldo, 2014; Yin, Price, & Rudnicki, 2013). Aged muscle fibers express elevated levels of FGF2 transcript and protein, while SCs and interstitial cells did not undergo age-dependent changes in FGF2, suggesting that muscle fibers are the main source of FGF2 in aged muscle (Chakkalakal et al., 2012). Increased FGF2 in fibers was shown to cause a break in SC quiescence. Addition of FGF2 was sufficient to increase cycling cells when added to isolated adult muscle SCs in vitro, and downregulation of FGF2 signaling by administrating an inhibitor of FGFR1 systemically abrogated the number of cycling SCs in aged mice in vivo (Chakkalakal et al., 2012). Thus, signaling from aberrant expression of FGF2 in the muscle fiber to FGFR1 in the aged SCs enforces proliferation.

The transforming growth factor beta (TGF-β) ligands perform multiple cellular functions in both adult and developing embryos (Siegel & Massague, 2003). TGF-β ligands promote phosphorylation of SMAD complex to transactivate downstream target genes (Feng & Derynck, 2005). TGF-β1 is expressed at low levels in adult myofibers, and is highly expressed during early stages of muscle regeneration in the newly formed myofibers as well as other cell types (McLennan, 1993; Murakami et al., 1999; Noirez et al., 2006). In aged mice, circulating TGF-β1 levels increase and correlate with increased phosphorylated SMAD3, and a delay in cell cycle entry in vitro (Beggs et al., 2004; Carlson et al., 2009, 2008). Neutralizing TGF-β signaling is shown to improve aged muscle regeneration, suggesting that aging phenotypes can be alleviated by dampening TGF-β activity (Carlson et al., 2008). In addition to the direct effect on SCs, a recent study demonstrated that TGF-β signaling promotes survival of FAPs in regenerating muscle (Lemos et al., 2015). Thus, increased TGF-β from the niche (myofibers and FAPs) has an inhibitory role in muscle repair, limiting SC proliferation and promoting fibrosis.

Notch signaling plays important roles in maintaining muscle progenitors during muscle development as well as adult regeneration (Schuster-Gossler, Cordes, & Gossler, 2007; Vasyutina et al., 2007). In aged muscle, Notch ligand Delta1 level was decreased in myofibers, and SCs fail to induce Notch signaling during regeneration (Conboy et al., 2003). Stimulating Notch signaling by injection of extracellular fragments of Notch1 or through providing young serum via heterochronic parabiosis was able to increase Delta1 levels and promote proliferation of aged SCs (Conboy et al., 2005, 2003). What causes the aged niche to dysregulate Notch signaling is currently not known. Taken together, alteration of FGF, TGF, and Notch signaling pathways is detrimental to the maintenance, survival, and proliferation of aged muscle SCs. Thus, these studies highlight the importance of the niche for muscle regeneration.

3.4 Cell-Autonomous Changes in Niche Sensing

Many ECM proteins form active ligand–receptor signaling complexes (Gattazzo et al., 2014; Kim, Turnbull, & Guimond, 2011). Therefore, changes in any receptor complexes will alter the cells signaling response to the niche. Fibronectin and integrin family members act together to activate a downstream effector, focal adhesion kinase (FAK) (Lukjanenko et al., 2016), which is important for SC adherence to the matrix (Lukjanenko et al., 2016). In comparison to adult, aged SCs and their progeny have lower levels of fibronectin, integrin-β1 (Itgβ1), and FAK that lead to a decreased cell adhesion and increased cell death (Lukjanenko et al., 2016; Rozo, Li, & Fan, 2016). Although SCs are not the major contributor of FN in regenerating muscle tissue, these data suggest that aged SCs may limit their regeneration capacity through altered niche sensing.

Primary myoblasts cultured on fibronectin promoted a decrease in p38 MAPK and ERK signaling pathways, which are two of the main signaling pathways that are upregulated in aged muscle SCs. Culturing aged muscle SCs on fibronectin also improved adherence that was sufficient to promote cell proliferation and decrease cell death. Moreover, treatment of aged SCs with p38 inhibitors in vitro improved cell adhesion, suggesting that a fibronectin–p38 axis regulates SC adhesion (Lukjanenko et al., 2016).

The importance of Itgβ1 signaling in aged SCs is highlighted by its regulation of FGF signaling (Rozo et al., 2016). Aged muscle SCs have lower activity levels of Itgβ1 compared to adult SCs (Rozo et al., 2016), and activation of Itgβ1 in aged SCs using Itgβ1-activating antibody, TS2/16 (Hintermann, Bilban, Sharabi, & Quaranta, 2001), was able to improve muscle regeneration in aged mice. Activating Itgβ1 signaling via TS2/16 treatment combined with Fgf2 treatment enhanced Fgf2 signaling and SC proliferation in vitro (Rozo et al., 2016). Together this suggests that the loss of Itgβ1 signaling in the aged SCs decreases the sensitivity to Fgf2 from the niche. Thus, changes in ECM niche composition and niche sensing by the aged SC limit SC function and its muscle regenerative capacity. Whether changes in the ECM composition within the niche cause a decrease of ECM sensing in aged SCs needs to be determined.

4. CELL-INTRINSIC CHANGES DURING AGING IMPACT STEM CELL FUNCTION

Stem cell transplantation studies have shown that aged muscle SCs engrafted into a preinjured adult host have reduced potential to repopulate the niche and differentiate (Chakkalakal et al., 2012; Cosgrove et al., 2014; Price et al., 2014; Sousa-Victor et al., 2014). Moreover, using serial transplant assays, aged SC defects persist through multiple rounds of regeneration in a young host, suggesting that the young environment cannot rejuvenate self-renewed aged donor cells (Cosgrove et al., 2014).

Many signaling pathways that play essential roles in SC maintenance and myogenesis are deregulated in aged muscle SCs. These pathways include the p38 MAPK signaling pathway, and JAK–STAT and RTK signaling. Dysregulation of these pathways causes aged SCs to display phenotypes observed across many somatic cell types, including DNA damage, increased oxidative stress, imbalance of protein homeostasis, and terminal senescence. In the following sections, we will discuss how intrinsic cell signaling pathways alter muscle SCs in an aged context.

4.1 SCs From Aged Mice Possess a Cell-Autonomous Defect in Self-Renewal via Altered p38 MAPK

p38, a subgroup of an evolutionarily conserved mitogen-activated protein kinase (MAPK) pathway, was initially described as a responder to environmental stress conditions (Cuenda & Rousseau, 2007). However, p38 is also activated by nonstress stimuli to regulate a wide range of cellular response in various cell types (Cuenda & Rousseau, 2007). p38 is rapidly activated as SCs transition from quiescence to an activated state and regulates proliferation and differentiation (Bernet et al., 2014; Hausburg et al., 2015; Jones et al., 2005; Segales, Perdiguero, & Munoz-Canoves, 2016; Troy et al., 2012).

p38 signaling pathway is constitutively active in aged muscle SCs. This is shown by increased phosphorylation of p38 and expression of its downstream target MAPKAPK-2 (MK-2) as well as transcriptome KEGG (Kyoto Encyclopedia of Genes and Genomes) pathway analysis (Bernet et al., 2014; Cosgrove et al., 2014; Price et al., 2014). Consistent with a role of p38 in cell cycle exit and differentiation, reducing p38 activity in aged SCs via small-molecule inhibitors or siRNA knockdown increases SC proliferation in vitro (Bernet et al., 2014; Cosgrove et al., 2014). Moreover, the aged SCs proliferate in a less committed state (higher Pax7 and lower myogenin expression), and expressed low levels of the conserved cellular senescence marker, p16ink4a (Bernet et al., 2014). Asymmetric activation of p38 in SC daughters is a key regulator of SC self-renewal (Troy et al., 2012). Symmetric activation of p38 in aged SC daughters leads to symmetric differentiation and a loss of self-renewal. Pharmacological inhibition of p38 was able to partially restore the proliferative and self-renewal potential of aged SCs (Bernet et al., 2014). Thus, aberrant expression of p38 leads to various aging phenotypes of aged muscle SCs, and these cell-intrinsic changes can be reversed using pharmacological approaches.

FGF2 is capable of activating p38 (Jones, Fedorov, Rosenthal, & Olwin, 2001; Jones et al., 2005), suggesting that upregulated FGF2 in the aged niche may promote activation of p38 in quiescent SCs. FGF2 signals via FGFR1 in SCs; we found that 6 weeks administration of FGFR1 inhibitor to aged mice rescued many of the aged SC phenotypes in vivo, and during activation in high mitogen in vitro (Chakkalakal et al., 2012). Interestingly, under high mitogen conditions aged SCs are not able to fully activate FGFR1; however, transfection of a ligand-independent constitutive active form of FGFR1 was able to rescue the asymmetric expression of p38 and proliferation of aged SCs in vitro (Bernet et al., 2014). Together this suggests that chronic elevation of FGF2 from the aged niche leads to sustained activation of the FGFR1–p38 MAPK signaling pathway. This is followed by desensitization of the FGFR1 receptor (Bernet et al., 2014), rendering it incapable of robust activation required for rapid proliferation during muscle injury. While we have discussed the role of FGF2 as a potential activator of p38 in aged SCs, there are likely multiple factors that remain to be determined.

4.2 Increase in JAK/STAT Signaling Causes Aging in Satellite Cells

The JAK/STAT signaling pathway, activated by cytokines and growth factors, is a well-conserved pathway from *Drosophila* to mammals (Rawlings, Rosler, & Harrison, 2004), and previously has been described to play a role in SC proliferation (Toth et al., 2011). KEGG pathway analysis reveals that JAK/STAT signaling pathway is highly upregulated in aged muscle SC (Price et al., 2014). Furthermore, increased phosphorylation of STAT3 in aged SC shows that aged quiescent SCs are actively signaling through JAK/STAT pathway (McKay et al., 2013).

Two independent studies have demonstrated that inhibition of JAK/STAT signaling pathway, through either genetic or pharmacological inhibition in aged muscle, promoted cell proliferation and prevented cell differentiation (Price et al., 2014; Tierney et al., 2014). Moreover, regeneration capacity of aged muscle SCs can be improved by transient downregulation of JAK/STAT signaling during early stages of SC activation process (Price et al., 2014; Tierney et al., 2014). Mechanistically, increased JAK/STAT signaling impaired SC function by promoting asymmetric division that is biased to generate committed daughter cells (Price et al., 2014), or by regulating MyoD expression to promote differentiation at the expense of cell proliferation (Tierney et al., 2014). Thus, increases in intracellular JAK/STAT signaling contribute to aged muscle SC phenotypes, and aging-associated impairment can be reversed by transient regulation of JAK/STAT pathway activity.

4.3 Spry1 Is Required for Satellite Cell Quiescence and Long-Term Satellite Cell Maintenance

Sprouty1 (Spry1) is an intracellular inhibitor of RTK signaling, expressed in quiescent SCs, that decreases during proliferation and is reestablished as cycling SCs return to a quiescent state (Shea et al., 2010). Loss-of-function studies show that Spry1 is dispensable for quiescence in uninjured muscle, but required for self-renewal of the SC pool in response to muscle injury

(Shea et al., 2010). While some RTK ligands activate, FGF2 inhibits Spry1 expression (Gross, Bassit, Benezra, & Licht, 2001). The aged niche expresses high FGF2 levels that inhibit Spry1 expression and increase proliferation in SCs in vivo (Chakkalakal et al., 2012). Chronic stimulation of the FGF2-Spry1 axis leads to a depletion of the SC pool and an exaggeration of the aged muscle regeneration phenotypes. In contrast forced expression of Spry1 in aged SCs in vivo preserves the SC pool (Chakkalakal et al., 2012). Taken together, approaches that reinforce quiescence during normal tissue turnover may prevent SCs loss and enhance their response to injury during aging.

4.4 p16-Mediated Senescence Limits Satellite Cell Function

The accumulation of cellular stressors and damage leads to the upregulation of INK4A/ARF family of cell cycle inhibitors that drives a cell into an irreversible arrest, known as senescence. On one hand this would diminish the number of cells that could contribute to tissue repair; in addition, senescent cells secrete factors that hinder the function of cells within the community (Acosta et al., 2013; Campisi, 2013). However, recent studies suggest that secreted factors from senescent cells can positively regulate embryonic development, wound healing, and stem cell regenerative capacity (Chiche et al., 2017; Ritschka et al., 2017). Thus, senescence appears to play complex physiological roles in aging and tissue repair.

During aging, SCs display multiple indices of cellular damage including DNA damage and increased reactive oxygen species (ROS) levels (Sousa-Victor et al., 2014). In addition, there is a progressive accumulation of senescent SCs, based on well-described phenotypic markers such as SA-βgal and p16ink4a (Sousa-Victor et al., 2014). Moreover, components of the PRC1 complex that normally repress the INK4A locus are reduced in geriatric SCs, leading to derepression of INK4A locus and the expression of p16ink4a. Loss-of-function studies demonstrate that downregulation of p16ink4a in SCs from geriatric mice improved proliferation and differentiation under the setting of transplantation (Sousa-Victor et al., 2014). Therefore, inhibiting prosenescent factors can enhance aged SC function.

Since stem cells are the building blocks of tissues, the accumulation of senescent cells would eventually limit tissue and the whole organismal function. Supporting this idea, one recent study generated a transgenic mouse that ablates p16ink4a-expressing cells by expressing FK506-binding protein–caspase 8 protein driven by an INK4A-minimal promoter. Clearance of p16+ senescent cells extended median life span (Baker et al., 2016).

Elimination of p16+ cells decreased the expression of senescence markers and inflammation genes, suggesting a prevention of aging-associated phenotypes in multiple tissues (Baker et al., 2016). Thus, senescent cells are one of the main drivers of tissue and organismal aging.

5. CHANGES TO THE SATELLITE CELL EPIGENOME DURING AGING IMPACT STEM CELL FUNCTION

Chromatin is a highly compact nuclear structure composed of genomic DNA, protein, and RNA. Epigenetic changes are the molecular modifications to the chromatin state triggered by environmental cues and allow changes in the transcriptional landscape (Avgustinova & Benitah, 2016; Marks et al., 2012). Generally, a structurally loose chromatin promotes gene transcription (euchromatin), whereas a tightly compact chromatin (heterochromatin) is associated with repressive gene transcription and genome stability (Avgustinova & Benitah, 2016; Marks et al., 2012; Zhou, Goren, & Bernstein, 2011). Age-associated epigenome changes could be an underlying cause to the global transcriptional changes that are seen in aged muscle SCs and their progenitors. Transcriptome changes of aged muscle SCs include decreased antioxidant genes, altered myogenic differentiation genes, and decreased protein-folding genes (Baj et al., 2005; Bernet et al., 2014; Bortoli et al., 2003; Charge, Brack, & Hughes, 2002; Liu et al., 2013; Pietrangelo et al., 2009; Price et al., 2014). In this section, we will discuss the growing body of evidence suggesting that age-dependent changes in the epigenome (posttranslational histone modification and DNA methylation patterns) of muscle SCs and how this impacts muscle stem cell function.

5.1 Recessive Epigenome Changes in Aged Satellite Cells at the Global Level

In eukaryotic chromatic, the genomic DNA is wrapped around four core histones, H2A, H2B, H3, and H4, to form nucleosomes (Avgustinova & Benitah, 2016). The amino-terminal of the histone tails is subjected to a wide range of posttranslational modifications that allow a permissive state (e.g., H3K4me3) or a repressive state (e.g., H3K27me3, H3K9me2) of the chromatin (Avgustinova & Benitah, 2016). The overall chromatin structure of adult muscle SCs is in a permissive state (H3K4me3 markers at transcription start sites; TSSs) in quiescence (Liu et al., 2013; Schworer et al., 2016). During early activation in vivo, muscle SCs retain their H3K4me3 marks and acquire repressive marks, such as H3K27me3 (Liu et al.,

2013), thus resulting in a downregulation of quiescence genes in activated cells. Interestingly, quiescent SCs show a large body of bivalent marks on the chromosome containing both permissive and repressive marks, indicative of a poised state (Bernstein et al., 2006; Marks et al., 2012; Mikkelsen et al., 2007).

During aging, quiescent muscle SCs maintain permissive marks and acquire more repressive marks, including H3K27me3 (Liu et al., 2013) and H3K9me2 (Schworer et al., 2016). Thus, the chromatin becomes more bivalent. Of note, the repressive marks greatly increase within the intergenic regions on the chromosome, suggesting that aged cells lose their overall transcription potential (Liu et al., 2013). In particular, histone genes become bivalent and their transcript levels decline (Liu et al., 2013). A decrease in histone levels is another line of evidence supporting the notion that aging is associated with reduced transcript potential in SCs.

Human primary myoblasts from elderly patients have an overrepresentation of hypermethylated CpG islands throughout the genome, indicating chromatin compaction and gene silencing (Bigot et al., 2015). Hypermethylated genes include SPRY1 and NOTCH1, which are required for muscle SC quiescence. Treating aged SCs with a DNA demethylation agent rescued SPRY1 expression and increased the number of quiescent SCs (Bigot et al., 2015). Therefore, aging drives a change in DNA methylation patterns, resulting in loss of key quiescence genes and self-renewal potential. Similar to human SCs, *Spry1* levels are lower in aged murine SCs. While the Spry1 decline in murine cells was ascribed to increased FGF2 activity, it is interesting to consider whether RTK signaling can impact DNA methylation, as observed in ESCs (Choi et al., 2017).

5.2 Epigenetic Regulation of Individual Genes

Although histone-encoding genes accumulate repressive chromatin marks in quiescent SCs during aging, these genes are upregulated during SC activation, suggesting that there is a different mechanism to allow gene transcription in activated aged muscle SC (Liu et al., 2013). Consistent with this premise, activated old SCs display a more permissive chromatin state (increase in H3 and H4 acetylation motifs and decrease in H3K27me3 motifs) than activated young SCs (Schworer et al., 2016). Homeodomain transcription factor 9 (Hoxa9), a master regulator of development, gene locus is one example that showed aberrant gene expression in aged muscle SC during activation (Schworer et al., 2016). Knockdown of Hoxa9 gene, or its upstream H3K4me3 demethylase, Mll1 (mixed lineage leukemia), rescued impaired muscle regeneration in aged tissue and stem cell self-renewal capacity of SCs on fibers to the level of young SCs in vitro. Induction of Hoxa9 gene in young muscle SCs suppressed their proliferative capacity, mimicking aged SCs. Consistent with this observation, the transcriptome of Hoxa9 overexpressing young SCs was overrepresented with proaging signaling pathways including MAPK, TGF-β, Wnt, and JAK/STAT signaling (Schworer et al., 2016). Thus, aged satellite cells have altered epigenetic response to activating stimuli that limits tissue regeneration.

6. METABOLISM AND AGING IN SATELLITE CELLS

The relatively low metabolic rate of quiescence muscle SCs indicates that these cells are dormant. SCs have a compact cytoplasmic volume, relatively few active mitochondria, and less accumulation of ROS compared to their progeny (Latil et al., 2012; Lopez-Otin, Blasco, Partridge, Serrano, & Kroemer, 2013; Montarras, L'Honore, & Buckingham, 2013). For quiescent SCs to activate and progress through myogenesis, a drastic metabolic change is needed. It has been proposed that multiple cellular processes are utilized by dormant quiescent muscle SCs to increase metabolism rate during satellite cell activation, including an increase in mitochondrial activity (Rodgers et al., 2014), induction of autophagy (Tang & Rando, 2014), and increased glycolysis (Ryall et al., 2015). Dysregulation of cellular metabolism is a common feature seen in aged organisms as well as muscle SCs (Garcia-Prat & Munoz-Canoves, 2017; Sun, Youle, & Finkel, 2016). Since evoking metabolic changes is a crucial step in SC activation, age-associated changes in metabolism can be a major hindrance to limit regeneration capacity of aged SCs. In this section, we will highlight recent findings on dysregulation of mitochondria activity, autophagy in aged SCs, and two conserved longevity pathways that rejuvenate SC functions.

6.1 Mitochondrial Activity

Mitochondria, the energy-generating power plant in cells, play crucial roles in aging regulation. One of the widely observed changes across different organisms is that mitochondrial function declines in multiple cell types with aging (Garcia-Prat & Munoz-Canoves, 2017; Sun et al., 2016). Analysis of muscle SCs reveals that aging causes a decrease in mitochondrial membrane potential and oxygen consumption, and elevated ROS generation (Garcia-Prat et al., 2016; Zhang et al., 2006). At the transcription level, KEGG analysis indicated downregulation in TCA cycle and oxidative phosphorylation (Bernet et al., 2014; Liu et al., 2013; Price et al., 2014; Zhang et al., 2006).

Consistent with the inefficient mitochondrial activity, aged muscle SCs have diminished concentrations of ATP and NAD+, suggesting that cellular energy levels are low (Zhang et al., 2006).

Interventions that increase mitochondrial stress response pathway, such as mitochondrial unfolded protein response (UPR^{mito}), is one of the evolutionarily conserved longevity pathways (Sun et al., 2016). A recent study has proposed that increase life span by UPR^{mito} induction is through improving stem cell function (Zhang et al., 2006). Restoration of NAD+ levels by nicotinamide riboside treatment reduced DNA damage marks (γH2AX) and induced UPR^{mito} (Zhang et al., 2006). Consistent with these observations, there was a rescue in loss of muscle SC number and muscle strength in homeostasis, and an improved muscle regeneration capacity of aged SCs (Zhang et al., 2006). Thus, improvement in mitochondria function at the cellular level leads to longer life span at the organismal level.

6.2 Autophagy

Clearance of damaged organelles and macromolecules is crucial to maintain healthy cells. Autophagy is a regulated self-degenerative process to remove damaged macromolecules or organelles (He & Klionsky, 2009; Rubinsztein, Marino, & Kroemer, 2011). An autophagosome is formed by incorporating damaged proteins and organelles, and it fuses with lysosomes to degrade macromolecules and organelles via acid hydrolases. A decline in autophagic activity is observed in aged hematopoietic stem cells (Ho et al., 2017) and muscle SCs (Garcia-Prat et al., 2016) in mammals, and in other tissues across different species (Lopez-Otin et al., 2013). Thus, the decline in autophagic capacity can be considered a hallmark of aging.

The decrease in autophagic flux in aged muscle SCs leads to a reduction in stem cell transplantation capacity (Garcia-Prat et al., 2016). A genetic ablation of autophagy-related gene 7 (Atg7), a key regulator of autophagosome formation, specifically in muscle SCs, negatively impacted mitochondria as shown by high ROS levels and low mitochondrial activity (Garcia-Prat et al., 2016). This suggests that one of the consequences of diminished autophagy in aged muscle SCs is increased mitochondrial damage. Furthermore, antioxidant treatment partially reversed the increase in ROS levels and increased DNA damage markers and epigenetic upregulation of p16ink4a expression in aged muscle SCs as well as adult Atg7-mutant SCs. Moreover, antioxidant treatment to aged SCs partially restored their engraftment potential under transplantation conditions (Garcia-Prat et al., 2016). However, short-term antioxidant treatment may not be sufficient to restore damaged DNA that has been accumulated during a lifetime of aged muscle SCs, hence the partial rescue phenotype. Thus, activation of cellular processes to prevent proteotoxicity in muscle SCs is crucial for their function. It is also possible that the decline in protein homeostasis is tightly linked to cellular senescence.

One of the first interventions to extend life span in mammals is through caloric restriction (CR) (Kenyon, 2010; Lopez-Otin et al., 2013). At the organismal level in mammals, CR is beneficial by relaying the onset of cardiovascular diseases, cancer, and mortality (Anderson & Weindruch, 2012). Short-term CR in young mice improves muscle stem cell function via modulating mitochondrial biogenesis and inducing known longevity genes. Muscle SCs isolated from young CR mice show faster proliferation in vitro, increase in SC numbers under homeostasis in vivo, and improved regeneration potential in injury models and SC engraftment (Cerletti, Jang, Finley, Haigis, & Wagers, 2012). However, the effect of CR treatment has not been directly examined in aged mice under regenerative conditions (Cerletti et al., 2012). Since CR is known to be effective independent of the developmental stage or the age of animals in worms (Smith et al., 2008), flies (Mair, Goymer, Pletcher, & Partridge, 2003), and mice (Dhahbi, Kim, Mote, Beaver, & Spindler, 2004), CR may have similar metabolic changes in aged muscle SCs to improve stem cell function.

Downregulation of mammalian target of rapamycin (mTOR), a master regulator of nutrition sensing and growth promotion, extends life span longevity in multiple model organisms (Johnson, Rabinovitch, & Kaeberlein, 2013; Kenyon, 2010). Treatment of mice with rapamycin, a direct inhibitor of the mTOR complex, extends life span or health span across phyla (Johnson et al., 2013; Kenyon, 2010). Interestingly, CR mice show downregulation of mTOR activity, suggesting the two longevity interventions share similar outcomes (Johnson et al., 2013). However, other reports suggest that rapamycin treatment affects the metabolism of animals, but in a distinct way from CR, e.g., CR makes the animals more sensitive to insulin, whereas rapamycin treatment causes mild insulin resistance (Miller et al., 2014). In aged SCs, rapamycin treatment rejuvenates muscle stem cell function, in part, by rescuing autophagic flux to adult levels (Garcia-Prat et al., 2016). This rejuvenation partially rescues the poor engraftment potential of aged muscle SCs similar to a level seen in aged SCs overexpressing Atg7. Thus, conserved pathways that increase life span in animals seem to benefit organisms at a cellular level by promoting stem cell function and tissue regeneration.

7. CONCLUDING REMARKS

The decline in SC function during aging causes significant limitations for muscle regeneration. As our knowledge of SC heterogeneity evolves, it is apparent that age-dependent effects on subsets, rather than a uniform response across the whole population, may explain the widely described decline in SC function. It remains to be determined whether certain subsets are specialized to respond to certain cues. This may explain why SC function under muscle injury is deemed to be under cell nonautonomous control, while cell transplant is under cell-autonomous regulation. Identification of the highly potent subsets of SCs is necessary to realize the transplantation potential for therapy.

The molecular changes in the environment and within the SC during aging that cause defective SC function and impaired muscle regeneration are becoming realized. In the near future, these fundamental findings will hopefully translate into meaningful gains in humans.

ACKNOWLEDGMENTS

This work was supported by NIH grants (R01 AR060868, R01 AR061002) to Andrew S. Brack and Human Frontier Science Project (LT000781/2016-L) to Ara B. Hwang.

REFERENCES

Acosta, J. C., Banito, A., Wuestefeld, T., Georgilis, A., Janich, P., Morton, J. P., et al. (2013). A complex secretory program orchestrated by the inflammasome controls paracrine senescence. *Nature Cell Biology*, *15*, 978–990.

Anderson, R. M., & Weindruch, R. (2012). The caloric restriction paradigm: Implications for healthy human aging. *American Journal of Human Biology*, *24*, 101–106.

Avgustinova, A., & Benitah, S. A. (2016). Epigenetic control of adult stem cell function. *Nature Reviews. Molecular Cell Biology*, *17*, 643–658.

Baj, A., Bettaccini, A. A., Casalone, R., Sala, A., Cherubino, P., & Toniolo, A. Q. (2005). Culture of skeletal myoblasts from human donors aged over 40 years: Dynamics of cell growth and expression of differentiation markers. *Journal of Translational Medicine*, *3*, 21.

Baker, D. J., Childs, B. G., Durik, M., Wijers, M. E., Sieben, C. J., Zhong, J., et al. (2016). Naturally occurring p16(Ink4a)-positive cells shorten healthy lifespan. *Nature*, *530*, 184–189.

Beggs, M. L., Nagarajan, R., Taylor-Jones, J. M., Nolen, G., Macnicol, M., & Peterson, C. A. (2004). Alterations in the TGFbeta signaling pathway in myogenic progenitors with age. *Aging Cell*, *3*, 353–361.

Bentzinger, C. F., Wang, Y. X., Dumont, N. A., & Rudnicki, M. A. (2013). Cellular dynamics in the muscle satellite cell niche. *EMBO Reports*, *14*, 1062–1072.

Bernet, J. D., Doles, J. D., Hall, J. K., Kelly Tanaka, K., Carter, T. A., & Olwin, B. B. (2014). p38 MAPK signaling underlies a cell-autonomous loss of stem cell self-renewal in skeletal muscle of aged mice. *Nature Medicine*, *20*, 265–271.

Bernstein, B. E., Mikkelsen, T. S., Xie, X., Kamal, M., Huebert, D. J., Cuff, J., et al. (2006). A bivalent chromatin structure marks key developmental genes in embryonic stem cells. *Cell*, *125*, 315–326.

Bigot, A., Duddy, W. J., Ouandaogo, Z. G., Negroni, E., Mariot, V., Ghimbovschi, S., et al. (2015). Age-associated methylation suppresses SPRY1, leading to a failure of re-quiescence and loss of the reserve stem cell pool in elderly muscle. *Cell Reports*, *13*, 1172–1182.

Blau, H. M., Cosgrove, B. D., & Ho, A. T. (2015). The central role of muscle stem cells in regenerative failure with aging. *Nature Medicine*, *21*, 854–862.

Bortoli, S., Renault, V., Eveno, E., Auffray, C., Butler-Browne, G., & Pietu, G. (2003). Gene expression profiling of human satellite cells during muscular aging using cDNA arrays. *Gene*, *321*, 145–154.

Brack, A. S., Bildsoe, H., & Hughes, S. M. (2005). Evidence that satellite cell decrement contributes to preferential decline in nuclear number from large fibres during murine age-related muscle atrophy. *Journal of Cell Science*, *118*, 4813–4821.

Brack, A. S., Conboy, M. J., Roy, S., Lee, M., Kuo, C. J., Keller, C., et al. (2007). Increased Wnt signaling during aging alters muscle stem cell fate and increases fibrosis. *Science*, *317*, 807–810.

Brack, A. S., & Munoz-Canoves, P. (2016). The ins and outs of muscle stem cell aging. *Skeletal Muscle*, *6*(1).

Brack, A. S., & Rando, T. A. (2012). Tissue-specific stem cells: Lessons from the skeletal muscle satellite cell. *Cell Stem Cell*, *10*, 504–514.

Campisi, J. (2013). Aging, cellular senescence, and cancer. *Annual Review of Physiology*, *75*, 685–705.

Carlson, M. E., Conboy, M. J., Hsu, M., Barchas, L., Jeong, J., Agrawal, A., et al. (2009). Relative roles of TGF-beta1 and Wnt in the systemic regulation and aging of satellite cell responses. *Aging Cell*, *8*, 676–689.

Carlson, M. E., Hsu, M., & Conboy, I. M. (2008). Imbalance between pSmad3 and Notch induces CDK inhibitors in old muscle stem cells. *Nature*, *454*, 528–532.

Cerletti, M., Jang, Y. C., Finley, L. W., Haigis, M. C., & Wagers, A. J. (2012). Short-term calorie restriction enhances skeletal muscle stem cell function. *Cell Stem Cell*, *10*, 515–519.

Chakkalakal, J. V., Christensen, J., Xiang, W., Tierney, M. T., Boscolo, F. S., Sacco, A., et al. (2014). Early forming label-retaining muscle stem cells require p27kip1 for maintenance of the primitive state. *Development*, *141*, 1649–1659.

Chakkalakal, J. V., Jones, K. M., Basson, M. A., & Brack, A. S. (2012). The aged niche disrupts muscle stem cell quiescence. *Nature*, *490*, 355–360.

Charge, S. B., Brack, A. S., & Hughes, S. M. (2002). Aging-related satellite cell differentiation defect occurs prematurely after Ski-induced muscle hypertrophy. *American Journal of Physiology. Cell Physiology*, *283*, C1228–1241.

Cheung, T. H., & Rando, T. A. (2013). Molecular regulation of stem cell quiescence. *Nature Reviews. Molecular Cell Biology*, *14*, 329–340.

Chiche, A., Le Roux, I., von Joest, M., Sakai, H., Aguin, S. B., Cazin, C., et al. (2017). Injury-induced senescence enables in vivo reprogramming in skeletal muscle. *Cell Stem Cell*, *20*, 407–414.e404.

Choi, J., Huebner, A. J., Clement, K., Walsh, R. M., Savol, A., Lin, K., et al. (2017). Prolonged Mek1/2 suppression impairs the developmental potential of embryonic stem cells. *Nature*, *548*, 219–223.

Collins, C. A., Zammit, P. S., Ruiz, A. P., Morgan, J. E., & Partridge, T. A. (2007). A population of myogenic stem cells that survives skeletal muscle aging. *Stem Cells*, *25*, 885–894.

Conboy, M. J., Conboy, I. M., & Rando, T. A. (2013). Heterochronic parabiosis: Historical perspective and methodological considerations for studies of aging and longevity. *Aging Cell*, *12*, 525–530.

Conboy, I. M., Conboy, M. J., Smythe, G. M., & Rando, T. A. (2003). Notch-mediated restoration of regenerative potential to aged muscle. *Science*, *302*, 1575–1577.

Conboy, I. M., Conboy, M. J., Wagers, A. J., Girma, E. R., Weissman, I. L., & Rando, T. A. (2005). Rejuvenation of aged progenitor cells by exposure to a young systemic environment. *Nature*, *433*, 760–764.

Cosgrove, B. D., Gilbert, P. M., Porpiglia, E., Mourkioti, F., Lee, S. P., Corbel, S. Y., et al. (2014). Rejuvenation of the muscle stem cell population restores strength to injured aged muscles. *Nature Medicine*, *20*, 255–264.

Cuenda, A., & Rousseau, S. (2007). p38 MAP-kinases pathway regulation, function and role in human diseases. *Biochimica et Biophysica Acta*, *1773*, 1358–1375.

Dhahbi, J. M., Kim, H. J., Mote, P. L., Beaver, R. J., & Spindler, S. R. (2004). Temporal linkage between the phenotypic and genomic responses to caloric restriction. *Proceedings of the National Academy of Sciences of the United States of America*, *101*, 5524–5529.

Egerman, M. A., Cadena, S. M., Gilbert, J. A., Meyer, A., Nelson, H. N., Swalley, S. E., et al. (2015). GDF11 increases with age and inhibits skeletal muscle regeneration. *Cell Metabolism*, *22*, 164–174.

Elabd, C., Cousin, W., Upadhyayula, P., Chen, R. Y., Chooljian, M. S., Li, J., et al. (2014). Oxytocin is an age-specific circulating hormone that is necessary for muscle maintenance and regeneration. *Nature Communications*, *5*, 4082.

Feng, X. H., & Derynck, R. (2005). Specificity and versatility in TGF-beta signaling through Smads. *Annual Review of Cell and Developmental Biology*, *21*, 659–693.

Fiore, D., Judson, R. N., Low, M., Lee, S., Zhang, E., Hopkins, C., et al. (2016). Pharmacological blockage of fibro/adipogenic progenitor expansion and suppression of regenerative fibrogenesis is associated with impaired skeletal muscle regeneration. *Stem Cell Research*, *17*, 161–169.

Fry, C. S., Kirby, T. J., Kosmac, K., McCarthy, J. J., & Peterson, C. A. (2017). Myogenic progenitor cells control extracellular matrix production by fibroblasts during skeletal muscle hypertrophy. *Cell Stem Cell*, *20*, 56–69.

Fry, C. S., Lee, J. D., Mula, J., Kirby, T. J., Jackson, J. R., Liu, F., et al. (2015). Inducible depletion of satellite cells in adult, sedentary mice impairs muscle regenerative capacity without affecting sarcopenia. *Nature Medicine*, *21*, 76–80.

Garcia-Prat, L., Martinez-Vicente, M., Perdiguero, E., Ortet, L., Rodriguez-Ubreva, J., Rebollo, E., et al. (2016). Autophagy maintains stemness by preventing senescence. *Nature*, *529*, 37–42.

Garcia-Prat, L., & Munoz-Canoves, P. (2017). Aging, metabolism and stem cells: Spotlight on muscle stem cells. *Molecular and Cellular Endocrinology*, *445*, 109–117.

Gattazzo, F., Urciuolo, A., & Bonaldo, P. (2014). Extracellular matrix: A dynamic microenvironment for stem cell niche. *Biochimica et Biophysica Acta*, *1840*, 2506–2519.

Gilbert, P. M., Havenstrite, K. L., Magnusson, K. E., Sacco, A., Leonardi, N. A., Kraft, P., et al. (2010). Substrate elasticity regulates skeletal muscle stem cell self-renewal in culture. *Science*, *329*, 1078–1081.

Gross, I., Bassit, B., Benezra, M., & Licht, J. D. (2001). Mammalian sprouty proteins inhibit cell growth and differentiation by preventing ras activation. *The Journal of Biological Chemistry*, *276*, 46460–46468.

Hausburg, M. A., Doles, J. D., Clement, S. L., Cadwallader, A. B., Hall, M. N., Blackshear, P. J., et al. (2015). Post-transcriptional regulation of satellite cell quiescence by TTP-mediated mRNA decay. *Elife*, *4*, e03390.
He, C., & Klionsky, D. J. (2009). Regulation mechanisms and signaling pathways of autophagy. *Annual Review of Genetics*, *43*, 67–93.
Hintermann, E., Bilban, M., Sharabi, A., & Quaranta, V. (2001). Inhibitory role of alpha 6 beta 4-associated erbB-2 and phosphoinositide 3-kinase in keratinocyte haptotactic migration dependent on alpha 3 beta 1 integrin. *The Journal of Cell Biology*, *153*, 465–478.
Ho, T. T., Warr, M. R., Adelman, E. R., Lansinger, O. M., Flach, J., Verovskaya, E. V., et al. (2017). Autophagy maintains the metabolism and function of young and old stem cells. *Nature*, *543*, 205–210.
Joe, A. W., Yi, L., Natarajan, A., Le Grand, F., So, L., Wang, J., et al. (2010). Muscle injury activates resident fibro/adipogenic progenitors that facilitate myogenesis. *Nature Cell Biology*, *12*, 153–163.
Johnson, S. C., Rabinovitch, P. S., & Kaeberlein, M. (2013). mTOR is a key modulator of ageing and age-related disease. *Nature*, *493*, 338–345.
Jones, N. C., Fedorov, Y. V., Rosenthal, R. S., & Olwin, B. B. (2001). ERK1/2 is required for myoblast proliferation but is dispensable for muscle gene expression and cell fusion. *Journal of Cellular Physiology*, *186*, 104–115.
Jones, N. C., Tyner, K. J., Nibarger, L., Stanley, H. M., Cornelison, D. D., Fedorov, Y. V., et al. (2005). The p38alpha/beta MAPK functions as a molecular switch to activate the quiescent satellite cell. *The Journal of Cell Biology*, *169*, 105–116.
Kenyon, C. J. (2010). The genetics of ageing. *Nature*, *464*, 504–512.
Kim, S. H., Turnbull, J., & Guimond, S. (2011). Extracellular matrix and cell signalling: The dynamic cooperation of integrin, proteoglycan and growth factor receptor. *The Journal of Endocrinology*, *209*, 139–151.
Lacraz, G., Rouleau, A. J., Couture, V., Sollrald, T., Drouin, G., Veillette, N., et al. (2015). Increased stiffness in aged skeletal muscle impairs muscle progenitor cell proliferative activity. *PLoS One*, *10*, e0136217.
Latil, M., Rocheteau, P., Chatre, L., Sanulli, S., Memet, S., Ricchetti, M., et al. (2012). Skeletal muscle stem cells adopt a dormant cell state post mortem and retain regenerative capacity. *Nature Communications*, *3*, 903.
Lemos, D. R., Babaeijandaghi, F., Low, M., Chang, C. K., Lee, S. T., Fiore, D., et al. (2015). Nilotinib reduces muscle fibrosis in chronic muscle injury by promoting TNF-mediated apoptosis of fibro/adipogenic progenitors. *Nature Medicine*, *21*, 786–794.
Liu, L., Cheung, T. H., Charville, G. W., Hurgo, B. M., Leavitt, T., Shih, J., et al. (2013). Chromatin modifications as determinants of muscle stem cell quiescence and chronological aging. *Cell Reports*, *4*, 189–204.
Lopez-Otin, C., Blasco, M. A., Partridge, L., Serrano, M., & Kroemer, G. (2013). The hallmarks of aging. *Cell*, *153*, 1194–1217.
Lukjanenko, L., Jung, M. J., Hegde, N., Perruisseau-Carrier, C., Migliavacca, E., Rozo, M., et al. (2016). Loss of fibronectin from the aged stem cell niche affects the regenerative capacity of skeletal muscle in mice. *Nature Medicine*, *22*, 897–905.
Lv, H., Li, L., Sun, M., Zhang, Y., Chen, L., Rong, Y., et al. (2015). Mechanism of regulation of stem cell differentiation by matrix stiffness. *Stem Cell Research & Therapy*, *6*, 103.
Mair, W., Goymer, P., Pletcher, S. D., & Partridge, L. (2003). Demography of dietary restriction and death in Drosophila. *Science*, *301*, 1731–1733.
Mann, C. J., Perdiguero, E., Kharraz, Y., Aguilar, S., Pessina, P., Serrano, A. L., et al. (2011). Aberrant repair and fibrosis development in skeletal muscle. *Skeletal Muscle*, *1*, 21.

Marks, H., Kalkan, T., Menafra, R., Denissov, S., Jones, K., Hofemeister, H., et al. (2012). The transcriptional and epigenomic foundations of ground state pluripotency. *Cell, 149*, 590–604.
McKay, B. R., Ogborn, D. I., Baker, J. M., Toth, K. G., Tarnopolsky, M. A., & Parise, G. (2013). Elevated SOCS3 and altered IL-6 signaling is associated with age-related human muscle stem cell dysfunction. *American Journal of Physiology. Cell Physiology, 304*, C717–728.
McLennan, I. S. (1993). Localisation of transforming growth factor beta 1 in developing muscles: Implications for connective tissue and fiber type pattern formation. *Developmental Dynamics, 197*, 281–290.
Mikkelsen, T. S., Ku, M., Jaffe, D. B., Issac, B., Lieberman, E., Giannoukos, G., et al. (2007). Genome-wide maps of chromatin state in pluripotent and lineage-committed cells. *Nature, 448*, 553–560.
Miller, R. A., Harrison, D. E., Astle, C. M., Fernandez, E., Flurkey, K., Han, M., et al. (2014). Rapamycin-mediated lifespan increase in mice is dose and sex dependent and metabolically distinct from dietary restriction. *Aging Cell, 13*, 468–477.
Montarras, D., L'Honore, A., & Buckingham, M. (2013). Lying low but ready for action: The quiescent muscle satellite cell. *The FEBS Journal, 280*, 4036–4050.
Mueller, A. A., van Velthoven, C. T., Fukumoto, K. D., Cheung, T. H., & Rando, T. A. (2016). Intronic polyadenylation of PDGFRalpha in resident stem cells attenuates muscle fibrosis. *Nature, 540*, 276–279.
Murakami, N., McLennan, I. S., Nonaka, I., Koishi, K., Baker, C., & Hammond-Tooke, G. (1999). Transforming growth factor-beta2 is elevated in skeletal muscle disorders. *Muscle & Nerve, 22*, 889–898.
Naito, A. T., Sumida, T., Nomura, S., Liu, M. L., Higo, T., Nakagawa, A., et al. (2012). Complement C1q activates canonical Wnt signaling and promotes aging-related phenotypes. *Cell, 149*, 1298–1313.
Noirez, P., Torres, S., Cebrian, J., Agbulut, O., Peltzer, J., Butler-Browne, G., et al. (2006). TGF-beta1 favors the development of fast type identity during soleus muscle regeneration. *Journal of Muscle Research and Cell Motility, 27*, 1–8.
Paliwal, P., Pishesha, N., Wijaya, D., & Conboy, I. M. (2012). Age dependent increase in the levels of osteopontin inhibits skeletal muscle regeneration. *Aging (Albany NY), 4*, 553–566.
Pietrangelo, T., Puglielli, C., Mancinelli, R., Beccafico, S., Fano, G., & Fulle, S. (2009). Molecular basis of the myogenic profile of aged human skeletal muscle satellite cells during differentiation. *Experimental Gerontology, 44*, 523–531.
Price, F. D., von Maltzahn, J., Bentzinger, C. F., Dumont, N. A., Yin, H., Chang, N. C., et al. (2014). Inhibition of JAK-STAT signaling stimulates adult satellite cell function. *Nature Medicine, 20*, 1174–1181.
Rawlings, J. S., Rosler, K. M., & Harrison, D. A. (2004). The JAK/STAT signaling pathway. *Journal of Cell Science, 117*, 1281–1283.
Ritschka, B., Storer, M., Mas, A., Heinzmann, F., Ortells, M. C., Morton, J. P., et al. (2017). The senescence-associated secretory phenotype induces cellular plasticity and tissue regeneration. *Genes & Development, 31*, 172–183.
Rodgers, J. T., King, K. Y., Brett, J. O., Cromie, M. J., Charville, G. W., Maguire, K. K., et al. (2014). mTORC1 controls the adaptive transition of quiescent stem cells from G0 to G(Aler)t. *Nature, 510*, 393–396.
Rozo, M., Li, L., & Fan, C. M. (2016). Targeting beta1-integrin signaling enhances regeneration in aged and dystrophic muscle in mice. *Nature Medicine, 22*, 889–896.
Rubinsztein, D. C., Marino, G., & Kroemer, G. (2011). Autophagy and aging. *Cell, 146*, 682–695.

Ryall, J. G., Dell'Orso, S., Derfoul, A., Juan, A., Zare, H., Feng, X., et al. (2015). The NAD (+)-dependent SIRT1 deacetylase translates a metabolic switch into regulatory epigenetics in skeletal muscle stem cells. *Cell Stem Cell, 16*, 171–183.
Schuster-Gossler, K., Cordes, R., & Gossler, A. (2007). Premature myogenic differentiation and depletion of progenitor cells cause severe muscle hypotrophy in Delta1 mutants. *Proceedings of the National Academy of Sciences of the United States of America, 104*, 537–542.
Schworer, S., Becker, F., Feller, C., Baig, A. H., Kober, U., Henze, H., et al. (2016). Epigenetic stress responses induce muscle stem-cell ageing by Hoxa9 developmental signals. *Nature, 540*, 428–432.
Segales, J., Perdiguero, E., & Munoz-Canoves, P. (2016). Regulation of muscle stem cell functions: A focus on the p38 MAPK signaling pathway. *Frontiers in Cell and Development Biology, 4*, 91.
Shea, K. L., Xiang, W., LaPorta, V. S., Licht, J. D., Keller, C., Basson, M. A., et al. (2010). Sprouty1 regulates reversible quiescence of a self-renewing adult muscle stem cell pool during regeneration. *Cell Stem Cell, 6*, 117–129.
Shefer, G., Van de Mark, D. P., Richardson, J. B., & Yablonka-Reuveni, Z. (2006). Satellite-cell pool size does matter: Defining the myogenic potency of aging skeletal muscle. *Developmental Biology, 294*, 50–66.
Siegel, P. M., & Massague, J. (2003). Cytostatic and apoptotic actions of TGF-beta in homeostasis and cancer. *Nature Reviews. Cancer, 3*, 807–821.
Sinha, M., Jang, Y. C., Oh, J., Khong, D., Wu, E. Y., Manohar, R., et al. (2014). Restoring systemic GDF11 levels reverses age-related dysfunction in mouse skeletal muscle. *Science, 344*, 649–652.
Smith, E. D., Kaeberlein, T. L., Lydum, B. T., Sager, J., Welton, K. L., Kennedy, B. K., et al. (2008). Age- and calorie-independent life span extension from dietary restriction by bacterial deprivation in Caenorhabditis elegans. *BMC Developmental Biology, 8*, 49.
Sousa-Victor, P., Gutarra, S., Garcia-Prat, L., Rodriguez-Ubreva, J., Ortet, L., Ruiz-Bonilla, V., et al. (2014). Geriatric muscle stem cells switch reversible quiescence into senescence. *Nature, 506*, 316–321.
Sun, N., Youle, R. J., & Finkel, T. (2016). The mitochondrial basis of aging. *Molecular Cell, 61*, 654–666.
Tang, A. H., & Rando, T. A. (2014). Induction of autophagy supports the bioenergetic demands of quiescent muscle stem cell activation. *The EMBO Journal, 33*, 2782–2797.
Tierney, M. T., Aydogdu, T., Sala, D., Malecova, B., Gatto, S., Puri, P. L., et al. (2014). STAT3 signaling controls satellite cell expansion and skeletal muscle repair. *Nature Medicine, 20*, 1182–1186.
Toth, K. G., McKay, B. R., De Lisio, M., Little, J. P., Tarnopolsky, M. A., & Parise, G. (2011). IL-6 induced STAT3 signalling is associated with the proliferation of human muscle satellite cells following acute muscle damage. *PLoS One, 6*, e17392.
Troy, A., Cadwallader, A. B., Fedorov, Y., Tyner, K., Tanaka, K. K., & Olwin, B. B. (2012). Coordination of satellite cell activation and self-renewal by Par-complex-dependent asymmetric activation of p38alpha/beta MAPK. *Cell Stem Cell, 11*, 541–553.
Uezumi, A., Fukada, S., Yamamoto, N., Takeda, S., & Tsuchida, K. (2010). Mesenchymal progenitors distinct from satellite cells contribute to ectopic fat cell formation in skeletal muscle. *Nature Cell Biology, 12*, 143–152.
Uezumi, A., Ito, T., Morikawa, D., Shimizu, N., Yoneda, T., Segawa, M., et al. (2011). Fibrosis and adipogenesis originate from a common mesenchymal progenitor in skeletal muscle. *Journal of Cell Science, 124*, 3654–3664.

Vasyutina, E., Lenhard, D. C., Wende, H., Erdmann, B., Epstein, J. A., & Birchmeier, C. (2007). RBP-J (Rbpsuh) is essential to maintain muscle progenitor cells and to generate satellite cells. *Proceedings of the National Academy of Sciences of the United States of America, 104*, 4443–4448.

Wood, L. K., Kayupov, E., Gumucio, J. P., Mendias, C. L., Claflin, D. R., & Brooks, S. V. (2014). Intrinsic stiffness of extracellular matrix increases with age in skeletal muscles of mice. *Journal of Applied Physiology, 117*, 363–369.

Yin, H., Price, F., & Rudnicki, M. A. (2013). Satellite cells and the muscle stem cell niche. *Physiological Reviews, 93*, 23–67.

Zhang, J., Grindley, J. C., Yin, T., Jayasinghe, S., He, X. C., Ross, J. T., et al. (2006). PTEN maintains haematopoietic stem cells and acts in lineage choice and leukaemia prevention. *Nature, 441*, 518–522.

Zhou, V. W., Goren, A., & Bernstein, B. E. (2011). Charting histone modifications and the functional organization of mammalian genomes. *Nature Reviews. Genetics, 12*, 7–18.

FSC
www.fsc.org
MIX
Papier | Fördert gute Waldnutzung
FSC® C083411

Kolibri 360 GmbH
Ferdinand-Jühlke-Straße 7
99095 Erfurt, Deutschland
produktsicherheit@kolibri360.de